The Latin American Studies Book Series

The Latin American Studies Book Series promotes quality scientific research focusing on Latin American countries. The series accepts disciplinary and interdisciplinary titles related to geographical, environmental, cultural, economic, political, urban and health-related research dedicated to Latin America. The series publishes comprehensive monographs, edited volumes and textbooks refereed by a region or country expert specialized in Latin American studies.

The series aims to raise the profile of Latin American studies, showcasing important works developed focusing on the region. It is aimed at researchers, students, and everyone interested in Latin American topics.

Submit a proposal: Proposals for the series will be considered by the Series Advisory Board. A book proposal form can be obtained from the Publisher, Andrea Brody-Barre (andrea.brody-barre@springer.com).

Victor A. Ramos

History of the Geological Sciences in Argentina

200 Years of Accomplishments

Springer

Victor A. Ramos
Instituto de Estudios Andinos "Don Pablo Groeber"
University of Buenos Aires - Conicet
Buenos Aires, Argentina

ISSN 2366-3421 ISSN 2366-343X (electronic)
The Latin American Studies Book Series
ISBN 978-3-031-83138-6 ISBN 978-3-031-83139-3 (eBook)
https://doi.org/10.1007/978-3-031-83139-3

Translation from the Spanish language edition: "Historia de la Geología Argentina: una crónica de más de dos siglos" by Victor A. Ramos, © Instituto de Geología y Recursos Minerales—SEGEMAR 2023. Published by Instituto de Geología y Recursos Minerales, Servicio Geológico Minero Argentino. All Rights Reserved.

This Springer imprint is published by the registered company Springer Nature Switzerland AG
The registered company address is: Gewerbestrasse 11, 6330 Cham, Switzerland

Foreword

The book that the reader has in his hands is the first comprehensive, experiential and encyclopedic work on the history of Argentine geology. It is a far-reaching, leading, passionate work by someone who was a first-hand witness to a substantial part of that rich history. Different partial essays followed one another over time with Franco Pastore, Pablo Groeber, Juan Olsacher, Guillermo Del Corro, Ángel Borrello, Osvaldo Bracaccini, Telasco García Castellanos, Marcelo Yrigoyen, Alberto Riccardi, Florencio Aceñolaza, Guillermo Ottone and many more.

Previous authors, almost all of them starting with the figure of the French geologist and disciple of George Cuvier: Alcide d'Orbigny. Many of the authors were unaware of or did not believe the geological and paleontological sketches carried out by religious people, especially Jesuits, in colonial times. Such was the case of the discovery of fossil bones attributed to "antediluvian human giants", which later turned out to belong to megatheriums or mastodons. José Guevara, José Sánchez Labrador, Thomas Falkner, Antonio Sepp; as well as Félix de Azara, Antón Helms, Dámaso Larrañaga, Tadeo Haenke and many others gave rise to little-known pre-Dorbignian geology. Geology is hidden in the pages of natural history books, which represent a quarry of thousands of pages, which require entire lives to analyze them, as the encyclopedic Jesuit father Guillermo Furlong Cardiff once attempted.

Dr. Victor A. Ramos, author of this book, is a widely recognized geologist at national and international level, and today the greatest expert in the geology and tectonics of the Andes. One of his great successes was the creation of the Andean Tectonics Laboratory, later converted into IDEAN (Institute of Andean Geology), which took the name of a giant of Argentine geology such as Dr. Pablo Groeber. This center is today a world reference for geological disciplines. Dr. Ramos is also a fervent scholar of the history of science in Argentina. His recent book on "The University in the Manzana de las Luces" (Eudeba, 2023), a feat of 300 years of history, is an excellent letter of introduction that exempts us from further comments. As are his biographies of great natural scientists such as Larrañaga, Haenke or Humboldt. Or the biographies of dozens of important Argentine geologists, many of them included in this book. Or the impressive work he did in gathering the biographies of the corresponding members, foreign geologists, in a special volume of the Argentine

Geological Association. Just as he did with *La Mujer en la Geología* of Argentina or with the teaching of geology in the country's universities. Dr. Ramos has thus been a driving force in solving per se or in activating, through the virtuosity of dissuasion, different colleagues to take on the task of writing about the biographies of their elders, teachers and professors.

In this new book by Dr. Ramos, which I have the honor of writing the prologue to, a journey is made through Argentine space, from colonial times, showing how the exponential evolution of geological knowledge was until today. The three great milestones of the 19th century are analyzed with the arrival of the French scholars of Urquiza with Victor Martin de Moussy, Auguste Bravard and Alfred Du Graty; the Italians in Buenos Aires with Pellegrino Strobel and Juan Ramorino or the Germans of Sarmiento in Córdoba with Hermann Burmeister, Alfred Stelzner, Ludwig Brackebusch, Wilhelm Bodenbender and many others.

The Patagonian and Antarctic expeditions are analyzed, the arrival of well-trained foreign geologists and disciples of great figures, the first Argentine geologists, José María Sobral, Franco Pastore and Juan J. Nágera; the creation of specific geological institutions, the old geological school of YPF and its great geologists and achievements, the role of museums and the successive mine managements that changed the name but not the objective, the geology courses in different Argentine universities, the wide-ranging scientific institutions, the national oil, coal and uranium organizations, among many aspects that the reader can explore in a careful visit to the thorough index that accompanies the book. Each chapter is also accompanied by a select bibliography to assist readers who wish to delve deeper into certain topics. This is an eclectic and heterodox work, with a profound content and a thematic body different from other historical works.

The author discusses the paradigm shift represented by the geosyncline theory to plate tectonics, of which the author himself has been a witness. He is also someone who can attest to the rocks and outcrops described by the elders, which he knows firsthand from having traveled through them throughout the country. Armed with the rod of justice, he gives each geologist his value at the time when he had to act and interpret, under the new ideas and theories, how close or how far they were with their speculations. It is enough to mention his pioneering ideas on the allochthony of Patagonia or the Precordillera, which led to years of fertile discussions in national and international congresses. Ideas that also brought relevant American and European scientists to the country to understand the new mechanisms of plate tectonics on the ground.

As the teacher Horacio J. Harrington taught us, geologists pass, and the outcrops remain. "New wine in old wineskins" he reflected. New interpretations of old rocks. Rocks put a humbling brake on the geologist who interprets them, and it is an act of justice to recover the effort of all since all genuinely contributed with their studies to reveal the mysteries of rocks and minerals and to extract from them their best fruits for the greatness of the country. Oil and gas resources, conventional and non-conventional, metal and non-metal deposits, application rocks, underground water, soils and foundations, nuclear minerals and other substances of the mineral kingdom that today form the heritage of the Nation with valuable resources and reserves, were discovered by geologists, with their successes and their errors, with their ideas and their models, some false and others correct, but all are responsible for having honestly contributed to the country those billions of tons of resources that can be transformed into trillions of dollars in foreign currency. And history, always history, marks the rhythm of the ages and their discoveries. Beyond the political vicissitudes that generate notable discontinuities in the development of knowledge. Several of these discontinuities in moments of splendor and prosperity or failure and decline are clearly analyzed in this fundamental text that politicians and statesmen should read with relish. These disruptions led us to the failed year of 1930, the night of the long batons in 1966, the coups of 1955 and 1976, among other political episodes and how they affected the normal development of Argentine science in general and geology in particular.

Dr. Ramos' major contributions to the field of modern geology led him to be awarded the most important national and international distinctions, including the prestigious Leopoldo von Buch Prize (2023) from Germany for foreign geologists received by some of the greatest exponents of the discipline in the last century; as well as his appointment in numerous national and international academies and even to occupy the highest positions such as the Presidency of the National Academy of Exact, Physical and Natural Sciences. He is undoubtedly the best-known and most respected Argentine geologist in the world today and although he needs no introduction or apology, it is worth mentioning him for the new generations of geologists.

The book is written for a wide universe of readers of the history of science, historians and philosophers, geologists and miners, students of the different branches of knowledge in Earth Sciences, academics and researchers, in short, a comprehensive work that marks a new milestone in Argentine historiographical studies. A work that is destined to remain in time as are its emblematic ancestors "El Windhausen", "El Leanza", "El Turner", "El Caminos", "El Volkheimer" (which never saw the light of day due to palace intrigues) and now "El Ramos". This book is destined to play its role as an everlasting historical scientific work. A treatise of this type was needed, and we celebrate that it has finally been published.

Ricardo N. Alonso
Emeritus Professor
Universidad Nacional de Salta
Salta, Argentina

Preface

Writing the History of Argentine Geology had been on my mind for several years. The invitation from the Argentine Scientific Society on its 150th anniversary to write, following the tradition of previous anniversaries, the history and state of the discipline inspired me even more, but I could not meet the established deadlines. However, I did not abandon the project, and without any pressing deadlines, I began to write these lines.

One of the reasons that drove me was that the current generations did not know the great men of Argentine geology, nor had they heard of the first professors, most of them foreigners, who came to the country at the beginning of the twentieth century. I feel privileged to have had professors who were the intermediate generation between the great masters and our generation. They had daily contact with teachers such as Juan Keidel, Pablo Groeber, and Horacio Harrington, among others. All these professors are no longer with us, as are many of the colleagues with whom we have shared the teachings, they have left us. The stories I had received from these professors were disappearing due to the inexorable passage of time.

The extensive conversations I had with Osvaldo Bracaccini, both during the campaign and in the cabinet, brought me closer to figures of our geology with interesting anecdotes that allowed me to better appreciate attractive personalities such as Guillermo Bodenbender, Egidio Feruglio, Juan Brandmayr and many others whom I never got to know. His opinions, for example, on the "eternal" nature of Feruglio's works, especially his description of Patagonia, which according to him was because in his three excellent volumes, he never risked an interpretation.

Sharing years of teaching as a professor in La Plata with Edgardo Rolleri, Alfredo Cuerda, Mario Teruggi, Rosendo Pascual, opened me up to a world full of stories from the present time that were unknown to me. However, it was at the old National Geological Survey where for more than a decade I shared hours of work with Eduardo Holmberg, Juan C. M. Turner, Guillermo Furque, Raúl Dessanti, Emilio González Díaz, among others, who brought me closer to leading figures such as Pablo Groeber and Horacio Harrington. These experiences were complemented by Jorge Polanski's classes and his memories of Fossa Mancini and Groeber, or Hildebranda Castellaro's admiration for Harrington. The opinions of Félix González Bonorino, always circumspect, and those of Horacio Camacho, made us meet Franco Pastore and Edelmira Mórtola and their jealousy for teaching.

In the pleasant meetings of the Argentine Geological Association at the geologist's house, at that time when cell phones and social networks did not exist, I was able to meet and exchange ideas with many of the geologists who appear in this book. Listening to their presentations from the first Argentine geological congresses and reading their works, allowed me to appreciate first-hand their contributions.

The experiences of campaigns that lasted months, with long evenings drinking mate around a fire, or the days of bad weather that forced us to long gatherings in the camp in some corner of the mountain range, enriched me with stories and experiences lived by these pioneers of our geology.

Having traveled the Argentine-Chilean mountain range from Jujuy to Tierra del Fuego interacting with different geologists and institutions, has allowed me to learn not only from their rocks but also from the experience of many local referents. The field teachings of Domingo Jakúlica and his anecdotes in the Subandino, or of Arturo Corte in the glaciers of the Cordillera Principal, of Marcelo Yrigoyen of Aconcagua, of Pedro Criado Roque and his mobile Mendoza-Pampean belt, competed with those of Apolo Ortiz and the Eastern Precordillera, those of Andrés Lencinas and the Pachón region, and those of Juvenal Zambrano and the beginnings of knowledge of our offshore basins. The bonhomie of Cesar Vilela and his stories of the first campaigns in the Mining Direction and the profound experiences of Eduardo Holmberg, fourth and last Baron of Kaunitz, whose family memories went back to 1812. His frequent question *Did you know Ramos…?* anticipated a fascinating story about Juan J. Nágera, Agustín Riggi or any other pioneer of geology. These memories and concerns were and are shared over the years with Norberto Malumián, another profound expert on our history with whom we rescued the memory of the first Argentine geologist, a native of Gualeguaychú, Mr. Juan José Nágera. The formal details of Dr. Horacio Camacho, which indicated that he had been second behind Pastore, did not impress us, given the passionate personality of Nágera who claimed to be the first.

These experiences were the basis that led me to investigate distant histories, which are now accessible through digital sources in various institutions, which allow direct access to these original publications. The amount of information available creates another even bigger problem, trying to select the most relevant to our history. In this selection we have tried to be as fair and balanced as possible, knowing in advance that our prejudices and own experience would not make it totally objective.

I would like to express my gratitude to Eduardo Zappettini who gave me access to the files of the first geologists of the National Direction of Geology and Mines that allowed me to reconstruct part of the history. This is extended to Guido Rodrigues Migueres of the Faculty of Exact and Natural Sciences of the University of Buenos Aires, who allowed me to consult the files of personnel who taught at this institution. Our recognition also goes to Silvia Carrasquero for her searches and support in locating old documents at the National University of La Plata and to Sergio Santillana for providing numerous documents from the Argentine Antarctic Institute.

I would like to thank the staff of the Mariano Moreno National Library, the Luis Federico Leloir Library of the Faculty of Exact and Natural Sciences of the University of Buenos Aires, the Florentino Ameghino Library of the La Plata Museum, and the library service of the ETH in Zurich, Switzerland, for their diligence and kindness in providing access to unpublished documents and old publications. I would also like to thank the employees of the National General Archives of Argentina for their advice that allowed me to quickly locate the documentation held by the institution.

Numerous colleagues throughout the country have provided me with information and stories that have enriched this work. I can only thank them all for their generosity and hope that these lines have correctly interpreted those stories.

This invitation to prepare an English version for Springer's Latin American Studies series has allowed me to revise the Spanish version once again, adding information that the criticisms received requested and correcting some inaccuracies and errors. I hope that this more complete version will meet readers' expectations and contribute to our history of geology.

Buenos Aires, Argentina
October 2024

Victor A. Ramos

Contents

Abbreviations

AGA	Asociación Geológica Argentina
AGN	Archivo General de la Nación
ANC	Academia Nacional de Ciencias
DNA-IA	Dirección Nacional del Antártico-Instituto Antártico
ETH	Eidgenössische Technische Hochschule Zürich
FCEN	Facultad de Ciencias Exactas y Naturales
ICS-IUGS	International Commission on Stratigraphy, International Union of Geological Sciences
Jujuy IGM	Instituto de Geología y Minería de Jujuy
MACN	Museo Argentino de Ciencias Naturales
MNCN-CSIC	Museo Nacional de Ciencias Naturales, Consejo Superior de Investigaciones Científicas
Segemar	Servicio Geológico Minero Argentino
UNSJ	Universidad Nacional de San Juan
YCF	Yacimientos Carboníferos Fiscales

Chapter 1
Introduction

Abstract The first and last attempt to write a history of Argentine Geology was made by Franco Pastore in 1925. Subsequently, there were only brief syntheses or partial attempts on specific aspects such as the History of Mineralogy, Petroleum Geology, or in-depth regional historical studies. These regional studies were carried out on the occasion of the National Geology Congresses, which allowed the history of the host province of the meeting to be reconstructed, covering a large part of the national territory. Despite the large quantity and quality of the historical works carried out, we lack a work that would cover more comprehensively the History of Geological Sciences of Argentina. We hope this work will fulfill these objectives 100 years after Pastore's first pioneering attempt. We take this opportunity to acknowledge all those who have contributed with their studies to the purpose of reconstructing our geological history.

Keywords Pioneering efforts · Previous attempts · Local history · Specific histories · New attempt

The first attempt to write a *History of Argentine Geology* was due to Franco Pastore, who, at the initiative of the Argentine Scientific Society upon celebrating 50 years of existence, wrote *Our Mineralogy and Geology during the last fifty years* (*1872–1922*) which it was published a few years later (Pastore 1925).

Upon the one hundred and fiftieth anniversary of the Argentine Scientific Society, I was invited to write a synthesis of the History of Geological Sciences in 2022, paying special attention to a series of events that led to a strong development of geological studies. The first of them was the need to know our subsoil which, with the founding of *Yacimientos Petrolíferos Fiscales* (Fiscal Oil Fields), required in the first decades of the twentieth century to intensify these studies, with the consequent increase in vocations in this discipline. The second was the scientific revolution that occurred at the end of the sixties, which essentially needed a total revision of previous premises and interpretations, which led to plate tectonics. The mobilist ideas made it necessary to discard many of the prevailing conceptions in the mainstream and change them with new hypotheses and premises that led to the current situation.

V. A. Ramos, *History of the Geological Sciences in Argentina*, The Latin American Studies Book Series, https://doi.org/10.1007/978-3-031-83139-3_1

1.1 Background on the History of Geology

Different attempts have been made to bring together the history of Earth Sciences in Argentina in its most diverse aspects. Among them, the pioneering synthesis of Juan Valentín (1867–1897) stands out, a young German geologist who had arrived in Argentina in 1894 to join the La Plata Museum. The following year he was hired by the Public Museum of Buenos Aires as head of the Geology Section to participate in the Second National Census of 1895. On this occasion, he wrote his Geological Sketch of Argentina, which was the first attempt to gather geological data and the figures who contributed to the birth of Argentine geology (Valentín 1897). This work was later included in Latzina's Geographical Dictionary (1898). Dr. Valentín was sent by the director of the Public Museum Dr. Carlos Berg (1843–1912) to carry out a series of surveys on the Atlantic coast of northern Patagonia, dying due to an accident, falling into a ravine with his pockets full of fossils, while studying the coastal cliffs near Rawson in 1897. In just over three years he had made more than 14 contributions to Argentine geology (see reviews by Rolleri 2007; Aceñolaza 2011).

It is necessary to highlight the Geological sketch of the Argentine Republic presented by Maak (1870) to the Boston Society of Natural History, where he made one of the first syntheses of our geology. It describes the peculiar conditions of the country between the Andes Mountains and the Atlantic coast with extensive plains. He thanks Hermann Burmeister for having introduced him to the subject during his visit to Argentina, while he ventures into the origin of the Pampeano and the fossiliferous wealth of our Cenozoic. He mentions the presence of granites and gneisses along with melaphyres.

It is also worth mentioning the review on mining by Henry David Hóskold (1829–1904), who was Director General of Mines and Geology between 1887 and 1904, for the universal exhibition in Paris of 1889 where there are extraordinary references on the deposits and their geology (Hóskold 1889). However, the first institutional synthesis was carried out for the first centenary of the country in an extensive and complete work that considered the development of geology at the dawn of the 20th century. This work, The Argentine Republic in 1910 presented by Urien and Colombo (1910), brought together the geological knowledge achieved at that time and made a review of the main naturalists who contributed to the understanding of geology that was available at that time. Considering the depth of the geological analysis and its precision, it seems that it was based on internal reports specially prepared for this event, given the similarities with the one made by Engineer Enrique Martín Hermitte (1871–1955) for the year 1910 and that appeared shortly after under his authorship (Hermitte 1912) and as part of the Third National Census of 1914. In this last work, he recognizes the general contributions of Hans Keidel and Richard Stappenbeck and the partial ones of Walther Schiller and Guido Bonarelli (Hermitte 1914). Mr. Hermitte was director of the Mines and Geology Division starting in 1904 (Fig. 1.1). He promoted the search for water and oil exploration, culminating in its discovery on December 13, 1907, which is why Oil Day is celebrated in Argentina.

Fig. 1.1 Enrique Hermitte, Franco Pastore and Pablo Groeber pioneer authors in the history of the Argentine geology. *Source* Biblioteca Digital F.C.EN

Another synthesis with levels like the previous ones but more complete was carried out by Franco Pastore (1885–1958), the first geologist to graduate in the country, who received his doctorate in 1914. In his work on the Fiftieth Anniversary of the Argentine Scientific Society, the early contributions made by a plethora of German geologists in the National Academy of Sciences from Córdoba and the birth of the *Dirección de Geología y Minas* (Direction of Geology and Mines) stand out, leaving a great void in the previous works (Pastore 1925).

In the excellent description of Argentine Geology by Anselmo Windhausen (1882–1932), although there is a complete bibliography for the time, there is no review of the authors who contributed to the knowledge of it (Windhausen 1931). Years later in his book Mineralogy and Geology, Don Pablo Groeber (1885–1964) in 1938 updated in one of his chapters, in a precise and personal synthesis, of the development of geological sciences in Argentina and its main protagonists (Groeber 1938, p. 451–467).

With the beginning of the Argentine Geological Conferences in 1960, precursors of the current Argentine Geological Congresses, Juan Olsacher (1903–1964) was commissioned to carry out the first conference report on *One Hundred and Fifty Years of Geology in Argentina*, where he compiled, in a tight synthesis, the beginnings of our geology (Olsacher 1962) (Fig. 1.2). In the second conference report, Ángel Borrello in 1963 compiled the latest progress of the discipline (Borrello 1965).

Years later, an important reconstruction of our first steps was carried out at the First Symposium on the History of Science held in Córdoba in 1969 by Telasco García Castellanos (1912–2008). On this occasion, the First Argentine Regional Geology Symposium had been held, organized by the National Academy of Sciences, whose results were published by the academy (see Riccardi 2020a). Shortly after, Del Corro in (1972) published a brief historical synthesis of geology in Argentina. Since the First Symposium on the History of Geology was held, all regional geology symposiums had a synthesis of the history included this time in each province. This line culminated with the synthesis by Edgardo Rolleri (Fig. 1.3) and collaborators that was prepared

Fig. 1.2 Juan Olsacher and Ángel Borrello, authors of the first reports of the Argentine geological congresses and Telasco García Castellanos, a promoter of the history of Argentine geology. *Source* Segemar archives

for the last edition of Argentine Geology by Roberto Caminos in 1999 (Rolleri et al. 1999; Caminos 1999).

Partial syntheses have been made in the various reports on Geology and Natural Resources of the different provinces on the occasion of the Argentine geological congresses. In these syntheses of geology, a detailed analysis of the geological contributions and their protagonists was presented for each province. Among these, those by Rolleri and Criado Roque (1970) for the province of Mendoza stand out; those of Turner (1975) and Rolleri (1975) from the province of Buenos Aires; that of Rolleri et al. (1978) on the History of knowledge of Neuquén; that of Yrigoyen (1981), in his synthesis of the geology of San Luis; that of Núñez (1984) on the History of the province of Río Negro; the historical background of the province of Tucumán (Aceñolaza 1984, 2014); the History of the Geology of Mendoza by Rolleri (1993);

Fig. 1.3 Edgardo Rolleri, Horacio Camacho and Alberto Riccardi, experts in the history of Argentine geology. *Source* Segemar archives

Riccardi's Homage at the end of the century to the precursors of Argentine Geology (1999) in northwest Argentina; the Review of geographical and geological knowledge by Rolleri and de Barrio (2002) in Santa Cruz; the review by Rolleri et al. (2005) in the new report of the province of Buenos Aires; the geological studies in Jujuy by Aceñolaza and Alonso (2008), together with that of Ramos and Leanza (2011) on History of the evolution of geological knowledge of Neuquén. A new synthesis of the Geology of Tucumán served as a framework to update its history (Aceñolaza 2014), as well as that of Chubut (Ramos 2022). This set of partial works brings together a large amount of regional information on the history of the geology of these provinces to which the reader is referred for further details.

The fiftieth anniversary of the Argentine Geological Association led to the reconstruction of its history and that of the great masters of Argentine geology, published in the Journal of the Argentine Geological Association in 1995. A similar event allowed the Argentine Paleontological Association years later to present a work about its fiftieth anniversary that brings together the historical development of the association and its different specialties in its Fifteith Anniversary volume edited by Archangelsky et al. (2007). The history of other disciplines has been treated in isolation, such as the History of Argentine Mining, edited by Lavandaio and Catalano (2004a, b) and the complete treatise on the History of Mineralogy published by Sureda (2008). Added to these is the synthesis of Two Hundred Years of Earth Sciences in Argentina, on the occasion of the bicentennial of the May Revolution (Ramos 2011), the 50 years of plate tectonics (Ramos 2021a), among others.

Another set of works presents the history of different institutions, such as those of geological sciences at the University of Córdoba and the National Academy of Sciences by Telasco García Castellanos (1963, 1970, 1974, 1987, 2000); The Natural Sciences at the University of Buenos Aires by Horacio Camacho (1971, 1974, 1996); the one produced by the Argentine Mining Geological Service under the coordination of Eduardo Zappettini on the occasion of the centenary of this leading institution of Argentine Geology and which includes interesting unpublished documentation (Zappettini 2004); those of El Museo de La Plata on the advancement of geological knowledge since the end of the 19th century by Mario Teruggi (1977, 1988) and those of Alberto Riccardi (2008, 2016); the history of exploration at YPF (Anonymous 1972; Turic and Ferrari 2000); the 150 years of the National Academy of Sciences in Córdoba by Pedro Depetris (2019); that of Victor A. Ramos on the bicentennial of the creation of the Public Museum of Buenos Aires in 2012 and that of the Corresponding Members of the Argentine Geological Association, among others (Ramos 2012, 2018, 2021b).

It is evident in this tight synthesis of the background that there is a profuse specific bibliography on the history of geology in Argentina (see Riccardi 2020b). The challenge is to try to present the most important milestones of that history, the framework in which they were developed, and the references in each era that promoted the advancement of knowledge. These eras will be separated by both political and disciplinary discontinuities that have led to abrupt changes and accelerations in their knowledge.

References

Aceñolaza FG (1984) Antecedentes históricos. In: Aceñolaza FG, Bossi GE, Toselli A (eds) Geología de Tucumán. Colegio de Graduados en Ciencias Geológicas de Tucumán, pp 9–18, San Miguel de Tucumán

Aceñolaza FG (2011) Juan Valentin: un geólogo que supo resumir la geología argentina. Revista De La Asociación Geológica Argentina 68(3):315–321

Aceñolaza FG (2014) Estudios geológicos en la historia de Tucumán. In: Moyano S, Puchulu ME, Fernández DS, Vides ME, Nieva S, Aceñolaza GF (eds) Geología de Tucumán. Colegio de graduados en Ciencias Geológicas, pp 12–19, Tucumán

Aceñolaza FG, Alonso R (2008) Los estudios geológicos en Jujuy. In: Coira B, Zappettini EO (eds) Geología y recursos naturales de la provincia de Jujuy. 17° Congreso Geológico Argentino, Relatorio: 3–10, San Salvador de Jujuy

Anónimo (1972) Cincuentenario de YPF, Una empresa al servicio del país, 1922–1972. Departamento de Difusión de YPF, 192 p, Buenos Aires

Archangelsky S, Sánchez T, Tonni EP (2007) Asociación Paleontológica Argentina, Ameghiniana, 50º aniversario. Asociación Paleontológica Argentina, Publicación Especial 11:1–287

Borrello A (1965) La geología argentina a través de sus últimos quince años de desarrollo. Segundas Jornadas Geológicas Argentinas (Salta, 1963), Actas 1: 15–19, Buenos Aires

Camacho HH (1971) Las Ciencias Naturales en la Universidad de Buenos Aires. Estudio histórico. Eudeba, 150 p, Buenos Aires

Camacho HH (1974) Origen y evolución de la enseñanza de las Ciencias Geológicas en la Universidad de Buenos Aires. 5º Reunión Científica de la Comisión Internacional de la Historia de las Ciencias Geológicas, Comunicaciones Científicas, pp 123–132, Madrid

Camacho HH (1996) Origen y Evolución de la enseñanza de las Ciencias Geológicas en la Universidad de Buenos Aires. Cienc Invest 48(1–2):48–53

Caminos R (1999) Geología Argentina, Instituto de Geología y Recursos Minerales, Anales 29: 1–796 p, Buenos Aires

Del Corro G (1972) Breve historia de la geología argentina. Museo Argentino de Ciencias Naturales "Bernardino Rivadavia", Publicación de Extensión Cultural y Didáctica 18: 1–23, Buenos Aires

Depetris PJ (ed) (2019) La Academia Nacional de Ciencias, 150 Años acompañando la Ciencia Argentina. Academia Nacional de Ciencias, 400 p, Córdoba

García Castellanos T (1963) Evolución de la enseñanza de las Ciencias Exactas y Naturales en la Universidad de Córdoba desde su fundación hasta Sarmiento. Academia Nacional de Ciencias, Miscelánea 42: 1–33, Córdoba

García Castellanos T (1970) La Academia Nacional de Ciencias en Córdoba. Significado de su creación. Boletín de la Academia Nacional de Ciencias 48 (1–4):7–18, Córdoba

García Castellanos T (1974) Historia de la enseñanza de las Ciencias Geológicas en la Universidad Nacional de Córdoba, Argentina. IUGS International Committe on the History of Geological Sciences, 5° Scientific Meeting, Madrid, Papers, pp 133–150

García Castellanos T (1987) Breve historia de la Academia Nacional de Ciencias de Córdoba. Academia Nacional de Ciencias, Miscelánea 75: 1–39, Córdoba

García Castellanos T (2000) Historia de la enseñanza de la ciencia geológica en la Universidad Nacional de Córdoba, Argentina. Academia Nacional de Ciencias, Boletín 64: 5–16, Córdoba

Groeber P (1938) Mineralogía y Geología, Espasa Calpe Argentina SA, 492 p, Buenos Aires

Hermitte EM (1912) Memoria de la División de Minas, Geología e Hidrogeología correspondiente al año 1910. Anales Del Ministerio De Agricultura, Sección Geología, Mineralogía y Minería 7(2):1–129

Hermitte EM (1914) La Geología y Minería Argentina en 1914. Tercer Censo de la República Argentina: 406–494, Buenos Aires

Hóskold HD (1889) Memoria general y especial sobre las minas, metalurgia, leyes de minas, recursos, ventajas, etc., de la explotación de minas en la República Argentina, ilustrada

con mapas topográficos, geológicos, mineros, planos, secciones y vistas. Publicación oficial, Imprenta y Estereotipia del *Courrier de la Plata*, 600 p, Buenos Aires
Lavandaio E, Catalano E (2004ª) Historia de la Minería Argentina. Tomo 1. Instituto de Geología y Recursos Minerales, Anales 40: 1–314, Buenos Aires
Lavandaio E, Catalano E (2004b) Historia de la Minería Argentina. Tomo 2. Instituto de Geología y recursos minerales, Anales 40; 1–418, Buenos Aires
Maak GA (1870) Geological sketch of Argentine Republic. Proc Boston Soc Nat Hist 18:417–428, Boston
Nuñez E (1984) Historia del conocimiento geológico. In: Ramos VA (ed) Geología y recursos naturales de la provincia de Río Negro. 9º Congreso Geológico Argentino, Relatorio: 37–74, Buenos Aires
Olsacher J (1962) Cincuenta años de geología en la Argentina. 1º Jornadas Geológicas Argentina (San Juan, 1960), Relatorio 1: 21–36, Buenos Aires
Pastore F (1925) Evolución de las ciencias en la República Argentina. VI Nuestra Mineralogía y Geología durante los últimos cincuenta años (1872–1922). Cincuentenario de la Sociedad Científica Argentina, 47 p, Buenos Aires
Ramos VA (2011) Doscientos años de Ciencias de la Tierra en Argentina. Revista De La Asociación Geológica Argentina 68(3):392–406
Ramos VA (2012) Bernardino Rivadavia y las Ciencias Naturales. Revista del Museo Argentino de Ciencias Naturales, n.s. 14(2):213–222
Ramos VA (2018) El desarrollo de la Geología en la Universidad de Buenos Aires. In: Martino R, Ramos VA (eds) El nacimiento de la geología en la enseñanza universitaria, Revista Facultad de Ciencias Exactas, Físicas y Naturales 5(Suplemento 1): 23–32, Córdoba
Ramos VA (2021a) Fifty years of plate tectonics in the Central Andes. J S Am Earth Sci 105:102997
Ramos VA (2021b) Los Miembros Correspondientes de la Asociación Geológica Argentina. Asociación Geológica Argentina, Publicación Especial 15, 316 p, Buenos Aires
Ramos VA (2022) La evolución del conocimiento geológico en la provincia de Chubut. In Giacosa, R. (ed.) Geología y Recursos Naturales de la Provincia del Chubut, 21° Congreso Geológico Argentino, Relatorio: 3–33, Puerto Madryn
Ramos VA, Leanza H (2011) Historia de la evolución del conocimiento geológico. In: Leanza H, Arregui, C, Carbone O, Danieli JC, Vallés JM (eds) Geología y Recursos Naturales de la Provincia del Neuquén. 17º Congreso Geológico Argentino, Relatorio: 13–27, Neuquén
Riccardi AC (1999) Homenaje de fin de siglo a los precursores de la Geología Argentina. In: González Bonorino G, Omarini R, Viramonte J (eds) Geología del Noroeste Argentino, 14º Congreso Geológico Argentino (Salta), Relatorio 2: 173–187, Salta
Riccardi AC (2008) El Museo de la Plata en el avance del conocimiento geológico a fines del siglo XIX. In: Aceñolaza FG (ed) Los geólogos y la Geología en la historia argentina. InsuGeo, Serie Correlación Geológica 24: 109–126, Tucumán
Riccardi AC (2016) Las investigaciones geológicas del Museo de La Plata desde la época del centenario a la del sesquicentenario: 1906–1966. In: Riccardi AC (ed) Historia de la Geología en el Bicentenario de la Argentina. Revista del Museo de La Plata, vol 1, Número Especial: 228–256
Riccardi AC (2020a) Telasco García Castellanos: Promotor de la Geología y de las ciencias. Revista Facultad de Ciencias Exactas, Físicas y Naturales 7 (Suplemento 1): 95–114, Córdoba
Riccardi AC (2020b) Historiadores de la Geología Argentina. Revista Facultad de Ciencias Exactas, Físicas y Naturales 7 (Suplem. 1): 115–132, Córdoba
Rolleri EO (1975) Provincias geológicas bonaerenses. In: Geología de la provincia de Buenos Aires. 6° Congreso Geológico Argentino, Relatorio: 29–54, Buenos Aires
Rolleri EO (1993) Historia de la geología de Mendoza. In: Ramos VA (ed) Geología y Recursos Naturales de Mendoza. 12° Congreso Geológico Argentino y 2° Congreso de Exploración de Hidrocarburos, Relatorio: 1–10, Buenos Aires
Rolleri EO (2007) Rememoración de un olvidado ilustre: el Dr. Juan Valentin. In: Aceñolaza FG (ed) Los geólogos y la geología en la historia argentina. InsuGeo, Miscelánea 16:37–38, Tucumán

Rolleri EO, de Barrio R (2002) Reseña del conocimiento geográfico y geológico. In: Haller MJ (ed) Geología y Recursos Naturales de Santa Cruz. 15º Congreso Geológico Argentino, Relatorio: 13–32, Santa Cruz

Rolleri EO, Criado Roque P (1970) Geología de la provincia de Mendoza. 4º Jornadas Geológicas Argentinas, Actas 2: 1–60, Buenos Aires

Rolleri EO, Stipanicic PN, Ramos VA (1978) Evolución del conocimiento geológico. In Geología y Recursos Naturales del Neuquén. 7º Congreso Geológico Argentino, Relatorio: 9–22, Buenos Aires

Rolleri EO, Caballé M, Tessone M (1999) Datos para una historia de la geología argentina. In: Caminos R (ed) Geología Argentina. Anales del Instituto de Geología y Recursos Minerales 29(1):1–33, Buenos Aires

Rolleri EO, Caballé M, Tessone M (2005) Breve reseña histórica y apuntes sobre los avances del conocimiento geológico del territorio bonaerense. In: de Barrio RE, Etcheverry RO, Caballé MF, Llambías E (eds) Geología y Recursos Minerales de la Provincia de Buenos Aires. 16° Congreso Geológico Argentino (La Plata), Relatorio: 1–20, La Plata

Sureda RJ (2008) Historia de la mineralogía. InsuGeo, Serie de Correlación Geológica 23, 193 p, Tucumán

Teruggi ME (1988) Museo de La Plata 1888–1988, una centuria de honra. Fundación Museo de La Plata Francisco Pascasio Moreno, 1ª. ed., 2ª. ed., 3ª. ed. 1994, 157 p, La Plata

Teruggi ME (1977) Cien años de geología en el Museo de La Plata. In: Obra del Centenario del Museo de La Plata, 1. Reseña Histórica, pp 59–73, La Plata

Turic MA, Ferrari JC (2000) La exploración de petróleo y gas en la Argentina: el aporte de YPF. Repsol - YPF S.A. 178 p, Buenos Aires

Turner JCM (1975) Síntesis. In Geología de la provincia de Buenos Aires. 6º Congreso Geológico Argentino, Relatorio: 9–27, Buenos Aires

Urien CM, Colombo E (1910) La República Argentina en 1910. Estudio histórico, político, social y económico publicado bajo el auspicio de la H. Comisión del Centenario de la Independencia Argentina y de la Junta de Historia Numismática Americana 1(9) Geología: 372–403, Buenos Aires

Valentin J (1897) Bosquejo geológico de la Argentina. Gmo. Van Woerden & Cia., 50 p, Buenos Aires

Windhausen A (1931) Geología Argentina, segunda parte: Geología Histórica y Regional del Territorio Argentino. Ed. Jacobo Peuser, 646 p, Buenos Aires

Yrigoyen M (1981) Síntesis. In: Yrigoyen M (ed) Geología y Recursos Naturales de la Provincia de San Luis. 8º Congreso Geológico Argentino, Relatorio: 7–32, Buenos Aires

Zappettini EO (2004) Servicio Geológico Minero Argentino 1994–2004. 100 años al servicio del desarrollo nacional. Instituto de Geología y Recursos Minerales—Segemar, 104 p, Buenos Aires

Chapter 2
The First Observations

Abstract Establishing the exact date of the beginning of geological studies is difficult. The discovery of "sepulchers of giants" has been taken as such, as the large vertebrate fossils found in the Cenozoic deposits of the Pampas were called in the American colonies in the 16th century. This led to great discussions about their meaning and allowed progress in the knowledge of these remains. It was not until 1751 that the first precise description of these findings was made, preserved under thick layers of ash. Numerous other discoveries followed until an almost complete specimen of Megatherium was found in Lujan, studied in 1793 in Spain, and classified as such by Cuvier in 1804. It was one of the first large fossil mammals found in the world. The first precise geological descriptions of the lithology and structure of the outcropping rocks from the Sierras Pampeanas to Bolivia were made by the German geologist Anton Zacharias Helms, published in 1798 and later translated into English and French. These observations were followed by those of Tadeo Haenke, a naturalist born in Bohemia who, after participating in the Malaspina Expedition, settled in Cochabamba, part of the Viceroyalty of the Río de la Plata at that time. He published numerous contributions at the beginning of the nineteenth century, although many of his descriptions remained unpublished. The first specific geological study of the United Provinces of the Río de la Plata was carried out in 1819 by Dámaso Antonio Larrañaga, a self-taught Río de la Plata native. His Escritos were published a century later.

Keywords Giants tombs · Jesuits · Megatherium · Early naturalists · First study

2.1 Sepulchers of Giants

It is difficult to establish a starting point for the history of Argentine Geology. The first tangible evidence from the subsoil is the news of "sepulchers of giants" (*sepulcros de gigantes*) buried in the pampas and the amazement that these creatures aroused. Although they were known from various places in the American colonies since the sixteenth century, in Argentine territory their first news is related to the stories

V. A. Ramos, *History of the Geological Sciences in Argentina*, The Latin American Studies Book Series, https://doi.org/10.1007/978-3-031-83139-3_2

of Father José Guevara who recounts the discovery in 1740 of bones of giants on the banks of the Carcarañá River. This Jesuit mentions skulls and other bones and having seen a petrified tooth the size of a fist from this town. Remains of "sepulchers of giants" were also mentioned a few leagues downstream from the city of Córdoba.

However, the first reliable description of fossil remains, and their location are due to Thomas Falkner (1702–1784). This Jesuit, author of the famous Description of Patagonia and the Adjacent Parts of South America in 1774, was a medical doctor by profession (Fig. 2.1). He describes having found in 1751 on the banks of the Carcarañá River, a few kilometers from its mouth in the Paraná, the remains of bones and a shell with hexagonal joints.

Falkner in his anatomical analysis related part of these bones to the upper fraction of the armor of an armadillo, but almost three meters long (Falkner, 1774). Today we know that they correspond to the first glyptodont found on Argentine soil. Falkner mentions and correctly interprets an "immense cloud of volcanic ash" that fell in 1750 on the Pampas region while he was near Cape San Antonio at that time. He says that the pastures were covered in ashes on both sides of the Río de La Plata. He clarifies that "*the cause of all this was the eruption of a volcano near Mendoza, and the winds that blew those light ashes to the incredible distance of three hundred and more leagues*".

A few years later, in 1766, the discovery of other remains in the vicinity of Arrecifes by Captain Estevan Álvarez del Fierro, a sailor who was soon to leave

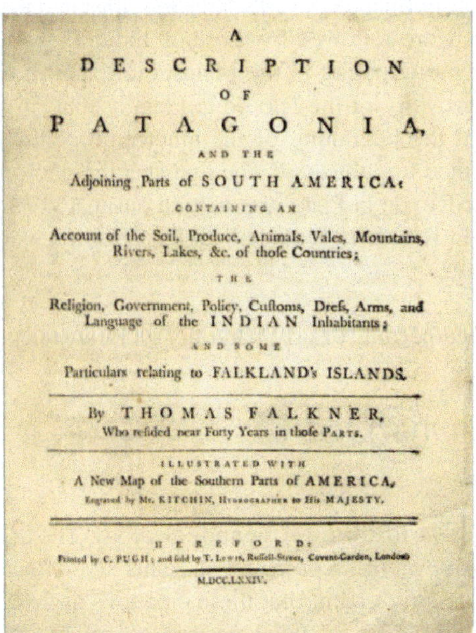

A
DESCRIPTION
OF
PATAGONIA,
AND THE
Adjoining Parts of SOUTH AMERICA:
CONTAINING AN
Account of the Soil, Produce, Animals, Vales, Mountains, Rivers, Lakes, &c. of thoſe Countries;
THE
Religion, Government, Policy, Cuſtoms, Dreſs, Arms, and Language of the INDIAN Inhabitants;
AND SOME
Particulars relating to FALKLAND's ISLANDS.
By THOMAS FALKNER,
Who reſided near Forty Years in thoſe Parts.
ILLUSTRATED WITH
A New Map of the Southern Parts of AMERICA,
Engraved by Mr. KITCHIN, Hydrographer to His MAJESTY.
HEREFORD:
Printed by C. PUGH; and ſold by T. Lewis, Ruſſell-Street, Covent-Garden, London.
M.DCC.LXXIV.

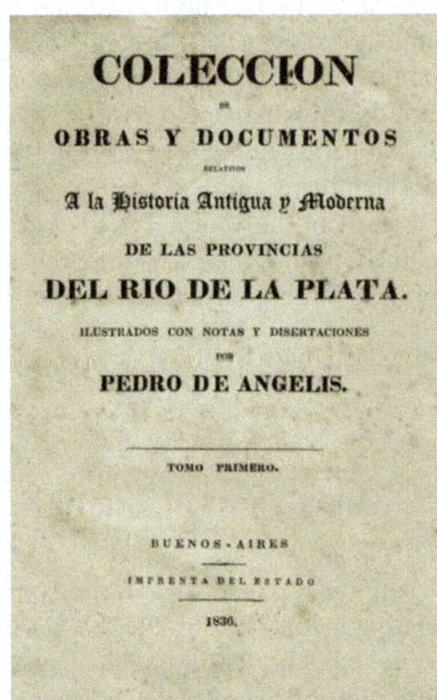

COLECCION
DE
OBRAS Y DOCUMENTOS
RELATIVOS
A la Historia Antigua y Moderna
DE LAS PROVINCIAS
DEL RIO DE LA PLATA.
ILUSTRADOS CON NOTAS Y DISERTACIONES
POR
PEDRO DE ANGELIS.
TOMO PRIMERO.
BUENOS-AIRES
IMPRENTA DEL ESTADO
1836.

Fig. 2.1 Facsimile of description of Patagonia by Thomas Falkner, Jesuit priest, and first version in Spanish published by Pedro de Angelis in 1838

for Spain, was recorded. According to Juan María Gutiérrez in 1866 in his work on "*Paleontology in the Spanish colonies*" (Gutiérrez 1866), Álvarez del Fierro presented a letter to the mayor of Buenos Aires informing him of the discovery of "sepulchers of giant-sized animals" and requested that expert people examine the place. Those sent by the mayor found the remains that were brought to Buenos Aires and before shipping them to Spain, doctors were asked to report whether or not they were human. One of the sworn doctors reported that the remains were human but extraordinarily gigantic. They were sent to Spain where they were examined by academics from the *Real Academia de Historia* who ruled that they were not human but rather appeared to be quadrupeds close to elephants (Fig. 2.2). A statement that we know today is correct given that the bones were from ancient mastodons (Cabrera 1930; Tonni et al. 2007; Pasquali and Tonni 2008; Orche and Amaré 2019a).

The most spectacular discovery was the Luján *Megatherium*, which in its time had a worldwide impact among experts and has been extensively analyzed over the years for its historical and paleontological importance. It was the first fossil mammal from South America described, and it was George Cuvier (1769–1832), the French naturalist, father of comparative anatomy and one of the great promoters of modern paleontology, who named it *Megatherium americanum* (Cuvier 1796, 1804).

These bone remains were found by the friar Manuel de Torres in the ravines of the Luján River in 1786, who requested help from the Viceroy Loreto so that he could extract it with the utmost care and draw the material found for its subsequent reconstruction (see details in Garriga 1796; Gutiérrrez 1835; Trelles 1882; Pasquali and Tonni 2008). The details of the excavation and the events that occurred with these remains until they were packed and sent to Spain have been recorded by various authors, with Viceroy Loreto actively participating in them. The Marquis of Loreto informs the king of the discovery in a detailed note accompanied by two drawings

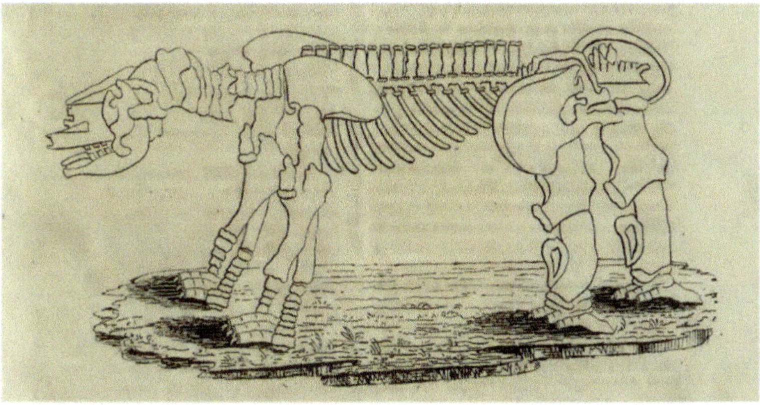

Fig. 2.2 The *Megatherium* drawn by Custodio Sa y Faría in Buenos Aires before sending it to Spain according to Gutiérrez (1835). Although the drawing is schematic, it accurately illustrates its shapes and dimensions. Gutiérrez already mentions that this animal rose on its hind feet to climb trees

made by the engineer José Custodio de Sa y Faría, a note where he observes that they do not correspond to humans and that they belong to an unknown large animal (Orche and Amaré 2019a).

The reconstruction of the *Megatherium* that arrived in the cabinet of the *Museo de Historia Natural de Madrid* in 1788 was carried out by Juan Bautista Bru (1740–1799), the cabinet's draftsman and taxidermist (Fig. 2.3). This reconstruction had certain errors in its morphological interpretation of the bones given the assumed quadrupedal posture, partly influenced by the in situ position of the almost complete skeleton found and drawn in Buenos Aires. When King Charles III, the enlightened monarch of the House of Bourbons, was able to observe the large dimensions of the received specimen, 5 m long by 3.5 m high, he was very impressed, as were the wise men and bureaucrats of the Court. Carlos III's instructions to the viceroy of Rio de la Plata became famous when he asked him to find out "*if in the district of Luján or in another of this viceroyalty, you can obtain a live animal, even if it is small, of the species of the said skeleton, by sending it alive if it could be, and, failing that, dissected and stuffed with straw…*" (Orche and Amaré 2019a, b, c).

Bru managed to assemble the skeleton and make a series of drawings along with a description of it in 1793, noting that he did not publish it due to a series of internal problems in the office of the *Museo de Historia Natural*. These notes and the respective drawings came into the hands of Cuvier who published a preliminary contribution in 1796 based on this information and recognizing that they corresponded to an unknown mammal, for which he created a new taxon, *Megatherium americanum* belonging to the toothless group. That same year, the Catalan engineer and naturalist Joseph Garriga, upset by the fact that foreigners were the first to report the existence of the skeleton, convinced Bru to sell him the text and the drawings that he had made.

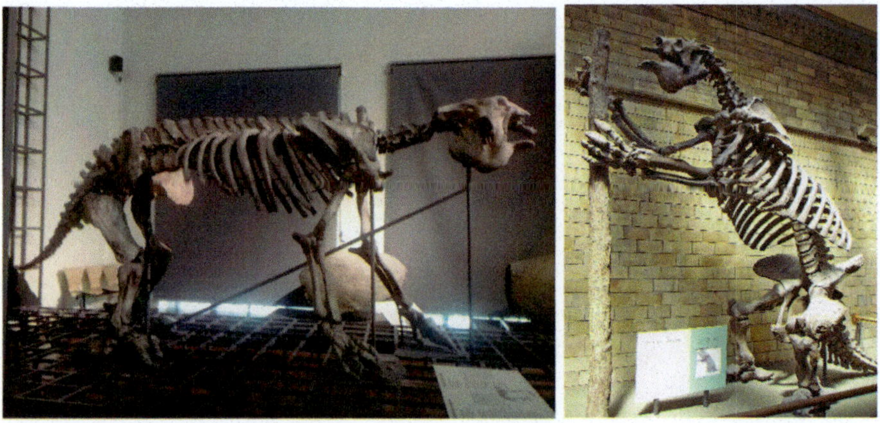

Fig. 2.3 The *Megatherium* that is exhibited with a quadruped posture, as it had been reconstructed by Juan Bautista Bru, except for the tail that was added later, in the *Museo de Historia Natural de Madrid* and the one that is exhibited in the Natural History Museum of London in a more natural upright position. *Source* Museo Nacional de Ciencias Naturales (MNCN-CSIC) and wikimedia

Fig. 2.4 George Cuvier (1769–1832) who unveiled the first example of the Luján *Megatherium* and William Clift (1775–1849) who studied the remains found on the banks of the Salado river, both in the province of Buenos Aires. *Source* Wikimedia commons and wellcome collection

With this material he published a pamphlet in 1796 with Bru's description, which also included the translation of Cuvier's article (Fig. 2.4). These new data led Cuvier in 1804 to publish in the *Annales du Muséum d' Histoire Naturelle* a new contribution on the *Megatherium*, where he corrected some initial errors and accompanied it with the French translation of Garriga's pamphlet made by Aimée Bonpland.

The Luján *Megatherium* has been an exceptional piece, being the first fossil vertebrate in history and the largest and rarest described until then. It was the first to be mounted with the posture it would supposedly have in life for public display.

Other remains of megatherids were found in 1832 on the banks of the Salado River in one of Rosas' ranches, who gave them as a gift to the English chargé d'affaires Woodbine Parish, who sent them to the United Kingdom. These remains were studied by William Clift who later made them known (Clift 1835). Although these remains were much more incomplete, they had preserved the tail that allowed the completion of the Luján *Megatherium* in the Museo De Historia Natural in Madrid years later.

2.2 The First Geological Observations

In the different syntheses that had been carried out since the 19th and 20th centuries on the History of Argentine Geology, there was a great absence, despite his pioneering and precise observations made throughout the viceroyalty of the Río de la Plata. It is up to Ricardo Alonso to rescue Antón Zacharias Helms (1750–1803) from oblivion, who has the merit of having made detailed observations from Buenos Aires to Bolivia in 1789 (Alonso 2000; Alonso and Egenhoff 2005).

Helms was born in Hamburg in 1750 and although little is known about the first decades of his life, he was considered a *Bergfachmann* (mountain professional). It is not known what studies he carried out, but the precision of his descriptions indicates that he was an expert who knew how to recognize the different types of rock, its mineralization and its structure (Helms 1807) (Fig. 2.5).

It is known that around 1780 he worked as a professional for the Polish crown in the mines of Krakow, becoming Director of Mines in Krakow. He participated in a mission organized by Baron Thaddeus von Nordenflicht to Schemnitz (Banská Stiavnika), today Slovakia, an important mining center of the Hungarian Empire to study the amalgamation system used in silver mines. Years later, the Baron of Nordenflicht, who had graduated as a Mining Engineer from the *Bergakademie Freiberg*, at the request of the King of Spain, traveled to the Viceroyalty of the Río de la Plata with a delegation of 15 miners, to improve the benefit of the silver mines of Potosi (Alonso and Egenhoff 2008). Helms participates in this expedition and 1788 he arrives in Buenos Aires from where he begins his way north. Throughout the journey through Córdoba, Tucumán, Salta, and Jujuy he writes a diary with precise details about the existing mines, the different minerals exploited, and the characteristics of the country rock of the deposits visited. He published these observations in German in

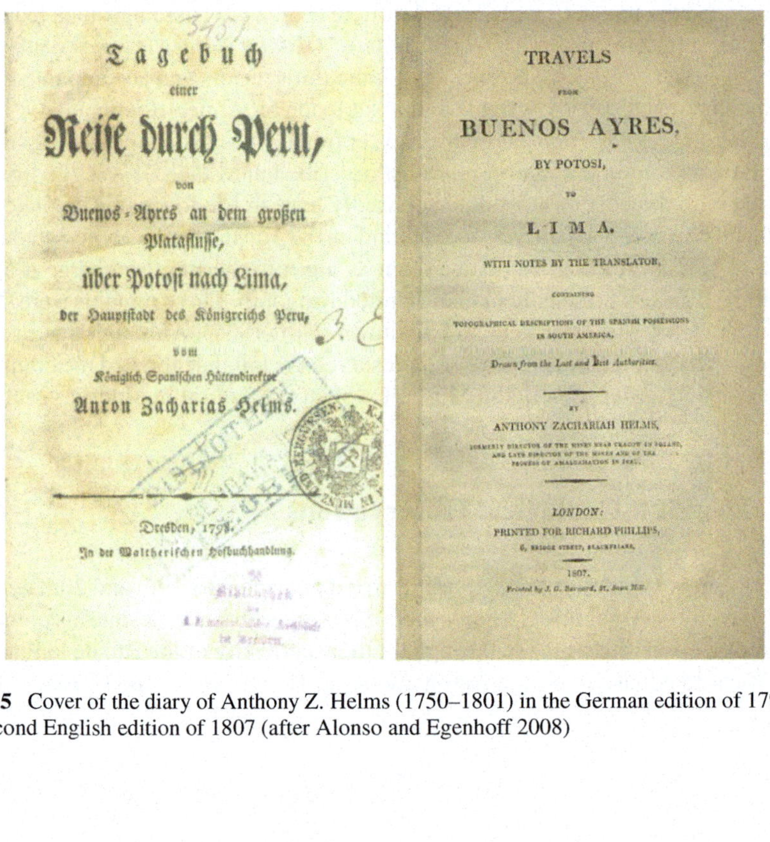
Tagebuch
einer
Reise durch Peru,
von
Buenos-Ayres an dem großen
Plataflusse,
über Potosi nach Lima,
der Hauptstadt des Königreichs Peru,
vom
Königlich Spanischen Hüttendirektor
Anton Zacharias Helms.
Dresden, 1798.
In der Waltherischen Hofbuchhandlung.

TRAVELS
FROM
BUENOS AYRES,
BY POTOSI,
TO
LIMA.
WITH NOTES BY THE TRANSLATOR,
CONTAINING
TOPOGRAPHICAL DESCRIPTIONS OF THE SPANISH POSSESSIONS
IN SOUTH AMERICA,
Drawn from the Last and Best Authorities.
BY
ANTHONY ZACHARIAH HELMS,
LONDON:
PRINTED FOR RICHARD PHILLIPS,
1807.

Fig. 2.5 Cover of the diary of Anthony Z. Helms (1750–1801) in the German edition of 1798 and the second English edition of 1807 (after Alonso and Egenhoff 2008)

1798 (Helms 1798) and years later two editions appeared in English in 1806 and 1807 and one in French in 1812. He describes the granites and gneisses of the basement of Córdoba, their metalliferous veins, precisely identifying their minerals and the alteration of feldspars to a porcelain clay (kaolin). He drew his attention to the extensive salt pans of Ambargasta and on his trip to Tucumán he distinguished a southern sector in Córdoba dominated by primitive granites, which in Tucumán give rise to partially weathered blue slates. Arriving in Salta he describes great thicknesses of ferruginous sandstones and limestones that rest on clayey slates. He recognizes these sedimentary rocks in different places in Salta, describes their course and inclination and upon reaching the *Quebrada de Humahuaca* he is impressed by the dimensions of the alluvial deposits of Volcán. In this region, he is struck by the strong deformation of the strata with abrupt changes in composition and alteration of their natural order due to faults that denote strong tectonics in this section of the Andes (Alonso and Egenhoff 2008).

Helms finally arrived in Potosí where he worked for three years, returning later to Europe where he published his travel diary, finally dying in Vienna in 1803. Helms was the first to describe precise geological observations along the road to Alto Peru, its structural characteristics, and associated mineralizations.

2.3 The First Naturalist of the Viceroyalty of the Río de la Plata

Thaddaeus Peregrinus Xaverius Haenke, better known as Tadeo Haenke (1761–1816), was one of the naturalists born in Bohemia who participated in the Malaspina Expedition, the first scientific expedition organized by the Spanish crown to its colonies between 1789 and 1794 (Fig. 2.6). This naturalist, after several years of traveling the world with that expedition, disembarked in Lima on the return trip and headed through Alto Peru to Cochabamba, where he settled until his death (Ramos and Alonso 2019).

He was one of the first naturalists who, leaving Buenos Aires, crossed the Andes at the latitude of Mendoza in 1790. He made observations at the southern end of the Sierras de Córdoba, ascending the Morro hill and the Punta de la Sierra de San Luis, places where he collected plants and minerals. He was the first to identify Tupungato as a volcano and describe the presence of hot waters in Villavicencio. He highlights that from the first foothills, evidence of earthquakes is observed in the valley, highlighting the position of the rocks, their strikes, their different types and mentions that there are many silver, copper and iron minerals, with very irregular exploitations. These mineral manifestations were those of the Paramillos of Uspallata.

His work, which comprises 7,000 handwritten pages in the Madrid archives, has been partially published by third parties and included in the "*Description of the Kingdom of Chile*" in 1942; in the "*Description of Peru*" in 1901 and in "*Tadeo Hanke, his work in the Andes and the Bolivian jungle*" in 1974. The French version of

Fig. 2.6 Tadeo Haenke as a young man (after Kuhnel, 1960) and a mature portrait preserved in the Town Hall of Cochabamba. *Source* Wikimedia and photo courtesy of Ramiro Suárez Soruco, Cochabamba

the "*Natural History of Cochabamba*" published as the second volume of "*Voyages dans l'Amérique Méridionale*" by Félix de Azara, which was included, as Azara clarifies, without permission from Haenke, given that "*the bohemian naturalist was in regions far from Europe and where it is impossible for him to have the results of his work printed*" (Walckenaer 1809).

Tadeo Haenke settled in Cochabamba in 1795 and continued receiving his salary from Spain as a naturalist for the expedition until 1810. His problems began that year when Viceroy Baltasar de Cisneros informed him from Buenos Aires that the *Suprema Junta de Sevilla* reported that he had been for 16 years "*unduly enjoying the salary assigned to him, (and that) the King our Lord D. Fernando VII has resolved... that Your Excellency immediately arrange for his return to this Peninsula*". Haenke answers the Viceroy in an illuminating letter in which he tells him how much he did for Spain in those years. That letter, the last known from Haenke preserved in the *Archivo General de la Nación* in Buenos Aires, allows us to reconstruct his contributions. Perhaps the most notable is having given the formula for the *Tarapacá* saltpeter mines dependent on Arequipa, currently Chilean, on how to benefit potassium nitrate. His method was used for decades and marked the beginning of wealth in that region because it made possible the commercial exploitation of potassium nitrate, which was essential for the manufacture of gunpowder.

Much of the material collected, both herbaria and samples of different minerals, was sent in 1799 to Buenos Aires, the head of the Viceroyalty of the Río de La Plata, in more than 40 drawers, along with an extensive report called "*Introduction to the Natural History of the province of Cochabamba and surrounding areas with their productions: Examined and described by Don Tadeo Haenke, partner of the Academies of Sciences of Vienna and Prague (1799)*". This manuscript with its original signature is preserved in the National Library of Buenos Aires.

This report presents a whole series of minerals described from different sectors of the viceroyalty of the Río de la Plata, from nitrates to copper minerals, gold, alum, etc. A fortuitous event caused his contributions, included in his substantial report to the crown of 1799, to be published in our country. In 1801, the *Telégrafo Mercantil*, the first newspaper in Buenos Aires, began to be published, and in those first years, there were numerous articles published that were extracted from Haenke's unpublished report of 1799 (Haenke 1801a, b, c, d; 1802a, b, c; 1807). These contributions are the first works published in Buenos Aires on mineralogy and its deposits. Something similar happened with the *Correo de Comercio* under the direction of Manuel Belgrano, which published three installments of its report on the "*Description of the Yuracares Indian Mountains*", the second of which appeared on May 25, 1810. These installments were extracted from Tadeo Haenke's report of 1799, but without citing his authorship. Finally, the report was published in its entirety by Paul Groussac in 1900 (Groussac 1900).

His ascent of the Misti volcano in the eighteenth century made him the first European and scientist to reach the highest summit of a mountain, before Humboldt himself. Many authors consider Tadeo Haenke the Humboldt of the southern sector of South America, due to his high academic preparation, his important explorations in the viceroyalties of Peru and the Río de la Plata and his contributions to the benefit of nitrate coming from the saltpeter mines. However, the fact that his reports remained largely unpublished during Haenke's lifetime and were only published many years after his death meant that he did not receive the recognition that his contributions would have deserved.

2.4 First Geological Study of the Río de la Plata

The first geological observations in the United Provinces of the Río de la Plata are generally associated with studies carried out by traveling naturalists, such as Alcide d'Orbigny between 1826 and 1833 and Charles Darwin between 1831 and 1836. However, a naturalist from the Río de la Plata province, Dámaso Antonio Larrañaga (1771–1848), had carried out the first geological study in 1819, a pioneering contribution that has remained almost unknown (see Larrañaga 1879) (Fig. 2.7). This Uruguayan savant has been one of the first naturalists from the South American continent who has dedicated a large part of his life to knowing the geology, flora and fauna, especially in its fossil records (Ramos 2020).

One of the main problems in delving into Larrañaga's scientific work has been that it has remained unpublished until several years after his death and is only fragmentarily known through partial publications until the publication of his *Escritos* (Writings) in 1922 and the complete paleographic version from his Natural History Diaries of 1808–1814 and 1813–1824 (Larrañaga 1922–1930, 1924, 2015, 2017).

His "*Geological memory on the formation of the Río de la Plata deduced from its fossil shells*", written in 1819, was a transcendental work for the history of Earth Sciences in the Viceroyalty of the Río de la Plata. Its merit is not only for having

Fig. 2.7 Dámaso Antonio Larrañaga (1771–1848) (Larrañaga 1930)

been the first geological study of Argentina and Uruguay, and especially of the Río de la Plata region, as Martínez (1998) highlights, but also for the precision of its observations and interpretations, both geological and paleontological, within the framework of the knowledge that was available at that time. In his prologue, he already establishes that "*In a country whose history is lost for a little more than three centuries, ... we have no other means than to resort to Geology, this modern science, which in such short days has made brilliant discoveries about the great events of our globe.*" "*Our country* (does not) *present…heights that in other countries have been so fruitful for geologists.*"

He describes that he has only found "*rocks that geologists call primordial*" on the left bank, and never alluvial deposits below them. For the first time it is recognized that "*gneiss forms the base of Montevideo, diabase is the material of its main hill*" and that "*reddish feldspar granite is what dominates the most*", to the east mica prevails more, while in the western one abounds in amphibole instead (Larrañaga 1894). These observations constitute the first description of the rocks of the basement of the Río de la Plata craton, together with his observations of "*micaceous schists… and even the calcareous stone are exempt from any organic body*".

Regarding marine sedimentary rocks, he states "*There are three only formations that I have found in these deposits, and they all seem to me to be of not very ancient origin*". It establishes its relative stratigraphy, the first corresponds to "*a calcareous gray composed of rounded quartz grains, some very shiny and diaphanous, others somewhat reddish and transparent, with an equal portion of shell fragments… The second formation is the one in which the* ***Mya labiata****, and to it belong all those* *that are currently found on both banks of the Guazú; that is, those of Ensenada, Riachuelo and Costa for San Isidro, Calera de los Padres de San Francisco and those on the opposite bank, which are those of Huérfanas, de las Vacas, and those of Santo Domingo Soriano… The third formation is that of the* ***Mytilus*** *or mussels; those that are around Montevideo and follow the coast towards Maldonado belong to it, and I do not know their limits*".

To plot his geological data, Larrañaga used the spherical chart of the Río de la Plata outlined by Felipe Bauzá, created by order of the King in 1789 and rectified in 1794 (Fig. 2.8).

Larrañaga not only classified the fossils found, but it raised important criticisms of the work of Maton (1809) published a few years earlier in the *Transactions of the Linnean Society of London* on the mollusks of the banks of the Río de la Plata. In his interpretations he makes correlations on both banks of the river, being able to relate the ravines of Ensenada de Barragán and San Isidro, with those of Santa Lucía on the opposite bank. Furthermore, he interprets how the waters of the Paraná and Uruguay rivers have made the sea retreat. The eclecticism that Larrañaga demonstrates in his *Escritos* far exceeds the dominant dogmas of his time. It is undoubtedly the first geological study carried out in the viceroyalty of the Río de La Plata, anticipating later traveling naturalists by several years.

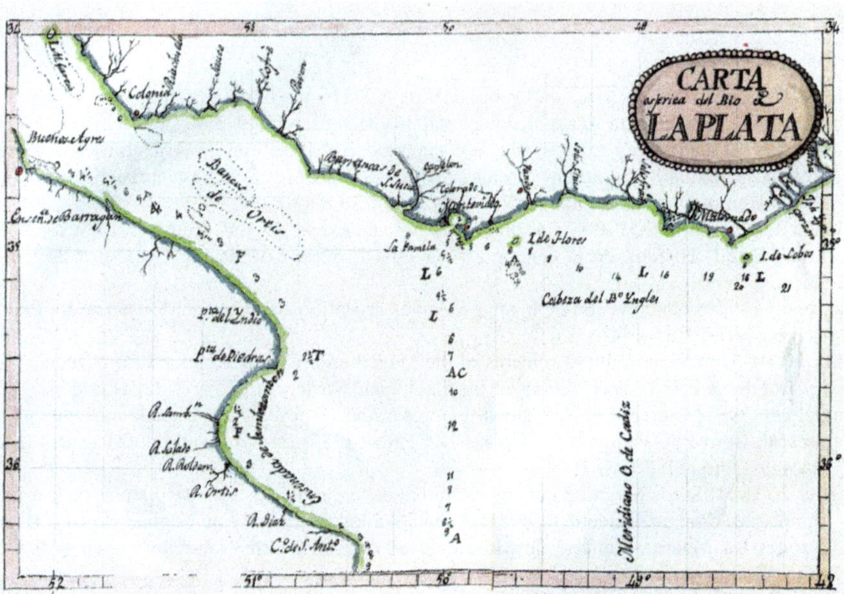

Fig. 2.8 Map made by Larrañaga that accompanies his "*Memory on the formation of the Río de la Plata deduced from its fossil shells*" of 1819 where it indicates the localities mentioned in the text and the existence of the ravines where he made the observations (Larrañaga, 1930, Plate. CXXX)

2.5 Final Considerations

We have seen that slowly, from the discovery of the first fossils with fanciful interpretations, to the specific observations of minerals and rocks, we arrived at the first geological study carried out in the Río de la Plata. The observations of Priest Dámaso Antonio Larrañaga stand out greatly if one considers the time in which they were written. In 1819, before Charles Lyell's *Principles of Geology* appeared in 1833, there was a self-taught River Plate wise man in these lands who alone used geology to unravel our past. This self-taught man compared his findings with those published in the London Transactions a few years earlier, which he criticized with solid taxonomic arguments. As Martínez (1998) points out, the rigor and updating that Larrañaga had in commenting and disagreeing with a work published by Maton (1809) is surprising. He draws attention to his concern to update himself and obtain these publications in an environment convulsed by revolutions and independence struggles, in complete intellectual solitude. Priest Larrañaga undoubtedly has a place among the pioneers of Río de la Plata Geology due to the value of his contributions at a time when this science was not known, nor the usefulness that it gave us years later in its different fields of application.

References

Alonso RN (2000) Las observaciones geológicas de A.Z. Helms (1789) en el interior de la República Argentina. Boletín de la Academia Nacional de Ciencias 64: 349–353, Córdoba

Alonso RN, Egenhoff SO (2005) The first geological observations in Argentina and southern Bolivia: The diary of Anthony Zacharias Helms (1788/1789). Neues Jahrbuch für Geologie und Paläontologie, Abhandlungen 236(1/2): 1–17, Stuttgart

Alonso RN, Egenhoff SO (2008) Las observaciones geológicas de A. Z. Helms en 1789. In: Aceñolaza FG (ed) Historia de la Geología Argentina I, Serie Correlación Geológica 24: 21–34, Tucumán

Cabrera A (1930) Una revisión de los mastodontes argentinos. Revista del Museo de La Plata 32 (tercera serie) 8:61–144

Clift W (1835) Some account of remains of the Megatherium sent to England from Buenos Aires by Woodbine Parish. Transactions of the Geological Society S2(3): 437–450, London

Cuvier G (1796) Notice sur le squelette d'une très grande espèce de quadrupède inconnue jusqu'a present, trouvé au Paraguay et dépose au Cabinet d'Histoire Naturelle de Madrid. Magasin Encyclopédique I: 303–310

Cuvier G (1804) Sur le Megatherium, autre animal de la famille des Paresseux, mais de la taille du rhinocéros, dont un squelette fossile presque complet est conservé au Cabinet Royal d'Histoire Naturelle à Madrid. Annales du Muséum National d'Histoire Naturelle 5: 376–400. (With translation by Aimé Bonpland from Bru's description)

Falkner T (1774) A Description of Patagonia and the adjoining parts of South America. Hereford, 144 p, London

Garriga J (ed) (1796) Descripción de un quadrúpedo muy corpulento y raro, que se conserva en el Real Gabinete de Historia Natural de Madrid. Imprenta Viuda de Ibarra, 16 p, Madrid

Groussac P (1900) Noticia de la vida y trabajos científicos de Tadeo Haenke. Anales de la Biblioteca, Tomo Primero: 17–193, Buenos Aires

Gutiérrez JM (1835) Megatherium (Animal desconocido). Museo Americano, Periódico ilustrado, Imprenta de la Libertad, pp 108–110, Buenos Aires
Gutiérrez JM (1866) La paleontología en las colonias españolas a mediados del siglo XVIII. Revista De Buenos Aires 11:100–114
Haenke T (1801a) Sal de Inglaterra, sal amarga en abundancia en estas provincias. Telégrafo Mercantil I, p 237. Buenos Aires
Haenke T (1801b) Nitro puro. Telégrafo Mercantil II (18) (4-IX-1801), p. 126–128, Buenos Aires
Haenke T (1801c) Cardenillo nativo recogido por los indios en los contornos de la laguna de Oruro. Telégrafo Mercantil I (33) (22-VII-1801), p 257, Buenos Aires
Haenke T (1801d) Oro pimiente del Perú. Telégrafo Mercantil I, p 34 (25-VII-1801), Buenos Aires
Haenke T (1802a) Sustancia mineral: Alumbre nativo o Cachina blanca. Telégrafo Mercantil V, pp1–3 (3-IX-1802), Buenos Aires
Haenke T (1802b) Vitriolo de Cobre, Vitriolo azul o Vitriolo de Chipre. Telégrafo Mercantil III (7) (14-II-1802), p 103, Buenos Aires
Haenke T (1802c) Minería. Mina de plata en barra, sin otro beneficio que cortarla a cincel llamada Mina Hedionda que se halla en el Partido de Lipes. Telégrafo Mercantil IV (13) (25-VII-1802), pp 225–226, Buenos Aires
Helms AZ (1798) Tagebusch einer Reise durch Peru, von Buenos Ayres an dem grossen Platastuffe, über Potosi nach Lima, der Haptstadt des Ronigreiches Peru. Dresden, Waltherrischen Losbuchhandlung, 300 p, Dresden
Helms AZ (1807) Travels from Buenos Ayres, by Potosi, to Lima. With notes by the translator, containing topographical descriptions of the Spanish possessions in South America. Printed for Richard Phillips, 92 p, London
Kühnel J (1960) Thaddeäus Haenke: Leben und Wirken eines Forschers, Lerche, 276 p, Munich
Larrañaga DA (1879) Memoria geológica sobre la formación del Río de la Plata deducida de sus conchas fósiles. Escrita en 1819. Revista de Ciencias, Artes y Letras, Boletín de las Universidades, Colegios y Escuelas de la República Argentina 3: 191–202, Buenos Aires
Larrañaga DA (1894) Memoria geológica sobre la formación del Río de La Plata deducida de sus conchas fósiles. Anales del Museo Nacional de Montevideo 5(1):3–11, Montevideo
Larrañaga DA (1922–1930) Escritos de Don Dámaso Antonio Larrañaga. Instituto Histórico y Geográfico del Uruguay, Imprenta Nacional, t. 1: 486 p. (1922); t. 2: 512 p. (1923); t. 3: 303 p. (1924); Atlas 1: 132 láms. (1927); Atlas 2: 131 láms. (1930), Montevideo
Larrañaga DA (1924) Reconquista de Buenos Aires. In Escritos de Don Dámaso Antonio Larrañaga. Instituto Histórico y Geográfico Del Uruguay, Imprenta Nacional t 3:107–110
Larrañaga DA (2015) Diario de Historia Natural 1813–1824. Biblioteca Artigas, Colección de Clásicos Uruguayos, Volumen fuera de serie 1, 745 p, Montevideo
Larrañaga DA (2017) Diario de Historia Natural 1808–1814. Biblioteca Artigas, Colección de Clásicos Uruguayos, Volumen fuera de serie 2, 160 p, Montevideo
Martínez S (1998) Los comienzos de la geología y paleontología en el Uruguay: la memoria sobre la reciente formación del Rio de la Plata, deducida de sus conchas fósiles por D.A. Larrañaga. Sociedad Uruguaya de Geología, 2° Congreso Uruguayo de Geología, Actas: 191–193, Montevideo
Maton WG (1809) Description of seven new species of testacea. Transactions of the Linnean Society of London 10: 325–332, tab. XXIV, London
Orche E, Amaré MP (2019a) El megaterio de Luján: una importante aportación hispana a la ciencia mundial: 1. Hallazgo, excavación y envío a España. In: Mansilla Plaza L, Mata Perelló JM (eds) El patrimonio geológico y minero. Identidad y motor de desarrollo, Instituto Geológico y Minero de España, Cuadernos del Museo Geominero 29: 531–546, Madrid
Orche E, Amaré MP (2019b) El megaterio de Luján: una importante aportación hispana a la ciencia mundial: 2. Estudio, montaje y repercusiones en el mundo científico. In: Mansilla Plaza L, Mata Perelló JM (eds) El patrimonio geológico y minero. Identidad y motor de desarrollo, Instituto Geológico y Minero de España, Cuadernos del Museo Geominero 29: 547–564, Madrid

Orche E, Amaré MP (2019c) El megaterio de Luján después del montaje de Juan Bautista Bru con especial atención a su iconografía. Siglos XVIII-XIX. In: Mansilla Plaza L, Mata Perelló JM (eds) El patrimonio geológico y minero. Identidad y motor de desarrollo, Instituto Geológico y Minero de España, Cuadernos del Museo Geominero 29: 565–582, Madrid

Pasquali RC, Tonni EP (2008) Los hallazgos de mamíferos fósiles durante el período colonial en el actual territorio de la Argentina. In: Aceñolaza FG (ed) Historia de la Geología Argentina. Serie Correlación Geológica 24: 35–42, Tucumán

Ramos VA (2020) El primer estudio geológico de las Provincias Unidas del Río de la Plata: Los aportes de Dámaso A. Larrañaga. Revista Facultad de Ciencias Exactas, Físicas y Naturales 7 (suplem. 1): 53–62, Córdoba

Ramos VA, Alonso RN (2019) Tadeo Haenke: primer naturalista del Virreinato del Río de la Plata. Anales de la Academia Nacional de Ciencias Exactas, Físicas y Naturales 70 (2018): 117–146, Buenos Aires

Tonni E, Pascuali R, Laza J (2007) Buscadores de fósiles. Los protagonistas de la paleontología de los vertebrados en la Argentina. Jorge Sarmiento Ed., 109 p, Buenos Aires

Trelles MR (1882) El padre Fray Manuel de Torres. Revista De La Biblioteca Pública 4:439–448

Walckenaer CA (1809) Voyages dans l'Amérique Méridionale par Don Félix de Azara. Tomo II:389–541

Chapter 3
The Traveling Naturalists

Abtract In the first decades of the ninteenth century, the country was visited by numerous foreign naturalists who travelled through the country on reconnaissance trips. Among them, Alcide d'Orbigny and Charles Darwin stood out for their pioneering observations and descriptions. Alcide d'Orbigny was sent by the *Muséum d'Histoire Naturelle* in Paris and after several years of stay in the southern sector of South America, he compiled a colossal work published in Paris. His contributions were mainly devoted to the Cenozoic deposits and the study of their abundant fauna. Charles Darwin carried out his studies along the Atlantic coast and made a complete crossing of the Argentine-Chilean Andes that was made known to the Geological Society of London. His monumental contribution in his book Geological Observations in South America remains a milestone in the understanding of the geological history of this region. These observations were fundamental for the development of his theory on the evolution of species, as he acknowledges was based on what he learned in those years of study in the visited region.

Keywords Foreign travelers · Cenozoic stratigraphy · Earthquakes · Uplifts · Fossil forest

The first decades of the nineteenth century were characterized by geological explorations carried out by notable European naturalists who made important observations, for example, Alcide d'Orbigny and Charles Darwin. However, there were fragmentary observations produced by other travelers such as Schmidtmeyer between 1820 and 1821, Miers in 1824, Temple in 1826, Domeyko in 1838, among many others. These travelers provided information that served as background and aroused the interest of new generations.

V. A. Ramos, *History of the Geological Sciences in Argentina*, The Latin American Studies Book Series, https://doi.org/10.1007/978-3-031-83139-3_3

3.1 Alcide d'Orbigny (1802–1857)

This young French naturalist was commissioned at the age of 24 by the *Muséum d'Histoire Naturelle* in Paris to undertake a scientific trip through South America that lasted eight years, between 1826 and 1833 (Fig. 3.1). He was in Buenos Aires several times. He studied the Paraná ravines, the Eastern Cordillera and the Altiplano of Bolivia, producing one of the first structural sections of the Andes. The scientific result of his expedition would be reflected in the nine volumes of his work *Voyage dans l'Amérique méridionale* that he would publish years later, dedicating a volume to Geology (d'Orbigny 1842).

His most important geological contributions were the Cenozoic stratigraphy of the Paraná ravines, which he studied during a year of stay in Corrientes, along with his descriptions of the Pampean deposits in the province of Buenos Aires and the sections exposed in the near Carmen de Patagones, where he stayed for nine months in 1829. On that occasion he analyzed the Tertiary sections, studying in detail the Cenozoic profiles of the coastal region. He coined the term "*Tertiaire Patagonien*" for the marine deposits that from Río Negro "*occupy almost all of Patagonia, on its eastern side, to the Strait of Magellan*". His studies included detailed documentation of Cenozoic marine fossils describing numerous taxa (d'Orbigny 1842).

In his catastrophic conception, he interpreted that the Tertiary seas retreated before the rise of the Andes Mountains (Fig. 3.2) and that the Pampean deposits were of marine origin. His contributions to paleontology and micropaleontology have been fundamental, as highlighted by Camacho (2001) and Tonni and Pasquali (2006).

He carried out important studies in Bolivia, presenting the first geological section of the Andes Mountains of Bolivia from the Pacific to the Eastern Cordillera, where

Fig. 3.1 Alcide d'Orbigny (1802–1857). *Source* d'Orbigny (1836) and wikimedia

Fig. 3.2 Geological section of the Bolivian Andes constructed by Alcide d'Orbigny (see d'Orbigny 1842 in Forbes 1861)

the Western Cordillera, the Altiplano and the Cordillera Real can be identified, paying special attention to the Illimani in the surroundings of La Paz (d'Orbigny 1842). He highlights in the profile the Corocoro mines, a famous sedimentary deposit of native copper.

3.2 Charles Darwin (1809–1882)

It is well established in the scientific community that Darwin's theory of the evolution of species is a milestone in the understanding of Life, and that this theory began to develop in Darwin's mind during his research and observations in South America, those that were consolidated after his return to Great Britain. However, his important ideas on the formation of mountains and his hypothesis on the mechanisms of subsidence of sedimentary basins are not so well known. These hypotheses were developed in his crossing of the Andes Mountains and in his observation of the characteristics of the Mesozoic sequences of Puente del Inca (Ramos 2009).

It is well documented that he considered himself a geologist, initially receiving a strong influence from the uniformitarian ideas of Charles Lyell. It is also important to note that after his Beagle voyage around the world, he contributed to improving Lyell's understanding and exerted a strong influence on his ideas. When we read that in 1838, he chose to characterize himself as "*I a geologist*" (Herbert, 2005), we should not be surprised that his monumental contribution to Geological Observations in South America remains a milestone in the understanding of geological history of this region (Darwin 1838, 1846).

At the age of 23, he began his trip around the world, spending more than two years in Argentina. He made fundamental observations about the relationship between volcanoes and earthquakes; the uplift of the Andes Mountains, being the first to identify it as an active orogen; the geology of the Tertiary ravines of the Atlantic coast; the glacial origin of the erratic blocks, among many others. His paleontological observations on the Atlantic coast allowed him to analyze the important transformations of past and present life that gave rise to his theory of evolution. Darwin's geological contributions in Argentina have been gathered and analyzed through a comprehensive description in a special issue of the Journal of the Argentine Geological Association (Aguirre Urreta et al. 2009) (Fig. 3.3).

Fig. 3.3 Darwin journeys in southern South America (Aguirre Urreta et al. 2009)

Darwin's geological contributions to our country were concentrated along the Atlantic coast and his landings that allowed him to cross the province between Patagones and Buenos Aires; his journey through the Paraná valley to Santa Fe and Bajada, as the city of Paraná was known in those years, and his attempts to go up the Santa Cruz River to its sources. When he was in Chile he crossed the Andes Mountains from Valparaíso to Mendoza, making his first detailed geological description and presenting two pioneering structural sections. In his expeditions, he coined terms such as the Pampean Formation, which he interpreted as the result of a slow sedimentation of a river system in contrast to the catastrophic ideas of d'Orbigny who associated it with a marine ingression produced by the sudden uplift of the Andes (Tonni and Pasquali 2006).

In Patagonia, he made the first description of the Jurassic porphyry series, today known as the Chon Aike Formation, along its coastal outcrops (Ramos 2011, 2022). He is highly surprised by the extension and extensive development of the Gravel Formation, which he calls the Patagonian Shingle Formation (Fig. 3.4), known today as the Patagonian Gravels. He considers these outcrops to be the largest gravel plain of all those known at that time and attributes an origin associated with the great transgression that occurred during the "*Great Universal Flood*" (Martínez et al. 2009).

However, where he makes fundamental contributions to Geology is in his description of the Andes coming from Chile (Giambiagi et al. 2009; Ramos 2009). He crosses through the Portillo pass to Mendoza and returns through the La Cumbre pass to Valparaíso (Fig. 3.5). Among his most notable observations is finding Jurassic marine deposits at more than 3,000 m high, which he links to the action of earthquakes. His interpretation was based on his observations of the upheavals produced by the Concepción earthquake, where he describes current marine deposits that in a few hours rose several meters from the seabed. Another milestone in that crossing is the discovery of what is today known as "*Darwin's Forest*", where he observed

Fig. 3.4 Charles Darwin in his youth and the typical Patagonian Gravels of the central plateau of Chubut. *Source* Wikimedia commons and Martínez et al. (2009)

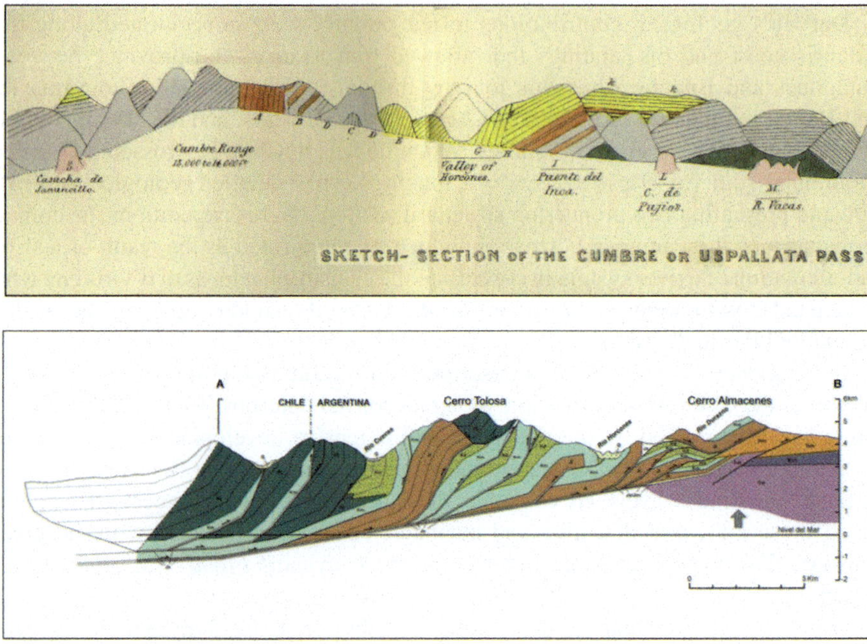

Fig. 3.5 Structural section surveyed by Darwin in 1835 (Darwin 1846) compared to the current structure according to Cegarra (1994)

dozens of trees in a growing position, silicified and covered by volcanic ash. It is also of interest to mention his geological observations in the Malvinas Islands.

3.3 Final Considerations

There were numerous naturalists who crossed the country making interesting geological observations. However, these were punctual or fragmentary, most of them obtained from rapid journeys through the national territory. This situation contrasts with the observations and studies carried out by d'Orbigny and Darwin, who have made the most important geological contributions in the first half of the nineteenth century. The success obtained by these two young travelers is that their observations were the result of several years of studies that allowed them to obtain a more appropriate regional framework in which to insert their descriptions, new hypotheses or proposals. Argentine Geology owes them both strong recognition for their contributions.

References

Aguirre Urreta B, Griffin M, Ramos VA (2009) Editores Invitados. Darwin En La Argentina, Revista De La Asociación Geológica Argentina 64(1):1–180

Camacho HH (2001) Las ciencias geológicas en la Argentina, hasta 1939. Saber y Tiempo 12:177–220

Cegarra MI (1994) La faja plegada y corrida de la cordillera principal entre Puente del Inca y Las Cuevas, Alta Cordillera de Mendoza, Argentina. Universidad de Buenos Aires, Facultad de Ciencias Exactas y Naturales, unpublished PhD Thesis, 297 p, Buenos Aires

Darwin CR (1838) On the connexion of certain volcanic phaenomena, and on the formation of mountain-chains and volcanos, as the effects of continental elevations. Proc Geol Soc 2:654–660, London

Darwin CR (1846) Geological observations on South America. Being the third part of the geology of the voyage of the Beagle, under the command of Capt. Fitzroy, R.N. during the years 1832–1836. Smith Elder and Co., 280 p, London

d'Orbigny A (1836) Voyage pittoresque dans les deux Amériques. Chez L. Tenré, Libraire-Éditeur, 864 p, Paris

d'Orbigny A (1842) Voyage dans l'Amerique Meridionale le Bresil, la Republique Orientale de l'Uruguay, la Republique Argentine, la Patagonie, la Republique du Chili, la Republique de Bolivia, la Republique du Perou, Execute Pendant les Annees 1826, 1827, 1828, 1829, 1830, 1831, 1832 et 1833. Tome Troisiene, Geologie. 290 p, Paris

Forbes D (1861) On the geology of Bolivia and Southern Peru. Q J Geol Soc 17:7–62, 3 plates, London

Giambiagi L, Tunik M, Ramos VA, Godoy E (2009) La Alta Cordillera de los Andes del centro de Argentina y Chile a lo largo de la transecta del Paso Piuquenes-Cordón del Portillo: Las observaciones pioneras de Darwin comparadas con la geología moderna. Revista De La Asociación Geológica Argentina 64(1):45–54

Herbert S (2005) Charles Darwin, Geologist. Cornell University Press, 485 p, Ithaca

Martínez OA, Rabassa J, Coronato A (2009) Charles Darwin and the first scientific observations on the Patagonian Shingle Formation (Rodados Patagónicos). Revista De La Asociación Geológica Argentina 64(1):90–100

Ramos VA (2009) Darwin en Puente del Inca: Observaciones sobre la formación de Puente del Inca y el levantamiento de montañas. Revista De La Asociación Geológica Argentina 64(1):170–179

Ramos VA (2011) Doscientos años de Ciencias de la Tierra en Argentina. Revista De La Asociación Geológica Argentina 68(3):392–406

Ramos VA (2022) La evolución del conocimiento geológico en la provincia de Chubut. In: Giacosa R (ed) Geología y Recursos Naturales de la Provincia del Chubut, 21° Congreso Geológico Argentino, Relatorio: 3–33, Puerto Madryn

Tonni EP, Pasquali RC (2006) Alcide d'Orbigny in Argentina: the beginning of stratigraphical studies and theories on the origin of the "pampean sediments." Earth Sci Hist 25(2):215–222

Chapter 4
Geology After Caseros

Abstract The national organization that followed the Battle of Caseros in 1852 led to the formation of the Argentine Confederation and brought great changes in geological studies. The opening of Natural History Museums in Paraná and Buenos Aires, both in charge of foreign naturalists, intensified the surveys throughout the country. In Paraná, the Belgian Marbais Du Graty stood out, a naturalist who promoted the mineral wealth of the Confederation at the universal exhibition in Paris. His investigations were continued by the French mining engineer Auguste Bravard who carried out important studies on the banks of the Paraná River and died in an earthquake in 1861 in Mendoza, where he had been commissioned to survey its mineral potential. The French naturalist Victor Martin de Moussy, traveled more than twenty thousand kilometers through the country, carrying out a complete *Description géographique et statistique de la Confédération Argentine*. The government's objectives were to publicize the country's mining potential and to encourage the settlement of European immigrants, so they actively participated in international exhibitions in Paris with important published works and samples of minerals, rocks, and fossils. When the government moved to Buenos Aires, another German-born naturalist, Hermann Burmeister, was hired as director of the Public Museum of Buenos Aires. This naturalist, in addition to participating with samples and publications in the Universal Exhibitions of Paris and Venice, began to edit the annals of the Museum, which quickly became a scientific journal widely recognized for the quality of its contributions. This period culminated with the studies of Francis Ignacio Rickard, an Irish mining engineer, who did meritorious work recognizing the mining potential of the central region of Argentina and published a study of wide international circulation on mining resources.

Keywords New museums · International exhibitions · Resource inventories · Mining potential · Physical description

The national organization that produced the Battle of Caseros on February 3, 1852, which brought about the end of the Rosas dictatorship, gave a new impetus to geology. Very little had happened in this field after the contributions of traveling naturalists.

V. A. Ramos, *History of the Geological Sciences in Argentina*, The Latin American Studies Book Series, https://doi.org/10.1007/978-3-031-83139-3_4

The Public Museum where Carlos Ferraris had developed an incipient Natural History cabinet for years had almost ceased to exist in 1840 with his return to Italy.

The profound change in geological studies that the Argentine Confederation brought in 1852 has been analyzed in detail by Florencio G. Aceñolaza (2008). The national territory had a population located in a strip north of the province of Buenos Aires and in the Mesopotamian region, and a corridor between Buenos Aires, Córdoba and the northern provinces and Greater Cuyo. Extensive regions of Patagonia were sparsely populated, and the Gran Chaco was "Indian territory" as Martín de Moussy identified it. The brand-new Argentine Confederation, which had the city of Paraná as its capital, had conflicts with many provinces, such as Buenos Aires, which was split until the battle of Pavón in 1861, when the constitution was reformed it was reintegrated into what has since been called the Argentina Republic.

4.1 The Paraná National Museum

Justo J. de Urquiza was responsible for promoting studies and surveys of the national territory, who attracted to the country prominent naturalists such as the Belgian Alfred Marbais Du Graty, the Frenchman Auguste Bravard, the doctor and geologist Víctor Martín de Moussy. and the German naturalist Hermann (Germán) Burmeister. All of them contributed to the geological knowledge of the country and to making known abroad the potential of natural resources and especially the mining resources, with their publications in famous European media.

4.1.1 Alfred Marbais Du Graty (1823–1891)

The first to offer his services to Governor Urquiza was the Belgian Marbais Du Graty (Fig. 4.1), a naturalist and geographer who graduated from the *École Militaire de Bruxelles* who from 1850 served as commander in the army of Entre Ríos participating in the battle of Caseros in 1852. Once the Urquiza government was consolidated, it promoted the creation of the National Museum of Paraná in 1854, which in its short history became a tireless promoter of knowledge of the mining riches of the Confederation. As soon as the museum was created, its director had the responsibility of guaranteeing the participation of the Argentine Confederation in the universal exhibition in Paris.

Du Graty spent long hours analyzing and describing the samples received from the different mining districts and sent the collection of minerals to France along with a "*Memory on the mining wealth of the Argentine Confederation*", where he described the interesting mining prospects that the country had. His samples of silver galena weighed "*300–400 pounds*" and received an honorable mention at the exhibition (Du Graty 1858).

Fig. 4.1 Portrait of Alfred Marbais Du Graty and facsimile of his main work (Du Graty 1858)

The museum did not have a physical reality and was used as an agent to disseminate abroad the image of an Argentina regenerated by new institutions and endowed with infinite riches (Podgorny 1997). Graty's main contribution has been his book on *La Confédération Argentine* published in 1858 where he strongly promotes the country's mineral wealth (Fig. 4.2). His map objectively shows the extent under the control of the Confederation and the territories that were under the dominion of different indigenous peoples. In Chap. 2, the book makes an exhaustive description of the main mining districts and their characteristics that allow for the first time to have an overall vision of the mineral resources of the Confederation. His main thesis was that the eastern slope of the Andes Mountains was at least as rich as its western slope.

In 1860, being a frontier commander in the Chaco, Du Graty requested leave from President Derqui and retired to Paraguay, traveling to Europe after two years and never returning. In his native land, he had extensive cultural and management activity, collaborating with the Argentine government, already as Baron Du Graty upon the death of his father. He died in Brussels in 1891.

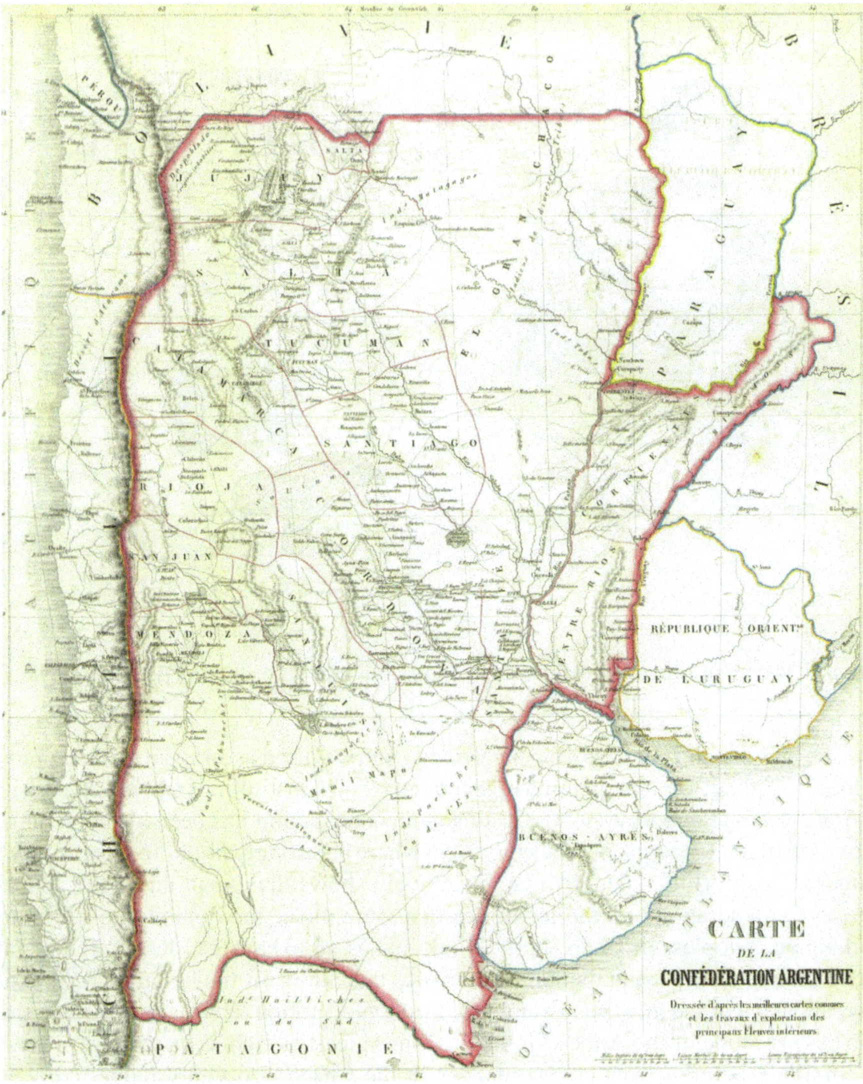

Fig. 4.2 Map of the Argentine Confederation compiled by Alfred Marbais Du Graty that illustrates his book. Note that Buenos Aires was not part of the Confederation, as well as the limits of the Gran Chaco and the limits of the provinces were very different from the current ones. He clarifies that much of the territory is in the hands of "indiens" from different tribes or is uninhabited. Bolivia still had access to the sea with its port of Cobija. The current Argentine Puna is listed as "Depopulated" (Du Graty 1858)

4.1.2 Auguste Bravard (1803–1861)

Bravard was a mining engineer who graduated in 1822 from the *École des Mines de Saint-Étienne* (Fig. 4.3), one of the most prestigious mining schools in France, standing out since graduation for his interest in geology. He worked for a few years in the exploitation of antimony and coal mines but quickly became interested in geology and exploring sedimentary sections in search of fossils. He produced several monographs with the results of his explorations in Auvergne and gathered collections of fossils that caught the attention of the British Museum in London, who offered him a fortune for them, but out of patriotism, he gave them at a lower price to the Jardin des Plants. Reading works such as those by Alcide d'Orbigny from 1826, Charles Darwin from 1835 and Woodbine Parish from 1838, led him to dream of exploring Argentina. Due to the political problems, he faced due to his republican and democratic ideas, which temporarily led him to prison, he decided to sell a collection of fossil vertebrates to the British Museum in 1852 to finance his trip to Buenos Aires (Podgorny 2001). Before embarking, he visits Charles Darwin in London, who advises him with his observations and recommendations. He finally arrived in Buenos Aires attracted by its fossil deposits in 1852, where he took up residence (Mege 1886).

He studied the ravines of the Río de la Plata and the Riachuelo where he extracted numerous remains of fossil vertebrates. In 1854 he sold again to the British Museum a collection made since his arrival in 1852. In 1856 he offered a new collection to the *Muséum d'Histoire Naturelle*, which he did not sell according to Podgorny (2001), and which was finally acquired for the Public Museum of Buenos Aires by Burmeister after his death (Castello 2014). He was the first to describe the discovery of marine

Fig. 4.3 Auguste Bravard (charcoal drawing by Cándido Villalobos de 1924). *Source* Biblioteca Digital F.C.E.N

fossils in the Belgrano ravines of the city of Buenos Aires in 1858, which were considered a classic locality of the Late Pleistocene ingression of the Belgranense sea. These deposits were exploited to produce lime and were known as the *Calera de los Franciscanos* (Franciscan lime pit) (Silveyra and Padula 2014).

The government hired him to travel to the Indian border and carry out geological observations in Sierra de la Ventana (Fig. 4.4). He reached the town of Bahía Blanca and produced the first geological map in Argentina. His "*Geological and topographic map of the surroundings of Bahía Blanca*" published in 1857 illustrates the general characteristics from the Sierras de la Ventana to the Atlantic coast, with special attention to its Cenozoic deposits (Bravard 1857a). He illustrates various stratigraphic profiles from Sierra de la Ventana to the coast, and sections of the Cenozoic deposits from Bahía Blanca to Monte Hermoso. On the geological map, he recognizes the different units, from metamorphic rocks (sic) in the Sierras de la Ventana, to the Pampean Quaternary deposits that he attributes to the results of atmospheric and terrestrial causes, discarding a marine or lacustrine origin as had been maintained by d' Orbigny and Darwin, since no alluvium features are observed (Bravard 1857b). Shortly afterward, Darwin wrote to Lyell saying that he had received two works from Bravard, *Geological Observations* (1857) and the *Monograph of the Tertiary Marine Terrains* (1858), highlighting that it seemed to him a "*strange doctrine that Pampean deposit is subaerial*" in his letter of February 14–15 in 1860. Various authors criticized Bravard's interpretation, such as Burmeister who in 1876 considered it an extravagant idea of the self-taught Bravard to assume an aeolian origin for the Pampean silts (Toledo 2022, p. 7), a hypothesis widely accepted today.

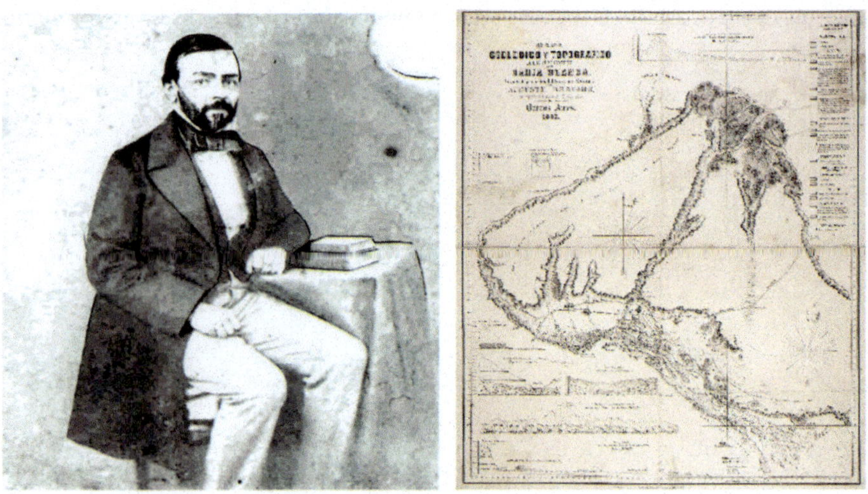

Fig. 4.4 Auguste Bravard and his geological map of the surroundings of Bahía Blanca. This first geological map made and published in the country contains precise stratigraphic sections of the Cenozoic deposits of the coast from Punta Alta to Monte Hermoso, as well as a geological section between Sierra de la Ventana and Puerto Nuevo (Borrello 1970). *Source* Museo de La Plata archives and Bravard (1857b)

Du Graty's resignation in 1857 from the direction of the National Museum provided the opportunity for Bravard to settle in Paraná starting in 1858, where he was hired as Inspector General of Mines and Director of the Museum. This allowed him to make several profiles in the Paraná ravines and publish his classic work on the marine deposits around the city of Paraná (Bravard 1858). In these studies, he demonstrated a good knowledge of the principles of Lyell's uniformitarianism by making an accurate reconstruction of the environment where he found abundant fossil remains (Tonni et al. 2006, 2008). Unlike Darwin and d'Orbigny, Bravard claimed that these fossil bones had not been transported and that the animals from which they came had lived and died in the place where their skeletons were found. This view has since been adopted by subsequent studies (Mege 1886).

In 1861, at the request of the Mendoza government, he was sent to explore this region, to evaluate the mining potential of the provinces of Mendoza and San Juan where he died during the great Mendoza earthquake.

His observations from previous days, especially those related to underground noises and marked variations in atmospheric pressure, made him alert about the possibility of an earthquake. He had interesting contacts with the local people, describing the unrest that existed in the population, who told him a series of notable changes and extraordinary noises that seemed to predict an earthquake, as he wrote in a letter to a friend in Paraná. Although he was not certain of the relationship between the barometric changes detected with the possibility of an earthquake, he decided to anticipate his plans and leave for Chile (Loos 1907; Fossa Mancini 1939). A Gibraltarian friend and colleague, Don Federico Benelische, recounted in 1887 the details of the earthquake and how Bravard died (Benelische 1887). Several days after the earthquake he found his friend in the ruins of the Cactus Hotel where he was staying, finding him buried in the rubble, crushed under a pool table where he sought refuge (Alonso 2016). This notable French geologist died in Mendoza on Ash Wednesday on March 20, 1861.

Upon his death, the collection of fossils that Bravard had been trying to sell to the *Museum d'Historie Naturelle* in Paris was deposited in a location at the *Café de Paris* in Buenos Aires. Burmeister, who claimed that the collection described in the Bravard Catalog was the best and most complete in the country in 1868, arranged for the government to acquire it for the Public Museum of Buenos Aires. The purchase of this collection was also based on the fact that he "*lost his life… carrying out instructions he had received from the National Government*" (Castello 2014).

4.2 The Physical Description of the Argentine Confederation

The Urquiza government had as one of its priorities to make known, mainly in Europe, the potential of the young nation's resources as a means of attracting interested parties for its development. To do this, he convened, hired and promoted the recognition of the territory by European scientific experts. We have seen Du Graty's description of *La Confédération Argentine* of 1858, which was followed by more complete works such as those of Martín de Moussy, and those of Burmeister and Rickard. These works in a certain way updated the pioneering work of the English diplomat Woodbine Parish (1796–1882), who had been a British representative in Buenos Aires between 1825 and 1832. Upon his return to London, he published "*Buenos Ayres and the Provinces of the Rio de la Plata from their discovery and conquest by the Spaniards to the establishment of their political Independence. With some account of their present state... and a description of the geology and fossil monsters of the pampas*" (Parish 1838), where he makes a detailed description of their physical characteristics.

The objective of the works promoted by Urquiza was to make known abroad the realities and benefits of the country to encourage European migration and encourage the settlement of the population in almost uninhabited regions despite their enormous resources. In a certain way, they tried to emulate the monumental work of the French naturalist Claudio Gay (1800–1873), hired by the Chilean government, who wrote the "*Physical and Political History of Chile*" in thirty volumes (Gay 1844–1871).

4.2.1 Victor Martin de Moussy (1810–1869)

This French naturalist who had studied Medicine in Strasbourg and Paris had a strong attraction to Natural Sciences (Fig. 4.5). He came to South America in 1841, emulating the steps of D'Orbigny and Bonpland, first to Brazil and then settling as a doctor in Uruguay (Leguizamón Pondal 1970). In 1855 he offered his services to General Urquiza who summoned him to carry out a comprehensive survey of Argentine territory.

He had the opportunity during his stay in the country to connect with Aimé Bonpland, every time this naturalist came from Santa Ana and exchange ideas and information.

With this purpose he traveled more than twenty thousand kilometers for four years, visiting the most remote places that allowed him to survey a large part of the country. His complete work *Description géographique et statistique de la Confédération Argentine* gives a precise description of its orography, hydrography, the main roads and the extent of the territories controlled by the native peoples with the name of the different tribes.

The work of Victor Martín de Moussy is the first comprehensive survey of Argentina. Although it was not written in its conclusions, it showed the Argentine oligarchy that it had a country with enormous potential, largely uninhabited by white, civilized and European men, with large areas of fertile and productive land in the hands of indigenous peoples.

Martin de Moussy traveled to Paris in 1859 to begin editing his work. The first two volumes appeared in 1860, the third in 1864, and a complete geographical atlas with which the contribution culminated in 1869 (Fig. 4.6).

Fig. 4.5 Victor Martín de Moussy (1810–1869). *Source* Wikipedia

It is interesting to note that, in his characterization of the orographic regions, in addition to a complete geographical description, he pays attention to the different aspects of Natural History. His knowledge allowed him to discuss with Aimé Bonpland and correspond with Alcide d'Orbigny on different aspects. He pays special attention to the geological characteristics describing the metamorphic rocks and granites of Tandil and the sedimentary rocks of the Ventana and Curamalal mountain ranges. He characterizes the petrography of the mountain ranges of Córdoba, San Luis and Pie de Palo, which to the north end of the Llanos de la Rioja. He makes a complete description of the Andes Mountains, its eastern foothills such as Payún and Cerro Nevado, and its most important known mineralizations. Although on some occasions he describes what he saw on his journey, most of the geological descriptions are based on previous authors such as Darwin, d'Orbigny, Parchape, etc. However, it is a great effort to make known the mining manifestations and their associated rocks (Martin de Moussy 1860). These analyses are complemented with a description of agricultural resources, communication routes and the population, with the aim of drawing the world's attention to the potential riches of the national territory (Leguizamón Pondal 1970; Zuretti 1970).

At the request of Juan María Gutiérrez, Urquiza retired from the presidency, Mitre summoned him to organize Argentine participation in the Paris Universal Exhibition of 1867, where he was appointed jury in a delegation of the Latin American countries. He died of apoplexy in 1869 in Paris at only 59 years old (Aceñolaza 2008).

Fig. 4.6 Facsimile of the Atlas of La Confédération Argentine by Martin de Moussy (1873) and an example of the maps that illustrate the work, where you can see the precise representation of orography and hydrography. Note the details of the towns concentrated along the main roads

4.2.2 *Hermann K. Burmeister (1807–1892)*

This German naturalist was a renowned professor at the University of Berlin where he taught Natural History and the University of Halle where he later taught Zoology. He had published works of high impact in Europe such as his "*History of Creation*" of 1843 (Fig. 4.7). For mainly political reasons he decided to travel to South America in 1850 recommended by Alexander von Humboldt.

After a short experience of almost two years in Brazil, interrupted by an accident, he returned to Germany. On a new trip, he arrived in Buenos Aires at the beginning of 1857, with a letter of recommendation for President Urquiza from Juan B. Alberdi, *chargé d'affaires* of the Confederation in Paris, where he had met him. He had been sent by the King of Prussia and recommended again by von Humboldt with the aim of studying the central region of the country. He was based in Paraná and Mendoza, toured various Argentine provinces, and crossed the Andes Mountain range to Copiapó, where he embarked on his return. Returning to Halle and motivated by academic difficulties he offered his services to the government. Minister Domingo F. Sarmiento was very interested in the offer and that is how Burmeister, at 55 years of age, resigned from the professorship he held in Germany to settle permanently in Argentina (Berg 1895; García Castellanos 1957; Biraben 1968; Leanza 1972; Carreras 2019).

His work as director of the Public Museum of Buenos Aires focused first on organizing and enriching its collections and then on making it known internationally through the Annals of the Public Museum of Buenos Aires and his own scientific works. This publication quickly achieved international prestige, a merit that corresponds almost exclusively to Burmeister, who wrote the articles, illustrated them as an excellent cartoonist, and edited them. These annals were the first highly prestigious scientific journal published by an Argentine institution. In addition, he published

Fig. 4.7 Burmeister in 1849 before he arrived in Argentina. Lithograph based on a daguerreotype from his book "*History of Creation*" (Leipzig, 1856 edition)

three fundamental works, the *Fossil Horses of the Pampas* in 1876, the *Description Physique de la République Argentine d'apres des observations personelles et étrangères* in four volumes published between 1876 and 1886 (Burmeister 2008), and the atlas *Vues pittoresques de la République Argentine* of 1881 edited by the Argentine Geographic Institute (Fig. 4.8). These three works represented us respectively at the Universal Exhibition in Paris in 1877 and 1889 and at the Geographical Exhibition in Venice in 1891, where he was awarded a gold medal.

In 1868 he offered to organize the Faculty of Exact, Physical, and Natural Sciences at the University of Córdoba, a task that Sarmiento accepted the following year and new professors from Germany began to be hired in 1870. Due to disagreements with the University and the new professors, he resigned the following year. In 1874 he accepted the appointment of Scientific Director of the Academy of Exact Sciences, to welcome new academics. The following year he resigned again due to continued problems and disagreements with the hired professors. A detailed analysis of these problems is developed by Carreras (2009) and Acosta (2015). Throughout his life, his sullen and authoritarian character brought him numerous problems, as highlighted by his biographers, who nevertheless recognized his talent as a tireless researcher (Berg 1895; Ulrich 1972).

The prestige achieved by the geological investigations carried out and promoted by Burmeister can be seen in the classic book, *The face of the Earth* by Eduard Suess (1904). He recognizes in his book that "*in carrying out a detailed examination of its structure we will begin with the south and the Argentine Republic, where under the leadership of Burmeister a school of researchers has grown and research is more advanced than in any other part of the great mountainous region that extends between the Caribbean Sea and Cape Horn*" (Zappettini 2004, p. 15).

Fig. 4.8 Vista de las barrancas de Paraná ilustrada por Burmeister (1876), donde se puede apreciar su talento como dibujante y su precisión en la caracterización geológica de las diferentes unidades

Various authors have analyzed his entomological, paleontological and zoological studies in general, as well as his time at the Public Museum of Buenos Aires and his management in the creation of the National Academy of Sciences in Córdoba. However, his geological contributions have not deserved detailed attention. Aceñolaza (2008) presents a synthesis of what was observed in his travels through the Sierras Pampeanas, and especially to the greywacke and Paleozoic limestone crossed by volcanic rocks of the Mendoza Precordillera. He once again found in Mendoza the remains of fossil trunks already observed by Darwin. In general, the descriptions of it are superficial, highlighting the high deformation of these rocks. Its geological contributions have been minor (Stipanicic and Cuerda 2004). The greatest merit of his work is the geographical description of the different itineraries carried out.

4.2.3 Francis Ignacio Rickard (1836–1906)

This mining engineer was born in Dublin, Ireland in 1836 and was known as Major Rickard, given his membership in the Corp of Royal Engineers and Miners (Fig. 4.9). After a six-year stay in Chile, he received in Valparaíso in 1862 an offer from the governor of San Juan Domingo F. Sarmiento to serve as Inspector General of Mines, which he quickly accepted.

During his exploration work on the Tontal mines, as well as those of the Sierra de la Huerta, he made contributions to the importance of these silver mines, their distribution and the grades of the ores, submitting a report to the authorities (Fig. 4.10). Later he publishes a description of the crossing of the Andes and his investigations in San Juan y Mendoza in London (Rickard 1863). With the support of the province of San Juan, he built one of the first beneficiation plants such as the Hilario mining establishment, which was active until at least 1866, smelting more than a hundred tons of silver ore (Michieli 2021).

When Sarmiento became president of the republic, he hired Rickard again as General Inspector of Mines of the Nation, extending his explorations and evaluation of the manifestations to other regions of the country (Rickard 1869a, b). In turn, at Sarmiento's request, he contributed to the first industrial exhibition held in Córdoba in 1871. He subsequently returned to the United Kingdom in 1872 where he added to his previous distinctions as a fellow of the Geological Society and the Royal Geographic Society, that of fellow of the Royal Society of Edinburgh. He dedicated his last years as a businessman to mining ventures in South Africa and Australia, until his death on the Isle of Wight, England, in 1906.

Fig. 4.9 Mayor Francis Ignacio Rickard dressed in military uniform in London and as a gaucho in 1867. *Source* Segemar archives

Fig. 4.10 Hilario mining establishment in 1864. Albumen on cardboard obtained by Ignacio Rickard. *Source* Segemar archives

4.3 Final Considerations

It can be partially concluded that the national organization produced after the battle of Caseros had as a priority making known and disseminating the mineral resources of the Argentine confederation, as can be seen in the works of Baron de Graty, Victor Martín de Moussy and Francis Rickard. The geological information produced by these authors is rather meager and was mainly based on the previous studies of Darwin and D'Orbigny. There were partial exceptions such as Auguste Bravard's studies in the Paraná ravines and in the Bahía Blanca region. The refoundation of the Public Museum of Buenos Aires carried out by Hermann Burmeister allowed important contributions to be made, but in the field of Paleontology. We must wait for the creation of the Natural Sciences degree at the University of Buenos Aires in 1865, as well as for the founding of the Academy of Sciences of Córdoba in 1869, to begin more detailed and larger geological studies.

References

Aceñolaza FG (2008) Estudios Geológicos de los años 1852–1868. Científicos y Exploradores en la Época de la Confederación Argentina. Serie Correlación Geológica 24: 71–84, Tucumán

Acosta LE (2015) Historia de la Zoología en la Universidad de Córdoba: los primeros años (1872–1916). Revista Facultad de Ciencias Exactas, Físicas y Naturales 2(1):75–95, Córdoba

Alonso RN (2016) La azarosa vida de Federico Benelische. Diario El Tribuno, 19 de diciembre de 2016, Salta

Benelische F (1887) 40 años de observaciones sobre la naturaleza y el hombre. Impreso en Buenos Aires, 174 p

Berg C (1895) Carlos Germán Conrado Burmeister. Reseña biográfica. Anales Del Museo Nacional De Buenos Aires 4:315–357

Biraben M (1968) Germán Burmeister. Su vida. Su obra. Ediciones Culturales Argentinas. 98 p, Buenos Aires.

Borrello AV (1970) Augusto Bravard, precursor de la cartografía geológica argentina. Boletín Academia Nacional De Ciencias Córdoba 48:455–460

Bravard A (1857a) Observaciones geológicas sobre diferentes terrenos de transporte en la hoya del Plata. Biblioteca del diario La Prensa, Buenos Aires

Bravard A (1857b) Geología de las pampas. En Territorio, Estado Físico del Territorio, Registro Estadístico del Estado de Buenos Aires, tomo primero. Imprenta de la Tribuna, 22 p

Bravard A (1858) Monografía de los terrenos marinos terciarios de las cercanías del Paraná. Imprenta del Registro Oficial, 107 p, Paraná. Reproducción facsimilar por el Congreso de la Nación en 1995 con prólogo de F.G. Aceñolaza

Burmeister G (2008) (1861) Viaje por los Estados del Plata. Academia Nacional de la Historia. V. 1: 492 p. y v 2: 512, Buenos Aires

Burmeister H (1876–1886) Description Physique de la République Argentine d´apres des observations personelles et étrangères. v1: 392 p. (1876), v2: 362 p. (1876) y v 3: 555 p. (1879), Buenos Ayres

Carreras S (2009) Una carrera científica entre Prusia y Argentina: el itinerario de Hermann Burmeister. Iberoamericana. América Latina-España-Portugal 9(33):89–101

Carreras S (2019) De los viajes de exploración a la experimentación genética. El papel de los científicos alemanes en la conformación de saberes transnacionales en Argentina, Chile y

Uruguay (siglo XIX a comienzos del XX). Revista Encuentros Latinoamericanos, segunda época 3(1):117–141
Castello HP (2014) Historia del Museo Público de Buenos Aires. Instituto De Investigaciones Históricas Manzana De Las Luces, Crónicas De Su Historia 11:1–30
Fossa Mancini E (1939) Terremotos y yacimientos de petróleo. Boletín de Informaciones Petroleras, 176:23–50
Du Graty AM (1858) La Confédération Argentine. Guillaumin et Cie. Editeurs, 371 p, París
García Castellanos T (1957) Carlos Germán Conrado Burmeister. Reseña Biográfica. Anales del Museo Nacional de Buenos Aires 4 (Serie 2)(1): 313–357, Buenos Aires
Gay C (1844–1871) Historia física y política de Chile: según documentos adquiridos en esta República durante doce años de residencia en ella. Museo de Historia Natural de Santiago. v. 1–8. Historia de Chile; v. 9–10. Documentos sobre la historia, la estadística y la geografía; v. 11–18. Botánica; v. 19–26. Zoología; v. 27–28 Agricultura; v. 29–30 Atlas, París
Leanza HA (1972) Aspectos biográficos de Germán Burmeister (1807–1892). Ameghiniana 29(4):390–392
Leguizamón Pondal M (1970) Victor Martín de Moussy: su vida y su obra. Boletín de la Academia Nacional de Ciencias 48: 573–582, Córdoba
Loos PA (1907) Estudios de sismologia. Los movimientos seísmicos de Mendoza. Anales del Ministerio de Agricultura, Sección Geología, Mineralogía y Minas 3(1):1–38
Martin de Moussy V (1860) Description géographique et statistique de la Confédération Argentine. Librairie de Firmin Didot Fréres, Fils et C[ia], Tome Premier, 582 p, Paris
Martin de Moussy V (1873) Description géographique et statistique de la Confédération Argentine. Librairie de Firmin Didot Fréres, Fils et C[ia], Deuxieme. Atlas, Paris
Mege F (1886) Un naturaliste issoirien, Auguste Bravard. Revue d'Auvergne, Societé d'emulation de l'Auvergne, 3: 189–217, Clermont-Ferrand
Michieli CT (2021) Hilario: estudio de las ruinas de una instalación metalúrgica promovida por Domingo F. Sarmiento. Teoría y Práctica de la Arqueología Histórica Latinoamericana Año X(12): 87–99, Rosario
Parish W (1838) Buenos Ayres, and The Provinces of the Rio de La Plata: Their present State, trade, and Debt. John Murray, 415 p, London
Podgorny I (1997) El Museo soy yo. Alfred Marcáis Du Graty en la Confederación Argentina. Ciencia Hoy 7(38):48–53
Podgorny I (2001) El camino de los fósiles: Las colecciones de mamíferos pampeanos en los museos franceses e ingleses del siglo XIX. Asclepio, Revista de historia de la medicina y de la ciencia 53(2):97–115, Madrid
Rickard FI (1863) A mining journey across the great andes: with explorations in the silver mining districts of the Provinces of San Juan and Mendoza, and a journey across the Pampas to Buenos Ayres, Smith, Elder & Co. London, 334 p
Rickard FI (1869a) Informe sobre los Distritos Minerales, Minas y establecimientos de la República Argentina 1868–1869. Ministerio del Interior, Imprenta, Litografía y Fundición de Tipos a Vapor, 174 p, Buenos Aires
Rickard FI (1869b) The mineral and other resources of the Argentina Republic in 1869. Longmans, Green and Co., 330 p, London
Silveira M, Padula H (2014) Valvas de moluscos y la calera de los franciscanos. In: Hernández de Lara O, Schávelzon D (eds) Casa del Virrey Liniers, Hallazgos arqueológicos, Patrimonio e Instituto histórico de la ciudad de Buenos Aires, Aspha ediciones, 131–136
Stipanicic PN, Cuerda AJ (2004) Pioneros de las Ciencias geológicas y afines de la Argentina y alguna de sus anécdotas. Academia Nacional de Ciencias, Miscelánea 102: 1–38, Córdoba
Suess E (1904) The face of the Earth (*Das Antliz der Erde*), Clarendon Press, 604 p, Oxford
Tonni EP, Pasquali RC, Laza JH (2006) Auguste Bravard en la Argentina: Su contribución al conocimiento geológico y paleontológico. Fundación Museo de La Plata, Museo 3(20):75–78, La Plata

Toledo M (2022) Typupiscis lujanensis (Ameghino 1874), asignación a Ancistrus cirrhosus (Valenciennes 1836) y su contexto histórico: la rivalidad Ameghino-Burmeister y el inicio de la fotografía científica en el Plata. Revista Mueso Ciencias Naturales 24(1):1–46

Tonni EP, Pasquali RC, Laza JH (2008) Auguste Bravard y su contribución al desarrollo de las ciencias de la tierra en la Argentina. Serie Correlación Geológica 24: 63–70, Tucumán

Ulrich W (1972) Hermann Burmeister, 1807 to 1892. Annu Rev Entomol 17:1–22

Zappettini EO (2004) Servicio Geológico Minero Argentino 1994–2004. 100 años al servicio del desarrollo nacional. Instituto de Geología y Recursos Minerales—Segemar, 104 p, Buenos Aires

Zuretti JC (1970) Victor Martín de Moussy: pionero de la geografía argentina. Boletín de la Academia Nacional de Ciencias 48: 593–597, Córdoba

Chapter 5
The Beginning of the Teaching of Geology in Argentina

Abstract Although the University of Buenos Aires was founded in 1821, it was not until 1865 that geology was taught in its classrooms. The hiring of Pellegrino Strobel, an Italian naturalist, allowed the first Geology classes to be given in the country. These courses were intended for engineering students. Strobel not only taught Geology classes but also organized the first geological expedition of an Argentine scientific institution to the Andes, which resulted in innovative publications. Upon his return to Italy, the classes continued with another Italian professor, Juan Ramorino, until his premature death in 1876. Shortly after, they continued with the first Argentine professor of geology, the engineer Eduardo Aguirre, who taught Geology and Mineralogy courses for 30 years. From 1907 onwards, the courses were taught by Engineer Enrique M. Hermitte, an Argentine who studied at the *École Nationale Superieure des Mines* in Paris. This marked a profound change, given that in those years Hermitte led the discovery of economically exploitable oil in Comodoro Rivadavia and Neuquén. His courses attracted the first students of geology and within a few years, starting in 1914, the first Argentine doctors in Natural Sciences with a geological orientation graduated at the University of Buenos Aires.

Keywords First classes · Italian professors · Buenos Aires university · Argentine professors · Oil discovery

The national organization after the battle of Caseros led to the resurgence of the University of Buenos Aires, since one of the first measures of Vicente López y Planes, the new governor of the province, was to restore the university's budget. The new rector José Barros Pazos (1852–1856) and his successor Antonio Cruz Obligado (1857–1861), began the recovery of the different faculties, but we must wait for the rectorship of Juan María Gutiérrez (1861–1873) for a true renaissance and reorganization of the university.

V. A. Ramos, *History of the Geological Sciences in Argentina*, The Latin American Studies Book Series, https://doi.org/10.1007/978-3-031-83139-3_5

5.1 The Creation of the Exact Sciences Department

The founding of the University of Buenos Aires in 1821 (Fig. 5.1) had begun with four departments, one of them Exact Sciences headed by Felipe Senillosa. On that occasion, courses in Mathematics, Physics and Chemistry were taught and physics and chemistry laboratories were established at the request of Bernardino Rivadavia. However, no Natural History or Mineralogy courses were taught despite the efforts made by Rivadavia to hire a professor in these disciplines (Ramos 2012). The Exact Sciences department disappeared in 1838, without being able to begin teaching Natural Sciences (Piñero and Bidau 1889). We must wait for the national organization after the battle of Caseros for its refoundation.

The appointment in 1861 of Juan María Gutiérrez as the new rector led to the refoundation of the Department of Exact Sciences, establishing as priorities for the new department that engineers, astronomers and geologists be trained. The need for engineers was evident, as well as for astronomers, given that they were the ones who had to contribute to putting in order the distribution and demarcation of the *estancias* in the province. The manifest need to train geologists came from his experience in exile in Chile, where he had learnt their importance in the development of mining, so significant for the progress of the neighboring country (Ramos 2016a, b). A few years earlier, the teaching of Natural Sciences had begun in the Americas, with the founding of the University of Chile in 1842 where Ignacio Domeyko (1802–1889) taught and

Fig. 5.1 The department of exact sciences was refounded by Juan María Gutiérrez in 1861 in the Manzana de las Luces where it operated until 1974. *Source* A.G.N. archives

the teachings of Antonio Raimoindi (1824–1890) at the University Mayor of San Marcos in Lima in 1851 (Ramos 2022). In North America, the Benjamin Silliman Chair stands out, as the first chair of Geology at the Yale School of Engineering inaugurated in 1852.

5.2 The Arrival of Foreign Professors

Rector Gutiérrez hired three Italian professors for pure and applied mathematics and for the teaching of geology, the latter falling to Pellegrino Strobel.

5.3 Pellegrino Strobel (1821–1895)

Dr. Pellegrino Strobel began the teaching of geology in our country with his classes taught starting on July 7, 1865 (Fig. 5.2). The new professor from the University of Parma had to give two courses per year, one on Geognosia and another on Applied Geognosia to the students of the new department (Camacho 1971, 1995). Strobel taught for two years, teaching his geology classes mainly for the engineering course, being a professor of the first classes of engineers received in the country (Ramos 2016b).

His coming to Argentina was largely motivated by his admiration for the explorations carried out in northern South America by Alexander von Humboldt (von Strobel 1998). This motivated him to request in his contract the opportunity to carry

Fig. 5.2 Pellegrino Strobel (1821–1895) during his stay in Buenos Aires between 1865 and 1867 (Ramos 2018)

out an expedition to the Andes Mountains, which he carried out in the summer of 1866.

In January of that year, he left Buenos Aires and examined the Sierras Pampeanas de San Luis where, in addition to his geological observations, he recognized the Inti Huasi cave and found the first stone objects polished by the hand of man. This finding, which was published the following year, is considered by Argentine archeology as the first study of this type and tribute is paid for having initiated this discipline in the country (Heider and Curtoni 2016).

Later he headed to Mendoza, crossing the first foothills of the Preandes, the name with which he identified the Precordillera. He crosses the mountain range through the *Paso de la Cumbre*, today known as Christ the Redeemer, to reach Curicó. He returns from there with a Chilean muleteer through the Planchón pass to cross to San Rafael and return to Mendoza (Strobel 1866). The geological findings and misadventures of this trip are related and described by Yrigoyen (1993). He basically had two problems on that trip, the compass breaking due to maneuvers by the mules, and that the muleteer, once they arrived at the Argentine side, did not know how to get to San Rafael. Despite these difficulties, he found the first Liassic fossil outcrops in Argentina in the Hermoso Valley, with abundant fossil fauna. His descent for the first time through the Las Leñas valley, named by Strobel, opened a new path towards Chile. His observations were recorded in a series of maps constructed by him without a compass and measuring the time it took the mules to travel. The relative precision of the maps was detailed analyzed by Yrigoyen (1996).

In the summer of 1867, he crossed the province of Buenos Aires, through the hills of Tandil and Ventana, passing by Bahía Blanca, and reaching Patagones. His observations were published in different media upon his return to Parma starting in 1867. His publications on different aspects that include geology, archaeology, flora and fauna among other disciplines have been compiled by von Strobel (1980) and Mezzadri (1998).

5.4 Juan Ramorino (1840–1876)

The pioneering stage initiated by Strobel, which lasted until 1867, continued with another Italian, Juan Ramorino until his early death in 1876 (Fig. 5.3). He, like Strobel, was hired for this purpose by the doctor Paolo Mantegazza, to whom some years before they had offered to become charge of the Natural History course. Juan Ramorino was born in Genoa, where he had received a Doctor of Science from the Royal University of this city in 1863, and he worked at the Museum of Natural History of Genoa until his coming to Argentina in 1866. In those first years he studied the Ligurian caves, where he found fossil human remains, revealing various works on cut and impact marks on bones of extinct fauna, plus the obvious association with lithics, as evidence of contemporary human activity with this fossil fauna.

Upon his arrival in Argentina, he contacted Florentino Ameghino, with whom he shares his interest in the origin of man. Ramorino recognized that Ameghino was the

Fig. 5.3 Dr. Juan Ramorino (1841–1876), Strobel's successor in the chair of geology in Buenos Aires, and his manual of rudiments of mineralogy from 1869 (Ramos 2016a)

first to point out in 1869 the presence of fossil man in Argentina contemporary with extinguished mammals, and in 1875 Ramorino invited him to present his findings at the Argentine Scientific Society (Camacho 1970a, b). Ramorino had a strong influence on Ameghino, being one of the few scientists of the time who supported his research and who paid attention to his findings in the Luján River valley (see Toledo 2009).

This professor began Natural History courses, with a strong emphasis on geological aspects in the preparatory courses at the university and a few years later at the National School of Buenos Aires until his early death at the age of 36. He wrote his "*Rudiments of Mineralogy*", which became the first manual of this discipline published in the country (Ramorino 1869). In this book, in addition to the basic concepts of mineralogy of that time, it describes what was understood as Natural Sciences, a much broader concept that contained Physics, Chemistry, Astronomy, Physical Geography and Natural History, which included Zoology, Botany and Mineralogy (Ramorino 1869, p. 3–4).

5.5 The First Argentine Professors

At the National College of Buenos Aires, upon the death of Ramorino, Isaac Larraín (1844–1924) replaced him in teaching the Natural History courses. His classes were published by the university in 1879, based on the compilation of his disciple Januario Escobedo (Fig. 5.4). This work became one of the first writings to escape from the European influence to put emphasis on the national, both in flora, fauna and gea (Ricardo N. Alonso, pers. comm., 2023).

Isaac Larraín was a professor of Castilian Grammar at the Colegio Nacional de Buenos Aires for many years, a discipline in which he stood out for his philosophical approaches (Lidget 2018), participating actively until at least 1884.

The chair of Geology and Mineralogy remained vacant in the preparatory studies at the University until a young engineer was appointed a few days before graduating, as interim professor ad honorem in 1878.

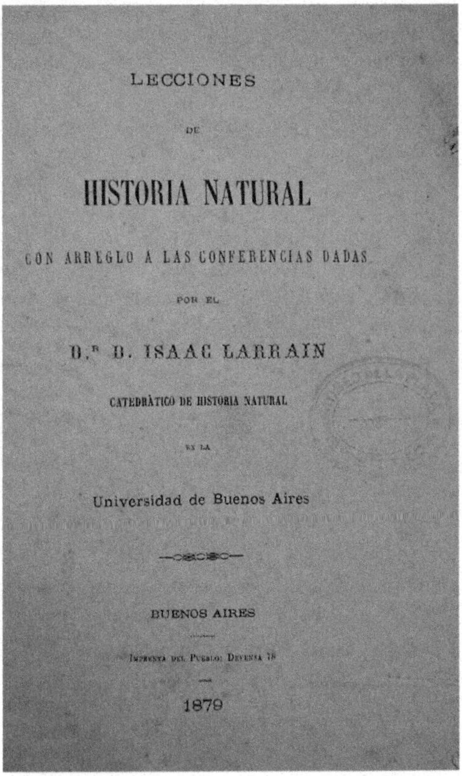

Fig. 5.4 Cover of Isaac Larraín's natural history lessons from 1879 (Larrain 1879). *Source* Biblioteca Florentino Ameghino, Museo de La Plata

5.6 Eduardo Aguirre (1857–1923)

Eduardo Aguirre had stood out as a brilliant student, actively collaborating in the Natural History Cabinet of the University (Fig. 5.5). He was a professor for more than 30 years in charge of the Mineralogy and Geology department, in addition to serving as dean of the Faculty, president of the National Academy of Exact, Physical and Natural Sciences and of the Argentine Scientific Society, among other important management tasks (Besio Moreno 1959; Camacho 1975; Hicken 1924).

Teaching and administration were no excuse for not carrying out pioneering geological studies in the province of Buenos Aires. With his students he surveyed the topography of the Sierra Baya and described the deposits of sedimentary and eruptive rocks, promoting their exploitation (Aguirre 1879). He studied the Ventania and Tandilia systems, unraveling their geological evolution and economic potential. He analyzed the artesian wells of the province of Buenos Aires (Aguirre 1882) and masterfully described the artesian origin of the El Balde water well in San Luis (Ramos 2016a).

Eduardo Aguirre, in addition to promoting the exploitation of rocks for application in the province of Buenos Aires, has been one of the outstanding Argentine scientists of the end of the century and in his honor since 1924 the main classroom of the

Fig. 5.5 Engineer Eduardo Aguirre (1857–1923) and his survey of the Sierra Baya preserved in the paleontology cabinet of the University of Buenos Aires. *Source* Biblioteca Digital F.C.E.N

Department of Geology of the University of Buenos Aires bears the name of *Aula Aguirre.*

5.7 Enrique M. Hermitte (1871–1955)

In the chair of Mineralogy and Geology of the doctorate in Natural Sciences at the University of Buenos Aires, Engineer Enrique M. Hermitte replaced Aguirre starting in 1907 (Fig. 5.6). This was a substantial change in the teaching of these disciplines. For the first time, an Argentine professor with a solid geological training and strong experience in the exploration and evaluation of the country's mineral and energy resources took charge of the teaching. Hermitte studied at the *l'École Nationale Superieure des Mines* in Paris between 1891 and 1894. He graduated as a Civil Mining Engineer, a title that he revalidated at the Faculty of Exact, Physical and Natural Sciences of Buenos Aires in 1901. He was a professor from 1907 to 1933 at the University of Buenos Aires.

Among his works, the *Argentine Geology and Mining* prepared in 1914 stands out, a synthesis work that was attached to the Third National Census, where Keidel, Stappenbeck, Schiller and Bonarelli, among others, collaborated (Hermitte 1915). This is complemented by his dissertation as president of the First Meeting of the Argentine Society of Natural Sciences in Tucumán in 1916 on the occasion of the Centennial of Independence where he spoke on the importance of geological studies,

Fig. 5.6 Enrique M. Hermitte (1871–1955) when he was a student at *the École Nationale Superieure des Mines* in Paris and at a mature age (Ramos 2016a)

particularly in the individualization of mineral deposits (Hermitte 1916). However, in addition to these excellent syntheses, his first achievement was to annul the provisions of the Mining Code that established that "*the state cannot exploit or dispose of mines*" (Hermitte 1921).

Eng. Hermitte was crucial in the conception of the decree that allowed the State to reserve the area of oil discovery in 1907 (Fig. 5.7). It was at Hermitte's suggestion that the national government would reserve the case by decree of December 15, 1907, at the time of discovery: "*The reporting of mining assets and granting of search permits in the port of Comodoro Rivadavia, Territory of Chubut, within a radius of 5 km leagues, in all directions, counting from that of the town, is prohibited.*" He was in charge of much of the activities for the discovery and development of oil in Comodoro Rivadavia and Neuquén until the creation of YPF in 1922 (Ramos 2016a).

His successes in the discovery of oil encouraged students to study Geology and he is responsible for the first graduates in the country such as Franco Pastore, Juan J. Nágera and the first woman Edelmira Mórtola, who received their PhDs in Natural Sciences with a geological thesis under his direction. Hermitte directed the thesis of Luciano Catalano, who although he received a doctorate in Chemistry in 1917, his thesis dealt with minerals from the Sierra de Pie de Palo, San Juan.

5.8 Final Considerations

The beginning of the teaching of Geology in Argentina marked sustained progress in the knowledge of our natural resources (Camacho 2007). The first Italian professors educated several generations of engineers who played an important role in the progress of the late nineteenth century in communications, water supply and in important tasks of development of our resources. Pellegrino Strobel stands out from this first part with the first crossing of the Andes by an Argentine institution whose results he published widely.

Among these first engineers was Eduardo Aguirre, who promoted the teaching of geology and the first geological investigations in the province of Buenos Aires. However, the fundamental change occurred with the return to the country of another Argentine with an outstanding education at that time, the mining engineer Enrique Hermitte, a graduate of a famous mining school in Paris. With his synthesis, he not only made known the potential of the resources of our mineral deposits (Hermitte 1904) but also fought to change the legislation, make explorations effective and provide the necessary logistical support. His actions allowed the discovery of oil in Comodoro Rivadavia in 1907 and in Plaza Huincul in 1917.

Fig. 5.7 Oil discovery well in Comodoro Rivadavia in December 1907. Transfer of discovery well No. 1 of the Neuquén basin to its drilling position in Plaza Huincul in 1917. Both wells were supervised by Hermitte. *Source* A.G.N, archives and Guevara (1927)

References

Aguirre E (1879) La geología de la Sierra Baya. Anales de la Sociedad Científica Argentina 8(2):34–45, Buenos Aires
Aguirre E (1882) Pozos artesianos en la provincia de Buenos Aires. Anales de la Sociedad Científica Argentina 13: 224–237, Buenos Aires
Besio Moreno N (1959) Ingeniero Civil Eduardo Aguirre. 1857–1923. Anales De La Academia Nacional De Ciencias Exactas, Físicas y Naturales 14:117–120
Camacho HH (1975) El ingeniero Eduardo Aguirre (1857–1923). Anales de La Academia Nacional de Ciencias Exactas. Físicas y Naturales (Número Del Centenario) 27:82–86
Camacho HH (1995) Origen y evolución de la enseñanza de las ciencias geológicas en la Universidad de Buenos Aires. Cienc Invest 48(1–2):48–53
Camacho HH (1970a) Reseña histórica de la enseñanza y los estudios de las Ciencias Naturales en la Universidad de Buenos Aires. Academia Nacional de Ciencias, Boletín 48(1–4):87–94, Córdoba
Camacho HH (1970b) El centenario de la obra "Rudimentos de Mineralogía" de Juan Ramorino y los primeros textos sudamericanos de Historia Natural. Academia Nacional de Ciencias, Boletín 48(1–4):489–493, Córdoba
Camacho HH (1971) Las Ciencias Naturales en La Universidad de Buenos Aires. Estudio Histórico, Eudeba, 150 p, Buenos Aires
Camacho HH (2007) La Contribución de la Dirección General de Minas, Geología e Hidrología de la Nación a la Formación de la Primera Generación de Geólogos Argentinos, y la Actuación del Ingeniero Enrique M. Hermite. Historia de la Geología Argentina, InsuGeo, Miscelánea 16: 41–42, Tucumán
Guevara C (1927) Historia del yacimiento fiscal de Plaza Huincul. Boletín de Informaciones Petroleras. Año IV, 40: 1175–1793, Buenos Aires
Heider G, Curtoni RP (2016) Investigaciones arqueológicas en la Provincia de San Luis: a 150 años de sus inicios, historia y perspectivas. Revista del Museo de Antropología 9(1):35–48, Córdoba
Hermitte EM (1904) Coal, petroleum and water in Argentine. In: The Saint Louis exposition. Ministry of Agriculture, Buenos Aires, pp 1–84
Hermitte EM (1915) Geología y Minería Argentinas en 1914. Contribución geológica al censo nacional de 1914, Compañía Sudamericana de Billetes de Banco, pp 1–112, Buenos Aires
Hermitte EM (1916) Importancia de los estudios geológicos particularmente en la individualización de los yacimientos minerales. Primera Reunión Nacional de la Sociedad Argentina de Ciencias Naturales, Sección Geología, Geografía y Geofísica, Disertación del Presidente, pp 13–26, Tucumán
Hermitte EM (1921) El estado de la cuestión petróleo precedido por el estudio de las cuestiones relacionadas con el aprovechamiento de los yacimientos de petróleo de la R. Argentina y en particular del de Comodoro Rivadavia. 1er. Congreso Nacional de Ingeniería, pp 2–49, Buenos Aires
Hicken C (1924) Eduardo Aguirre. Instituto De Botánica Darwinion, Darwiniana 1(3/4):183–186
Larraín I (1879) Lecciones de Historia Natural. Imprenta del Pueblo, 116 p, Buenos Aires
Lidgett E (2018) Una gramática escolar filosófica: reflexiones sobre el Curso gradual de gramática castellana de Isaac Larraín (1881). Boletín de la Sociedad española de historiografía lingüística. 12: 123–154, Valladolid
Mezzadri MG (1998) Bibliografía cronológica di Pellegrino Strobel. En Pellegrino Strobel (1821–1895) Omaggio nel centenário dela morte, Publicazioni del Museo di Storia Naturale, Universita di Parma, pp 51–67, Parma
Piñero N, Bidau EL (1889) Historia de la Universidad de Buenos Aires. Anales De La Universidad De Buenos Aires 3:1–542
Ramorino J (1869) Rudimentos de Mineralogía. Imprenta Americana, 107 p, Buenos Aires
Ramos VA (2012) Bernardino Rivadavia y las Ciencias Naturales. Revista del Museo Argentino de Ciencias Naturales, n.s. 14(2):213–222

Ramos VA (2016a) La primera clase de Exactas y el inicio de la enseñanza de la Geología. En Ramos, V.A. (ed. y coord.) 150 Años de Exactas, Eudeba, pp 2–49, Buenos Aires

Ramos VA (2016b) El inicio y desarrollo de la Geología en la Universidad de Buenos Aires. Revista Del Museo De La Plata, Número Especial 2016(1):217–227

Ramos VA (2018) El desarrollo de la Geología en la Universidad de Buenos Aires. In: Martino R, Ramos VA (eds) El nacimiento de la geología en la enseñanza universitaria, Revista Facultad de Ciencias Exactas, Físicas y Naturales 5(Suplemento 1): 23–32, Córdoba

Ramos VA (2022) Los inicios de la enseñanza de la geología en Latinoamérica. Pioneros de las geociencias, Revista Maya de Geociencias, Febrero: 48–56, México D.F.

Strobel P (1866) Excursión desde el Paso del Planchón, en los Andes Meridionales hasta San Rafael, en la Pampa del Sud, hecha en febrero de 1866. Relación Preliminar, Revista Farmacéutica 5(2):50–57, Buenos Aires

von Strobel V (1980) Bio-bibliografía del Naturalista Pellegrino von Strobel. Accademia Roveretana degli Agiati, Atti 228–229 VI, 18–19(B): 199–218, Rovereto

von Strobel V (1998) Prof. Dr. Pellegino Strobel, Naturwissenschaftler tra l'Europa ed il nuovo Mondo. En Pellegrino Strobel (1821–1895) Omaggio nel centenário dela morte, Publicazioni del Museo di Storia Naturale, Universita di Parma, 13–32, Parma

Toledo MJ (2009) Géoarchéologie de la transition Pléistocène-Holocène dans le nord-est pampéen (Buenos Aires, Argentine). Révision historique, stratigraphique et taphonomique. Perspectives pour le premier peuplement". Muséum National d'Histoire Naturelle, unpublished PhD Thesis, París

Yrigoyen MR (1993) Doctor Pellegrino Strobel: primer Profesor de Geología Aplicada a la Ingeniería. Asociación de Geología Aplicada a la Ingeniería, Actas 8: 229–248, Buenos Aires

Yrigoyen MR (1996) Dr. Pellegrino Strobel (1821–1895). Primer catedrático de Ciencias Naturales en la Universidad de Buenos Aires y descollante explorador de la Cordillera Andina. Ciencia e Investigación 49(1–2):48–56

Chapter 6
The National Academy of Sciences in Córdoba

Abstract In 1869, President Domingo F. Sarmiento promoted the creation of the National Academy of Sciences in Córdoba through Hermann Burmeister. This Academy included illustrious German scientists, among whom Alfred Stelzner stood out, a geologist who, during his short stay in the country, made fundamental contributions and founded the Mineralogy Museum of the University of Córdoba. He was a pioneer in the description of the Sierras Pampeanas, so called by him, and wrote about the Paleozoic strata of the Precordillera and the geological constitution of the Main Andes. His studies were continued by another German scientist, Ludwig Brackebusch, whose explorations were concentrated in the north of the country, where he identified the oil formation that years later would be an important source of oil production in northern Argentina. He produced the first geological map of the interior of the country. Their studies were continued by another German geologist, Guillermo Bodenbender, who not only carried out important research in the central Andean region but also for more than 30 years was a professor of geology at the University of Córdoba. It can be said that these three German scholars produced solid geological studies of the central region of the country, establishing its stratigraphy and structure. The National Academy of Sciences played an important role in financing the explorations and publishing the results obtained through its annals and scientific bulletins. These became an organ for the dissemination of geological studies not only of its academics but also of the contributions of the late nineteenth century.

Keywords Early research · Regional maps · Stelzner · Brackebusch · Bodenbender

The middle of the nineteenth century was a period of strong cultural resurgence. In those years, different schools of Natural Sciences teaching were emerging in the rest of the American continent. The University of Chile in 1842 was a pioneer through Ignacio Domeyko and Rodolfo Philippi, which was followed by the Universidad Mayor de San Marcos with Antonio Raimondi in 1851 and followed by the University of Buenos Aires in 1865 (Ramos 2022).

V. A. Ramos, *History of the Geological Sciences in Argentina*, The Latin American Studies Book Series, https://doi.org/10.1007/978-3-031-83139-3_6

In those years, two projects for the teaching and study of Natural Sciences were intertwined, one promoted by Juan María Gutiérrez and the other by Domingo F. Sarmiento supported by the director of the Public Museum of Buenos Aires, Germán Burmeister. In 1868, he proposed to Sarmiento a project for the development of Natural Sciences based on the existing University of Córdoba (Fernández 1970). Sarmiento, who shared these concerns, promulgated law 322/69 authorizing the hiring of foreign professors for these purposes. This resolution is taken as the founding act of the National Academy of Sciences. In 1872 the decree for the construction of the building was signed and the following year the Faculty of Physics and Mathematics opened its doors where the Academy dependent on the University of Córdoba would operate. In 1873, Germán Burmeister was appointed scientific director and his prerogatives and those of the academic professors were established (Depetris 2014; Martino 2018; Depetris and Maccioni 2019).

In 1874 another decree signed by Sarmiento established the functions of the Academy and its obligations to teach and research, making these results known in Minutes and Bulletins of the Academy of Exact Sciences. Before these regulations were approved, the first two academics arrived in Córdoba in 1871, followed by Alfred Stelzner, who was the Academy's first geologist.

The main geological contributions made by the National Academy of Sciences can be described by its three great geologists who succeeded one another over time, Alfred Stelzner, Ludwig Brackebusch, and Guillermo Bodenbender, whose careers will be seen in more detail.

6.1 Alfred Wilhelm Stelzner (1840–1895)

Stelzner was not only the first geologist to join the Academy, but also one of the best prepared (Fig. 6.1). He was a graduate of the *Bergakademie Freiberg*, one of the oldest institutions in Europe founded in 1765 dedicated to mining and geology, where Bernhard von Cotta (1808–1879) taught. He graduated in Dresden and obtained the highest degree at Freiberg in 1864, working at that Academy until 1870. At the request of Burmeister, he was hired by the Argentine government to teach and research at the Academy of Córdoba, where he arrived in April 1871 (Castellanos 1973; Toselli and Rossi 2008).

He founded the mineralogy museum that today bears his name, which he left with more than two thousand samples of rocks, minerals and fossils, in addition to the first microscopic preparations for petrographic analysis of local and foreign samples. He reveals a pioneering list of 30 species of Argentine minerals corresponding to pegmatites, metalliferous veins and crystalline limestones. His relations with Burmeister were difficult since his arrival, since he insisted on the priority of teaching, while the academics favored travel and research. Furthermore, Burmeister wanted to control what they published and where they published it, a situation that led to Stelzner's early resignation in 1874 (Tognetti 2004).

Fig. 6.1 Alfred W. Stelzner (1840–1895) a su llegada a la Academia de Ciencias Exactas en Córdoba y ya como profesor de la *Bergakademie Freiberg* a edad madura. *Source* A.N.C. archives

Despite having only been in Argentina for three years, his contribution to our geology has been enormous (Figs. 6.2, 6.3, 6.4 and 6.5). His *Beiträge zur Geologie und Palaeontologie der Argentinischen Republik* (Contributions to the geology and paleontology of the Argentine Republic) can be considered the first work with a solid description of our geology. He systematically approaches the description of the Sierras Pampeanas (*Pampinen Sierren*), thus baptized by Stelzner as a structure separate from the Andes, where he describes the igneous and metamorphic basement, its primitive slates, with their respective petrographic observations; He presents the first stratigraphic sections of the early Paleozoic sequences of the *Antecordillera* (today considered *Precordillera* following Strobel); he analyzes Triassic red sandstones and Jurassic sedimentary rocks and their fossils of the Main Andes; he presents the characteristics of ancient and recent eruptive rocks, with their petrographic descriptions and chemical analyses; He illustrates structural sections that show the tectonic characteristics between Santa Rosa de Los Andes in Chile and Mendoza, various areas of the Sierras Pampeanas and between the Precordillera and Pie de Palo (Stelzner 1885).

The collection of fossils collected by Stelzner is described in a series of monographs to his work made by Emanuel Kayser on Paleozoic invertebrates, by Hans Bruno Geinitz on plants and fish scales from the Rhaetian, and by Carl Gottsche on Jurassic marine fossils (Fig. 6.2). These paleontological studies were published in German in 1885, and appeared years later translated by Bodenbender in the annals of the Academy of Córdoba along with the geological part (Stelzner 1923–1924; Kayser 1925; Geinitz 1925; Gottsche 1925).

Fig. 6.2 Emanuel Kayser (1845–1927) studied the invertebrates of the lower Paleozoic (Kayser 1876); Carl Gottsche (1855–1909) did his doctorate with jurassic fossils in 1878 (Gottsche 1878), and Hans Bruno Geinitz (1814–1900), who was one of Stelzner's closest professors in Dresden, classified Triassic plants and fish scales (Geinitz 1876). *Source* cc-by-sa Deutsches Museum, Munich archives; geological survey of Norway archives, and wikimedia

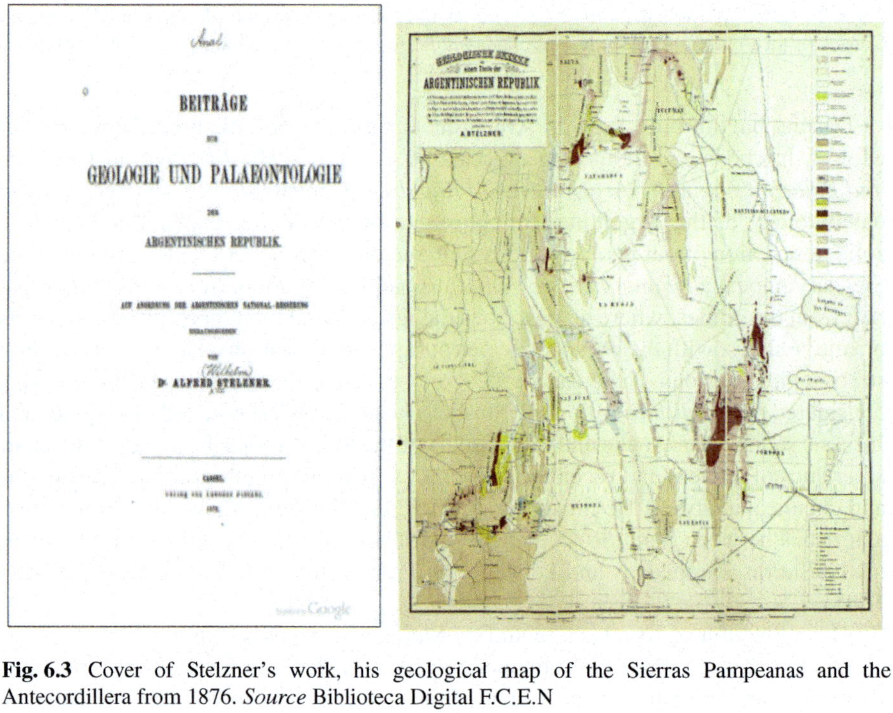

Fig. 6.3 Cover of Stelzner's work, his geological map of the Sierras Pampeanas and the Antecordillera from 1876. *Source* Biblioteca Digital F.C.E.N

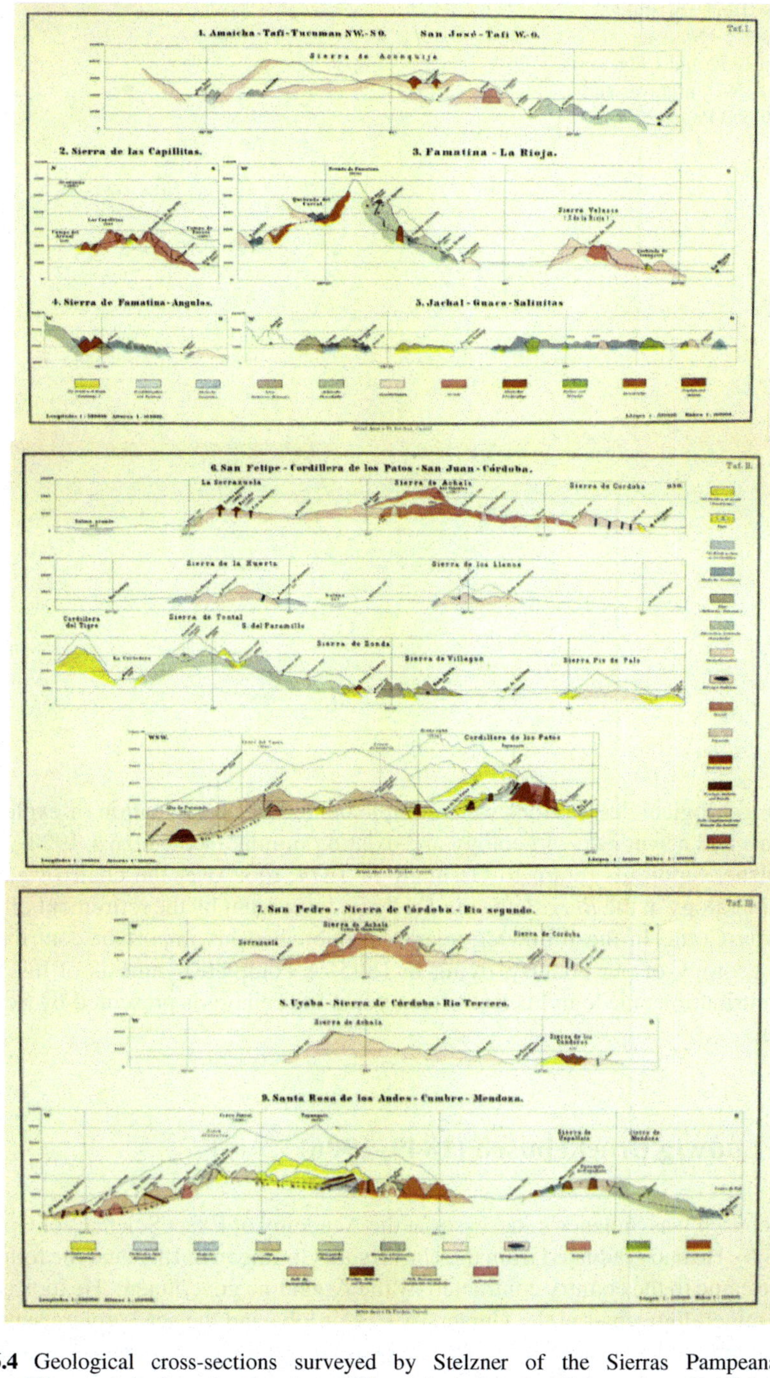

Fig. 6.4 Geological cross-sections surveyed by Stelzner of the Sierras Pampeanas, the Antecordillera and the frontal and main cordilleras, including the Chilean slope. Published in his famous work of 1886. *Source* Biblioteca Digital F.C.E.N

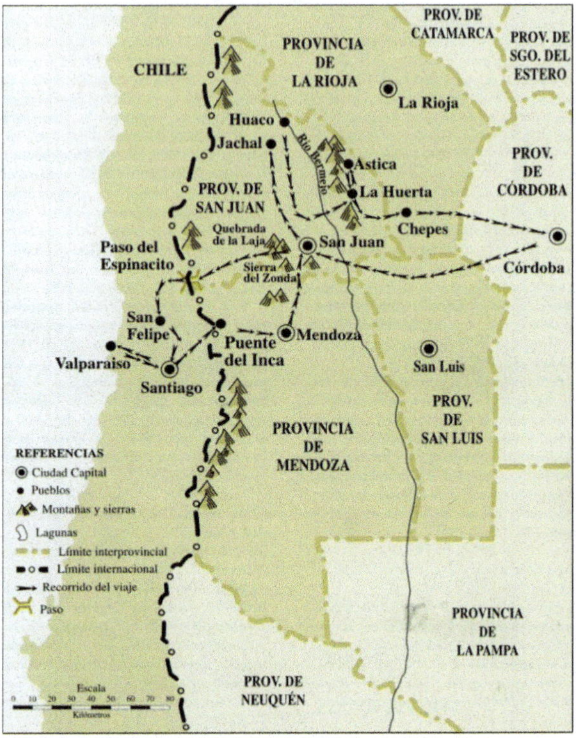

Fig. 6.5 Itinerary of the trips made by Stelzner between 1872 and 1874. *Source* A.N.C. archives and Tognetti and Page (2000)

The geological descriptions were complemented by an evaluation of exploitable minerals and a synthesis of geology and existing mining (see Stelzner 1923–1924).

Stelzner, upon his return to Germany in 1874, took over the chair of geology and mineralogy at the *Bergakademie Freiberg*, left vacant by the retirement of Bernhard von Cotta. In the following years, he published his important contributions to the geology of our country, dying in 1895. A complete synthesis of his works and contributions made in his last fifteen years in Freiberg is presented by Schmidt (1990).

6.2 Ludwig Brackebusch (1849–1906)

Brackebusch was Stelzner's successor at the Academy of Exact Sciences of Córdoba (Fig. 6.6). He had graduated from the Universität Göttingen in 1874 and the following year he came to the country contracted by the Argentine government. He took charge of the mineralogy chair at the University of Córdoba and the academy museum.

He continued Stelzner's work by carrying out extensive explorations starting in 1876, touring the south of the province of Córdoba and the north of San Luis, returning through Villa Mercedes. In 1881 he completely traveled through La Rioja and the

Fig. 6.6 Ludwig Brackebusch during his years in Córdoba. *Source* A.N.C. archives

northern foothills of San Juan, his most extensive trip being the one made to the Argentine northwest in 1883. On this occasion his tour consisted of leaving Tucumán, passing through the Sierra de la Candelaria, to reach San Antonio de los Cobres, from where it returns through Antofagasta, Fiambalá, Iglesia, San Juan and Mendoza to Córdoba (Tognetti 2004, Alonso 2020). These were long trips of almost four months each that served as a basis, first to finish the geological map of the province of Córdoba initiated by Stelzner and as a final goal his extraordinary Geological Map of the Interior of the Argentine Republic published later (Brackebusch 1891).

These field works, complemented with reconnaissance in northwest Argentina in the provinces of Salta and Jujuy, allowed him to carry out fundamental contributions such as his *Estudio de la formación petrolífera de Jujuy*, based on his observations in Laguna de la Brea and the recognition of the extension areal of the Salta System, as he called the Salta Group in his works (Brackebuscsh 1883). His work on gold, coal and the artesian wells of Catamarca, among many others, gave him fame as a complete geologist who was also in charge of evaluating mining resources and other applications in his geological studies, as can be seen in the complete list of publications presented by Alonso (2013). His scientific works alternated with vivid descriptions of his observations such as those gathered in *Por los Caminos del Norte*, where Gilberto Aceñolaza presents a compilation of his entertaining travel stories published on various occasions (see Brackebusch 1891; Stipanicic and Cuerda 2004). His numerous scientific works, both in German and Spanish, earned him recognition as one of the masters of Argentine geology (Bodenbender 1906; Turner 1970; García Castellanos 1981).

His geological work alternated with his administration and teaching tasks as professor and second dean of the Faculty of Exact, Physical and Natural Sciences, and director of the Museum of Mineralogy. His travels allowed him to increase the heritage with his collections (Tomasini and Alonso 2013), leaving the museum with more than 8,000 specimens. His mineralogical studies were reflected in his work on *Las especies minerales de la República Argentina* (Brackebusch 1879). His 13 years of work in Argentine geology culminated with his return to Germany, where he dedicated himself to publishing the map of the interior of the Argentine Republic at a scale of 1: 1,000,000, along with numerous publications in European magazines that were published in Spanish years later by the National Academy of Sciences (Fig. 6.7). He died in Germany in 1906.

6.3 Guillermo Bodenbender (1857–1941)

This German geologist was born in Meerholz in 1857 and had received his doctorate in Goetingen in 1884 (Fig. 6.8). At this university he met Brackebusch, who had invited him to come to Argentina, where he arrived in 1885 to work as an assistant professor in his chair and in the Mineralogy Museum of the Academia (Fig. 6.9). In 1889 he revalidated his doctorate degree at the National University of Córdoba. In 1891 he was appointed successor to Luis Brackebusch as professor of geology and mineralogy at the University and member of the Academy of Sciences in Córdoba (Miller 2005). He served as a professor for 30 years at the National University of Córdoba and was in charge of the museum until 1918.

His predecessors in the academy had done exceptional work; Following this line, Bodenbender is responsible for studying vast regions of the Cordillera Principal between San Juan and Neuquén, the Precordillera and the Sierras Pampeanas, especially the continental deposits of central and northwest Argentina.

He worked for almost 58 years in Córdoba except for a short period of time, when Engineer Enrique Hermitte took charge of the newly created Division of Mines, Geology and Hydrology of the Nation in 1904. He invited him to organize and take charge in Buenos Aires of the Geology Section task that he performed satisfactorily. In 1905 he carried out field studies in Orán, Salta, to determine the importance of coal in the Porongal region and later published the first geological map of the Direction of Mines at a scale of 1: 1,000,000 with its explanatory memory, which was the map compiled by Bodenbender of the province of Córdoba (Bodenbender 1906). Although in 1907 Juan Keidel took charge of the Geology Section, Bodenbender remained related to the Directorate of Mines as an *ad honorem* geologist until 1925 (Zappettini 2004).

Shortly after arriving in Córdoba in 1885, the *Sociedad Geográfica Argentina* entrusted him with an exploration in the summer of 1887–1888 south of Mendoza and Neuquén, a long campaign that he carried out with Dr. Federico Kurtz, a botanist at the academy (Bodenbender 1889). On that occasion he discovered numerous Jurassic and Cretaceous marine sequences, documented with abundant ammonites that gave rise to his classic work on *Sobre terreno Jurásico y Cretácico en los Andes Argentinos, entre los Ríos Diamante y Limay* (Jurassic and Cretaceous terrain in the Argentine Andes,

Fig. 6.7 Geological map of the interior of the Argentine Republic compiled by Brackebusch and published in Germany in 1891. The missing southwestern sector was surveyed but was not printed or distributed, apparently due to border boundary problems (pers. comm., Telasco Castellanos 1973). *Source* A.N.C. archives

between the Diamante and Limay Rivers, Bodenbender, 1892). That campaign was followed by others to the Main Cordillera of San Juan, making a new reconnaissance of the Espinacito Pass, collecting ammonites that A. Tornquist studied and made known as *El Dogger en el Paso del Espinacito.*

He studied the Precordillera of La Rioja, San Juan, and Mendoza, describing their Silurian and Devonian fossils. He discovered the Neopaleozoic continental deposits in the western sector with abundant fossil plants that Federico Kurtz studied. These works are complemented by the study of the Neopaleozoic and Triassic deposits of

Fig. 6.8 Wilhelm Bodenbender at 70 years old at his tribute from the National Academy of Sciences. *Source* A.N.C. archives

Fig. 6.9 Old building of the National Academy of Sciences built at the end of the nineteenth century in 1887 and what it looked like in 1927. *Source* García Castellanos (1987) and A.N.C. archives

the Sierras Pampeanas, where he defines, divides and makes known the stratigraphy of what he calls Paganzo Strata, in his work Geological Constitution of the southern part of La Rioja and bordering regions (Bodenbender 1911).

Bodenbender's contributions to Argentine geology and its mineral manifestations are numerous, which he published largely in Spanish in the bulletins of the Academy and in the annals of the Directorate of Mines. However, he also published his observations in German in prestigious magazines of the time (see complete list in Sparn 1927 and Hunicken 2005). Likewise, in recent years he translated from German some eleven classic works such as those by Stelzner and Gerth, in addition to various paleontological works such as those by Steuer, Behrendsen, Kayser, Jaworski, Krantz,

Tornquist, Gottsche and Geinitz published in the 1920s in the Academy bulletins that allowed these fundamental works to be disseminated in our country.

Despite his modesty, he received numerous tributes. At 60 years old, the German Scientific Association of Buenos Aires pays tribute to him (Keidel 1917). At the age of 70, the National Academy of Sciences dedicated a special volume to him in 1927 with the participation of the most prominent colleagues of the time, which brought together 19 contributions. When he turned 77, the University of Goettingen awarded him the title of *Doctor Honoris Causa* and at 80 he received the honor from the Argentine Scientific Society. He died surrounded by the affection of his colleagues and students in Córdoba in 1941 at the age of 84. It was once again up to Juan Keidel to publish an extensive appreciation of his scientific work in the pages of the Academy (Keidel 1941) and to Walther Schiller in the La Plata Museum Magazine (Schiller 1941).

6.4 Final Considerations

When the work carried out by Stelzner, Brackebusch and Bodenbender at the end of the 19th century as part of the research program of the National Academy of Sciences is observed as a whole, one can only admire the extraordinary effort made by these three geologists and the importance of their contributions. Before them, there was no geological map of our country, and from the first one made by Stelzner (1885) and substantially improved by Brackebusch (1891) of the interior of the Argentine Republic, there was the first 1,000,000-scale map of a part substantial of our country. The geological map of the province of Córdoba, initiated by Stelzner (1885), improved by Brackbusch and published in 1905 with contributions from Bodenbender, marked a milestone in the geological maps of our provinces.

The Patagonian region was not covered by the explorations programmed by the Academy. However, a scientific commission that accompanied the army participated in the so-called Desert Campaign in 1879. On that commission, there were several academics among whom was Adolfo Doering (1848–1925). This naturalist had arrived in the country in 1872 brought by Burmeister and actively participated in the Academy from 1878 (Acosta 2015). He was a chemistry professor and made important contributions as a geologist on the expedition. His best-known work is the Official Report of the scientific commission attached to the general staff of the expedition to the Río Negro (Patagonia) in its Geology section written by Doering (1882). In this work he describes the Tertiary and Quaternary deposits, making a detailed stratigraphy that served as the basis for the proposal made by Ameghino later (Tonni 2021).

The maps produced by these academics allow us to visualize the large extensions of our territory explored by these geologists that include the Argentine northwest, the central region of the Sierras Pampeanas, the Precordillera and the Main Cordillera, with pioneering contributions to its stratigraphy, petrography and paleontology, which have been unparalleled in anything previously done. It is with justice that they have been recognized as masters of our geology by the Argentine Geological Association.

All this activity recorded by the National Academy of Sciences, which was not only in the field of geology but encompassed the rest of the Natural Sciences, began to decline in 1890. Stelzner and Brackebusch had returned to Germany, and only Bodenbender remained. In our country. The financial situation of the Academy had a sharp decline in its income from the national treasury starting in 1890. This situation, along with a general malaise in relations with the National University of Córdoba, only improved in 1906 when the academies were separated from the management of the universities and began to have an exclusive advisory function.

This decline in the geological activity of the Academy was also observed in the teaching of Geology in Córdoba, where the lack of professors was accompanied by the students' little interest in pursuing this career. The situation begins to change with the arrival of Anselmo Windhausen in Córdoba, which will be analyzed in the following chapters.

References

Acosta LE (2015) Historia de la Zoología en la Universidad de Córdoba: los primeros años (1872–1916). Revista Facultad de Ciencias Exactas, Físicas y Naturales 2(1):75–95, Córdoba

Alonso RN (2013) Ludwig Brackebusch (1849–1906). Asociación Geológica Argentina. Grandes Geólogos de Argentina, 18 p, Buenos Aires

Alonso RN (2020) Personajes de la geología: Ludwig Brackebusch (1849–1906). Boletín Brackebuschiano. Geociencias y Sociedad. Asociación Geológica Argentina, agosto-septiembre 2020, 1: 7–18, Buenos Aires

Bodenbender G (1889) Expedición al Neuquén. Boletín Instituto Geográfico Argentino, 10:311–329, one map, Buenos Aires

Bodenbender G (1906) Necrología. Dr. Luis Brackebusch. Anales Sociedad Científica Argentina 62: 153–155, Buenos Aires

Bodenbender G (1911) Constitución geológica de la parte meridional de La Rioja y regiones limítrofes. Boletín de la Academia Nacional de Ciencias 19:5–220, Córdoba

Brackebusch L (1879) Las especies minerales de la República Argentina. Anales de la Sociedad Científica Argentina 7: 5–26; 49–60; 100–113; 193–205; 279–286 y 8: 5–17; 81–90; 123–134; 204–208, Buenos Aires

Brackebusch L (1883) Estudios sobre la formación petrolífera de Jujuy: Informe oficial. Boletín de la Academia Nacional de Ciencias 5: 137–184, Córdoba

Brackebusch L (1891) Mapa geológico del interior de la República Argentina. Construido sobre los datos existentes, y sus propias observaciones hechas durante los años 1875 hasta 1888. Escala 1:1.000.000. Gotha: Instituto Geográfico de Hellfahrt. Ed. Academia Nacional de Ciencias, Córdoba

Depetris PJ (2014) Las Ciencias de la Tierra en la FCEFyN: breve historia de la investigación científica. Revista De La Facultad De Ciencias Exactas Físicas y Naturales 1(1):99–111

Depetris PJ, Maccioni HJ (2019) La gestación de un proyecto: los viajes de Sarmiento (1831–1869) y sus consecuencias. In: Depetris PJ (ed) La Academia Nacional de Ciencias: 150 años acompañando la ciencia argentina. Academia Nacional de Ciencias de Córdoba, 400 p, Córdoba

Fernández J (1970) Intenciones paralelas en la génesis del estudio de las Ciencias de la Tierra en la Argentina: Sarmiento, Burmeister y Juan María Gutiérrez. Boletín de la Academia Nacional de Ciencias 48(1–4): 513–521, Córdoba

García Castellanos T (1973) Alfredo Guillermo Stelzner, 1840–1895. Boletín Academia Nacional de Ciencias, 50: 5–26, Córdoba

García Castellanos T (1981) Prólogo. In: Por los caminos del Norte. Colegio de Graduados en Ciencias Geológicas de Tucumán, 109 p, Tucumán

García Castellanos T (1987) Breve historia de la Academia Nacional de Ciencias de Córdoba. Academia Nacional de Ciencias, Miscelánea 75: 1–39, Córdoba

Geinitz HB (1876) Ueber rhätische Thier-und Pflzenreste in der argentinischen Provinzen La Rioja, San Juan und Mendoza. Palaeontographica, Suppl. III-2: 1–14

Geinitz HB (1925) Contribuciones a la Paleontología de la República Argentina. Sobre plantas y animales réticos en las provincias argentinas de La Rioja, San Juan y Mendoza. Traducido por G. Bodenbender, Actas de la Academia Nacional de Ciencias 8(3–4): 333–347, Córdoba

Gottsche C (1878) Ueber Jurasische Versteinerungen in der argentinische Cordillere. In Beiträge zur Geologie und Palaeontologie der Argentinischen Republik. Palaeontologischer Theil II, 1–50, 8 Tablas

Gottsche C (1925) Contribuciones a la Paleontología de la República Argentina. Sobre fósiles jurásicos de la Cordillera Argentina (Paso del Espinacito, prov. de S. Juan). Traducido por G. Bodenbender, Actas de la Academia Nacional de Ciencias 8(3–4): 231–296, Córdoba

Hünicken MA (2005) Doctor Guillermo Bodenbender, su obra geológica en Argentina. In: Aceñolaza FG et al (eds) Simposio Bodenbender, InsuGeo, Serie Correlación Geológica 19: 15–22, Tucumán

Kayser E (1876) Ueber primordiale und untersilurische Fossilien aus der argentinischen Republik. Palaeontographica Suppl. III, Lief. 2:1–33

Kayser E (1925) Contribuciones a la Paleontología de la República Argentina. Sobre fósiles primordiales e infrasilurianos. Traducido por G. Bodenbender, Actas Academia Nacional de Ciencias 8(3–4): 297–332, Córdoba

Keidel H (1917) Wilhelm Bodenbender. Zu seinem 60. Geburtstag—Zeitschrift des deutschen wissenschaftlichen Vereins 3: 202–210, Buenos Aires

Keidel J (1941) Guillermo Bodenbender, reseña de su obra científica. Boletín de la Academia Nacional de Ciencias 35(1–4): 269–308, Córdoba

Martino RD (2018) Breve reseña del nacimiento de la enseñanza en Ciencias Geológicas en la Universidad Nacional de Córdoba y de sus primeros 70 años (1870–1940). In: Ramos VA, Martino RD (eds) El nacimiento de la Geología en la enseñanza universitaria, Revista Facultad de Ciencias Exactas, Físicas y Naturales 5 (Suplemento 1): 17–22

Miller H (2005) Wilhelm Bodenbender: reseña de su vida. In Aceñolaza FG et al (eds) Simposio Bodenbender, InsuGeo, Serie Correlación Geológica 19: 11–14, Tucumán

Ramos VA (2022) Los inicios de la enseñanza de la geología en Latinoamérica. Pioneros de las geociencias, Revista Maya de Geociencias, Febrero, 48–56, México D.F

Schiller W (1941) Guillermo Bodenbender. Revista Museo de La Plata, Nueva Serie, Sección Oficial: 143–148

Schmidt P (1990) Alfred Wilhelm Stelzner, com destaque Quanto ao seu relacionamento com América do Sul. In: Lopes M, Figueiroa S (eds) O conhecimento geológico na America Latina, Questoes de Historia e Teoria, Unicamp. Campinas, pp 33–43

Sparn E (1927) Nómina de los trabajos publicados por Guillermo Bodenbender. Boletín de la Academia Nacional de Ciencias 30: 11–15, Córdoba

Stelzner A (1885) Beiträge zur Geologie und Palaeontologie der Argentinischen Republik. Ed. Theodor Fischer, 329 p, Cassel und Berlin

Stelzner A (1923–1924) Contribuciones a la Geología de la República Argentina, con la parte limítrofe de los Andes Chilenos entre los 32° y 33° S. Traducido por G. Bodenbender, Actas de la Academia Nacional de Ciencias 8: 1–227, Córdoba

Stipanicic PN, Cuerda AJ (2004) Pioneros de las Ciencias geológicas y afines de la Argentina y alguna de sus anécdotas. Academia Nacional de Ciencias, Miscelánea 102:1–38, Córdoba

Tognetti L (2004) La Academia Nacional de Ciencias en el siglo XIX. Los naturalistas, publicaciones y exploraciones. Historia de la Academia Nacional de Ciencias 2, 121 p, Córdoba

Tognetti L, Page C (2000) La Academia Nacional de Ciencias, Etapa fundacional, Siglo XIX. Historia de la Academia Nacional de Ciencias 1, 95 p, Córdoba

Tomasini A, Alonso RN (2013) La vida del minero en la República Argentina por Ludwig Brackebusch. Traducción del alemán, estudio preliminar, notas y comentarios. Mundo Gráfico Salta Editorial, 112 p, Salta

Tonni EP (2021) Adolfo Doering. Un científico multifacético entre dos siglos. Historia Natural, Fundación Azara, Tercera Serie 11(1):37–46

Toselli AJ, Rossi JN (2008) Alfred W. Stelzner ¿Por qué solo tres años en Argentina? En: Aceñolaza FG (ed) Historia de la Geología Argentina I, Serie Correlación Geológica 24: 91–102, Tucumán

Turner JCM (1970) Historia de la geología del noroeste argentino. In Primer Congreso Argentino de Historia de la Ciencia (1969), Boletín de la Academia Nacional de Ciencias 48:495–512, Córdoba

Zappettini EO (2004) Servicio Geológico Minero Argentino, 100 años al servicio del desarrollo nacional. Segemar, 104 p, Buenos Aires

Chapter 7
The Museum of Natural Sciences of la Plata

Abstract The preceding studies allowed the central and northern parts of the country to be known. However, large areas of Patagonia remained unknown, not only from a geological but also from a geographical point of view. Francisco P. Moreno was an enthusiastic explorer of the southern regions from 1873 onwards. He carried out several expeditions to Patagonia, collecting numerous geological samples and fossils through the Argentine Scientific Society, which allowed him to initially create a modest museum in his own house. With the creation of the city of La Plata, he proposed a plan to build a museum on par with the new city, which was approved by the authorities. In 1884, the La Plata Natural Sciences Museum was built under his direction, which became a leading institution not only for Argentina but also on a continental level. With the collaboration of Florentino Ameghino, his paleontological collections were expanded, and through hired personnel, important expeditions to Patagonia were initiated. During those years, the German geologist Rodolfo Hauthal took charge of the museum's Geology section, and Alcides Mercerat of the Paleontology section. The explorations were reinforced by the participation of Carl E. Burckhardt and Leo Wehrli, Swiss geologists who carried out research in the northern sector of the Patagonian Cordillera. In 1902, border problems with Chile led Francisco Moreno to take charge of the studies as an expert to collect the necessary information to establish the border on both sides of the Andes Mountains. The extensive studies carried out by a legion of topographers, naturalists, and geodesists made it possible to gather the necessary information to present at the arbitration, which implied a great advance in the geological knowledge of Patagonia.

Keywords Perito Moreno · La Plata Museum · Patagonian geology · German experts · Boundary problems

In the previous chapters, we analyzed the profound modification in the country after the battle of Caseros in 1852. While Justo J. Urquiza was president of the Argentine Confederation, geology had its epicenter in Paraná through various initiatives. At the beginning of the 60s, geological activity moved to Buenos Aires, which was split

V. A. Ramos, *History of the Geological Sciences in Argentina*, The Latin American Studies Book Series, https://doi.org/10.1007/978-3-031-83139-3_7

from the Confederation, through the initiatives of Juan María Gutiérrez as rector of the University of Buenos Aires, supported by Mitre amid the war with Paraguay.

Simultaneously with the creation of the Department of Exact Sciences, including natural sciences, and the beginning of the teaching of Geology at the University of Buenos Aires, the Public Museum was reactivated under the direction of Germán Burmeister. We had seen Burmeister's role, not only in the development and progress of the museum but in the creation of an Academy of Exact Sciences in Córdoba, through the support of Sarmiento.

The initiatives of the Academy of Sciences allowed the first regional geological surveys to be carried out and to produce extensive coverage in the central-northern sector of the country which, together with the petrographic and stratigraphic surveys, were a notable advance in knowledge. However, the Patagonian region did not have similar activity. The Academy of Sciences participated with its academics attached to the army in the so-called Conquest of the Desert, exclusively recognizing the Río Negro region (Doering 1882; Tonni 2021).

The great extensions of our Patagonia remained practically unknown, except for some specific surveys on the Atlantic coast and some expeditions carried out through the Argentine Scientific Society and private initiatives. This society, created in 1872, had received a strong push for its creation through the students from the Department of Exact Sciences of the University of Buenos Aires. Emilio Rosetti, one of the Italian professors brought by Juan María Gutiérrez, summoned the professors and students for the creation of the society, temporarily assuming the presidency, later delegated at the constitutive meeting at the National College of Buenos Aires, to the engineer Luis A. Huergo, the first graduate of the department and first president of the Argentine Scientific Society (Babini 1986). From its first years, the Society assisted the explorations of Patagonia through its scientific meetings and supported some expeditions within its possibilities.

However, in those years the internal region of Patagonia remained unknown, and it was up to Francisco P. Moreno (1852–1919) to begin the first explorations with funding provided by his father in 1873 (Fig. 7.1). In 1875 with the support of the Argentine Scientific Society (Babini 1986) organized an expedition from Carmen de Patagones, arriving at Lake Nahuel Huapi between the end of 1875 and the beginning of 1876, being the first civilized man to arrive (Riccardi 2021). Ramón Lista (1856–1897) also received financial support to carry out a new expedition to Patagonia in 1877. In successive expeditions Moreno, in some of them accompanied by Carlos Moyano (1854–1910), discovered the Argentino and San Martín lakes, collected numerous samples and described the extent of the Patagonian boulders and the Patagonian transgression (Moreno 1879). These last expeditions continued to be carried out through personal efforts and partially sponsored by the Argentine Scientific Society. His explorations in the southern part, especially the eastern slope of the Patagonian Cordillera, are reported in Moreno (1882). It is noteworthy that these explorations by Moreno in various regions of Patagonia were the only ones carried out in those years and that allowed the incorporation of knowledge of vast previously unexplored sectors (Riccardi 2008).

Fig. 7.1 Francisco P. Moreno (1852–1919) as a young explorer and at a mature age. *Source* A.N.C. archives and Museo de la Patagonia, Bariloche

In 1877 he donated the collections obtained from his explorations to make an Anthropological and Archaeological Museum of the province of Buenos Aires. Upon his return from a trip to museums in France and England in 1881, he presented a project for the construction of a new museum, which was crystallized in 1884 when the provincial authorities gave their support to build it in the city of La Plata, the brand-new capital of the province.

7.1 The Creation of the la Plata Museum

The foundation of the museum allowed Moreno not only to deposit his collections but, after the purchase of Florentino Ameghino's collection by the province, to increase the museum's collection and have Ameghino as vice director from 1886 (Riccardi 2011). Thanks to Ameghino's contribution, the museum became a reference center for the study of the fossil megafauna of South America (Fig. 7.2).

New expeditions began, such as those of Carlos Ameghino to Chubut between 1888 and 1889, which provided new collections of fossil remains of mammals and dinosaurs. The Santa Cruz region was explored by Clemente Onelli, where they explored the Tertiary coals and collected abundant remains of mammals. In 1887, the publications of the La Plata Museum began, with a work by Florentino Ameghino followed by several works by Alcides Mercerat (Fig. 7.3), in charge of the Paleontology section, who describes the collections made by Moreno in his first expeditions. These explorations were followed by others such as those carried out by Carlos Burmeister between 1891 and 1892, which widely covered the province of Santa

Fig. 7.2 The La Plata Museum built between 1884 and 1887, was officially inaugurated in 1888. Its researchers under the direction of Perito Moreno had an important role in the geological and topographical studies for the demarcation of the boundaries between Argentina and Chile. *Source* Museo de La Plata archives

Cruz (Riccardi 2008). The Cenozoic fossil invertebrates were studied by Fernando Lahille, a French ichthyologist invited by Moreno who arrived in the country in 1893, serving as head of the Zoology section until 1899 (Teruggi 1977).

At the same time, the creation in 1891 of the geology and mineralogy section of the Museum and the appointment of Rodolfo Hauthal as its head marked the beginning of the systematic exploration of vast sectors of the Andean Mountains and especially Patagonia. This German geologist made important observations in the region of Ventania and Tandilia, as well as in the foothills of the southern part of the Cordillera Principal de Mendoza where he described the outcrops of bituminous

Fig. 7.3 Rodolfo Hauthal (1854–1928), Alcides Mercerat (1869–1934), and Fernando Lahille (1861–1940), heads of the geology, paleontology and zoology sections of the Museum of La Plata. *Source* Leibniz institute for regional geography, archive for geography, Biblioteca Digital F.C.E.N. and Museo de La Plata archives

coals (Hauthal 1892, 1896a, b, 1901). In 1896, Moreno was appointed Expert Representative for Argentina in the boundary dispute with Chile and made the Museum's services available to the National Government. He led commissions of traveling topographic engineers and naturalists in the field that explored all the disputed regions that extended along the Andes Mountains (Riccardi 2008, 2021). The participation of the La Plata Museum to establish the limits between Argentina and Chile was fundamental given that Perito Moreno, in the next 10 years, together with Rodolfo Hauthal, Santiago Roth, Leo Wehrli and Carl Burckhardt, produced a notable advance in the Andean geology of Mendoza and Neuquén and the Patagonian Mountains.

Among these contributions stand out Hauthal's surveys between Lago Belgrano and Seno de la Última Esperanza, his first sections of Lago Blegrano and a complete geological map between Lago Argentino and Seno de la Última Esperanza, where he identified the main stratigraphic units of this sector of the Southern Patagonian Mountain Range (in Wilckens 1905). Another of his classic contributions was the distribution and classification of volcanic centers along the Argentine-Chilean border (Hauthal 1904). Hauthal's scientific production includes more than seventy contributions carried out from his sixteen years of work in Argentina (Hunicken 1970). Furthermore, this German geologist was an extraordinary mountain climber (Hernández n.d.), and he is responsible for the first ascents of numerous hills carried out between 1893 and 1899 (see Ramos 2023). He returned to Germany in 1906 to take charge of the Direction of the Roemer-Museum in Hildesheim, Lower Saxony.

In the central region, the research of two Swiss geologists stands out, Carl E. Burckhardt and Leo Wehrli who between 1897 and 1898 made two geological cross-sections of the Andes at the latitudes of Malargüe and Las Lajas in Neuquén (Burckhardt 1898; Wehrli and Burckhardt 1898) (Fig. 7.4).

Details of these transects can be found in Aguirre Urreta et al. (2020). Burckhardt later released a series of fundamental contributions on the geology of Neuquén published extensively in the annals and journal of the Museo de La Plata (Burckhardt 1900a, b; 1902a, b, c) (Fig. 7.5). Burckhardt returned to Europe in 1901 to finish studying the collected Mesozoic fossils that he later published (Burckhardt 1903) and did not return to the country. He spent his last years making important contributions to Mexican geology (Buxtorf 1935; Carrasquero 2016; Muellerried 1936).

Santiago Roth was a renowned paleontologist who had made important contributions to Pampas' geology and made previous incursions into the Patagonian Mountains when Moreno contacted him to replace Mercerat as head of the paleontology section in 1895, a position he held until 1924.

7.2 The Foundation of the National University of la Plata

Francisco P. Moreno, once the international arbitration for the border issue ended in 1902, remained in the direction of the La Plata Museum until 1906, resigning when it became dependent on the National University of La Plata (Riccardi 2008;

Fig. 7.4 Carl E. Burckhardt (1869–1935) and Leo Wehrli (1870–1954), geologists at the La Plata Museum between 1896 and 1900. *Source* Muellerried (1936) and ETH image archives

Fig. 7.5 The 1896 expedition to the south of Mendoza by Hauthal and Burckhardt commissioned by the Museo de La Plata (photo by Leo Wehrli). *Source* ETH-Bibliothek and wikimedia

Teruggi 1977). The incorporation of the Museum into the recently created university motivated Perito Moreno's definitive departure from this institution.

Starting in 1906, Santiago Roth was appointed professor of paleontology (Fig. 7.6), being the first in this discipline at the National University of La Plata. His geological and paleontological work was widely recognized by Machon (1925), Kraglievich (1925), Torres, (1927), Bond (1999), Scheinsohn (2021), among others.

Fig. 7.6 Santiago Roth (1850–1924) and his first exploration of Patagonia in 1892 (Machon 1925; Weigelt 1951)

He studied the Neocretaceous-Danian fossiliferous levels of General Roca, explored the mountain foothills between Esquel and Río Senguerr, found the marine Lower Jurassic of Piedra Pintada and described the Tertiary deposits of Río Frías currently in Chilean territory (Ramos 2023), collecting the faunas that allowed Ameghino to define the Friasense stage (Roth 1899, 1902, 1904, 1925).

Perito Moreno had invited Walther Schiller (1879–1944), at the suggestion of Gustav Steinmann, whose thesis student and disciple he was, to work as head of the Geology and Mineralogy Section of the Museum of La Plata. He replaced Hauthal in 1905, who had returned to Germany. In those early years, he organized and increased the collection of rocks and minerals, producing a complete catalog.

7.3 Final Considerations

The La Plata Museum continued to provide important services to the geological knowledge of the country after the departure of Perito Moreno, although it is necessary to highlight how much it accomplished during the years that it collaborated with the establishment of the limits with Chile. This work does not only include geological studies, but important topographic surveys were carried out on both sides of the mountain range that served to establish the line of high peaks and the interoceanic division of the waters, which in a large part does not coincide with the maximum height line of the mountain range (see details in Riccardi 2021).

The surveys directed and supervised by Perito Moreno and his efforts strongly enhanced the limited information available on Patagonian geology before 1870. These studies include Rodolfo Hauthal, Santiago Roth, Carl Burckhardt and Leo Wehrli, who surveyed the Patagonian Mountain Range from Pino Hachado to the

Seno de la Última Esperanza, producing geological maps, structural sections, stratigraphic profiles, and an important collection of fossils, not only in the Andean region but throughout extra-Andean Patagonia. A large part of these contributions has been treasured in the annals and bulletins of the Museum of La Plata, an institution that must be recognized for a great boost in the knowledge of Patagonian geology at the end of the 19th century and the beginning of the 20th, whose studies made it possible to safeguard our sovereignty before British arbitration.

References

Aguirre-Urreta B, Mescua JF, Vennari VV, Ramos VA (2019) Tras los pasos de Carl E. Burckhardt en los Andes Mendocinos, 5º Congreso Argentino de Historia de la Geología, Academia Nacional de Ciencias, Miscelánea 107: 21–23, Córdoba

Babini J (1986) Historia de la ciencia en la Argentina. Dimensión Argentina, Ediciones Solar, 273 p, Buenos Aires

Bond M (1999) Santiago Roth. Revista De La Fundación Museo De La Plata 2(13):33–37

Buckhardt C (1898) Rapport préliminaire sur une expédition géologique dans la région andine, situé entre Las Lajas (Argentine) et Curacautin (Chili). Revista Del Museo De La Plata 9:197–220

Burckhardt C (1900a) Profils géologiques transversaux de la cordillére argentino-chilienne. Stratigraphie et Tectonique. Anales del Museo de La Plata, Sección Mineralogía y Geología 1(2):1–136

Burckhardt C (1900b) Coupe Géologique de la Cordillére entre Las Lajas et Curacautin. Anales Del Museo De La Plata, Sección Mineralogía y Geología 1(3):1–102

Burckhardt C (1902a) Traces Géologiques d'un ancien Continent Pacifique. Revista Del Museo De La Plata 10:179–192

Burckhardt C (1902b) Le gisement supracrétacique de Roca (Rio Negro). Revista Del Museo De La Plata 10:209–222

Burckhardt C (1902c) Sur les fossiles marins du Lias de la Piedra Pintada avec quelques considérations sur l´age et l´importance du gisement. Revista Del Museo De La Plata 10:243–249

Burckhardt C (1903) Beiträge zur Kenntniss der Jura- und Kreideformation der Cordillere. Palaeontographica 50:1–144

Buxtorf (1935) Dr. Carl Emanuel Burckhardt, 1869–1935. Verhandlungen der Schweizerischen naturforschenden Gesellschaft 116: 425–435, 1 pl

Carrasquero SI (2016) Naturalistas suizos en el Museo de La Plata (Siglos XIX-XX). Revista Del Museo De La Plata, Número Especial 2016(1):55–60

Doering A (1882) Informe oficial de la comisión científica agregada al estado mayor general de la expedición al Río Negro (Patagonia) realizada en los meses de Abril, Mayo y Junio de 1879 bajo las órdenes del General D. Julio A. Roca. Entrega III Geología, Imprenta de Ostwald y Martínez, pp 298–530, Buenos Aires

Hauthal R (1892) Informe sobre el carbón de piedra de San Rafael (Prov. de Mendoza). Revista Del Museo De La Plata 4:101–113

Hauthal R (1896a) Notas sobre algunas observaciones geológicas en la Provincia de Mendoza. Revista Del Museo La Plata 7:69–96

Hauthal R (1896b) Contribución al estudio de la geología de la provincia de Buenos Aires I. Las Sierras entre Cabo Corrientes é Hinojo. Revista Del Museo De La Plata 7:477–489

Hauthal R (1901) Contribuciones al conocimiento de la geología de la provincia de Buenos Aires: I Excursión a la Sierra de la Ventana. II Apuntes geológicos de las Sierras de Olavarría. La Plata: UNLP, Facultad de Ciencias Físico-Matemáticas 1, 3 p, La Plata

Hauthal R (1904) Distribución de los centros volcánicos de la Rep. Argentina y Chile. Revista del Museo de La Plata 11: 177–192, La Plata
Hernández JH (n.d.) s/f. Hauthal. Geólogo y explorador alemán, nacionalizado argentino, su pasión por las ciencias naturales, acompañada por su afición a las excursiones y a las exploraciones a las montañas lo vincularon a participar con el Perito Moreno en la Comisión de Límites con Argentina y Chile, cuya labor fue clave para definir la frontera entre los dos países. Centro Cultural Argentino de Montaña, Enciclopedia, en prensa
Hünicken MA (1970) Labor científica del Doctor Rodolfo Hauthal. Su contribución al conocimiento geológico y geográfico de la Argentina. Primer Congreso Argentino de Historia de la Ciencia (1969), Boletín de la Academia Nacional de Ciencias, 48 (1–4): 531–550, Córdoba
Kraglievich L (1925) En memoria del Dr. Santiago Roth, geólogo y paleontólogo. Physis 7:412–417, Buenos Aires
Machon F (1925) Prof. Dr. Santiago Roth (1850–1924). Actes de la Société Helvétique des Sciences Naturelles 106(5): 35–41, Zurich
Moreno FP (1879) Viaje a la Patagonia austral emprendido bajo los auspicios del Gobierno Nacional. 1876–77, Buenos Aires
Moreno FP (1882) Patagonia, resto de un continente hoy sumergido. Anales De La Sociedad Científica Argentina 14:97–131
Muellerried FKG (1936) Carlos Burckhardt. 1869–1935. Boletín de la Sociedad Geológica Mexicana 9(1):23–29
Ramos VA (2023) Historia de la Geología Argentina. Una crónica de más de dos siglos. Instituto de Geología y Recursos Minerales, Servicio Geológico Minero Argentino, Anales 61. Asociación Geológica Argentina, Serie D, Publicación Especial 18, 316 p, Buenos Aires
Riccardi AC (2008) El Museo de La Plata en el Avance del Conocimiento Geológico a Fines del Siglo XIX. In: Aceñolaza FG (ed) Los geólogos y la geología en la historia argentina. InsuGeo, Serie Correlación Geológica 24: 109–126
Riccardi A (2011) The development of invertebrate paleontology in the La Plata Museum. Revista De La Asociación Geológica Argentina 68(3):380–391
Riccardi AC (2021) Francisco P. Moreno en el avance del conocimiento de la Patagonia. Ciencia e Investigación 71(3): 13–31, Buenos Aires
Roth S (1899) Reconocimiento de la región andina de la República Argentina. Apuntes sobre la Geología y la Paleontología de los territorios del Río Negro y Neuquén (Diciembre de 1895 a Junio de 1896). Revista Del Museo De La Plata 9:141–196
Roth S (1902) Le découverte du gisement de la Piedra Pintada. Revista Del Museo De La Plata 10:227–234
Roth S (1904) Noticias preliminares sobre nuevos mamíferos fósiles del Cretáceo superior y Terciario inferior de la Patagonia. Revista Del Museo La Plata 11:135–158
Roth S (1925) Investigaciones Geológicas en la Región Norte de la Patagonia durante los años 1897 a 1899. Revista Del Museo De La Plata 28:146–180
Scheinsohn V (2021) Kaspar Jacob "Santiago" Roth. In: Dávila L, Arenas P (eds) El americanismo germano en la antropología argentina de fines del siglo XIX al siglo XX, Editorial CICCUS-CLACSO, pp 101–130
Teruggi ME (1977) Cien años de geología en el Museo de La Plata. Obra del Centenario del Museo de La Plata. I, Reseña Histórica: 59–73
Tonni E (2021) Adolfo Doering. Un científico multifacético entre dos siglos. Historia Natural, Tercera Serie 11(1): 37–46, Buenos Aires
Torres LM (1927) Doctor Santiago Roth (1850–1924). Revista Del Museo De La Plata 30:165–169
Wehrli L, Burckhardt C (1898) Rapport préliminaire sur une expédition géologique dans la cordillere argentino-chilienne entre le 33° et 36° latitude sud. Revista Del Museo De La Plata 8:373–388
Weigelt G (1951) Santiago Roth, 1850–1924, ein Berner ald Wissenschaftlicher Pionier in Südamerika. Berner Zeitschrift für Geschichte und Heimatkunde 1951(1):19–39
Wilckens O (1905) Die Meeresablagerungen der Kreide- und Tertiär Formation in Patagonien. Neues Jahrbuch Für Mineralogie, Geologie und Paläontologie 21:98–195

Chapter 8
Geological Explorations at the End of the Nineteenth Century

Abstract Several foreign expeditions were launched at the end of the nineteenth century to the southern tip of South America and to Antarctica. One of the first was the Italian-Argentine southern scientific expedition of 1882 led by Giacomo Bove, which was followed by other expeditions. One of the reasons that guided these explorations was motivated by the unusual astronomical phenomenon known as the transit of Venus that could be observed in the Magellanic region. This led to an important French expedition to Cape Horn and another German one to the south of Chile and Argentina. At the same time, expeditions from Princeton University with John B. Hatcher explored the southern Patagonian Cordillera. The International Congress of Geography in London in 1895 promoted the exploration of the Antarctic Region, as one of the least-known regions of the world. This motivated some 16 expeditions, several of them in the Argentine Antarctic sector and the adjacent Patagonian regions. Thus, three Swedish expeditions were organized over time, successively covering Patagonia, Tierra del Fuego, and Antarctica. José María Sobral participated in the latter, and together with his companions, he spent the winter in Antarctica for two consecutive years. In addition, the Scottish expedition to Antarctica, carried out with logistical support from Argentina, founded the first observatory on Laurie Island, which has been operated by Argentine scientists without interruption since 1904. All these expeditions produced a large collection of data that promoted a profound advance in the geological knowledge of the southern areas of Patagonia and Antarctica.

Keywords Patagonian · French · Scottish · Swedish · Antarctic expeditions

In the first chapters, we analyzed the contribution of traveling scientists such as d'Orbigny and Darwin, and the geographical-geological descriptions made after the national organization, after the battle of Caseros, carried out by foreign naturalists hired by the national government. The latter, with some exceptions, were mere descriptions that tried to promote the geological-mining resources of the country (see Chap. 4).

V. A. Ramos, *History of the Geological Sciences in Argentina*, The Latin American Studies Book Series, https://doi.org/10.1007/978-3-031-83139-3_8

These first contributions were followed by geological studies carried out by the National Academy of Sciences of Córdoba and later by those carried out through the Museum of La Plata, partly related to the demarcation of the limits with Chile. These institutional efforts were partially contemporary with other activities carried out by foreign organizations attracted by specific problems and without a direct relationship with Argentine institutions (Riccardi 2015). We will try to present in chronological order these expeditions and the geological impact they had on the advancement of geological knowledge of the country.

8.1 The Italian Argentine Southern Scientific Expedition

Following an initiative by Giacomo Bove (1852–1887), an Italian sailor and explorer, the Argentine Geographic Institute became interested in organizing the first Antarctic expedition to the South Shetland Islands and Graham Land. Bove had accompanied Baron Adolf E. Nordenskiöld on his Swedish-Italian expedition to the Arctic as second commander on the Vega between 1878 and 1879, who was the first to cross the Arctic via the Northeast Passage from Europe to Asia. Baron Nordenskiöld was a geologist and uncle of another geologist, Otto Nordenskjöld, who encouraged the latter to undertake the Antarctic expedition.

The Geographic Institute through Estanislao Zeballos (1854–1923) interested public opinion in raising funds through a Cooperating Commission for the Expedition to the South Pole. Bove addressed the authorities offering to carry it out under full Argentine sponsorship (Bove 1881; Lista 1881). The intrepid Italian sailor, on a short previous visit to Buenos Aires in 1881, achieved the support of General Roca in those eight days and gave a conference to 400 people in the Italian Circle on the Antarctic Expedition, which was received with enthusiasm by the authorities and those present (Fig. 8.1). In May 1881, Roca informed him that "*the offer made by Lieutenant Bove, through the Argentine Geographic Institute, has been accepted and that, ... he will be given the direction of the exploration work that must be carried out on the southern coasts*" (Pérez 1950).

Finally, the expedition was made up of two Italian scientists, the geologist Domenico Lovisato (1842–1916) and the zoologist Decio Vinciguerra (1856–1934), plus the botanist Carlos Spegazzini (1858–1926), an Italian researcher at the Natural History Cabinet of the University of Buenos Aires. The expedition left in December 1881 and returned in September 1882. They left Buenos Aires in two ships, traveling along the Atlantic coast to Santa Cruz. They toured the Island of the States for 40 days and then sailed through the Beagle Channel to Ushuaia (Fig. 8.2). One of the ships went to Sloggett Bay and when they were leaving for Picton Island it was shipwrecked near the coast, being rescued by Allan Gardiner, an English missionary based in Ushuaia. They continued their journey to Ushuaia and Punta Arenas, where they began their return.

From a geological point of view, the main contributions have been the pioneering studies of Lovisato on the States Island (Lovisato 1883a, b). This geologist, who like

Fig. 8.1 Lieutenant Giacomo Bove, Italian sailor and explorer, and geologist Domenico Lovisato. *Source* Giodino (2013) and Fabre et al. (2017)

Fig. 8.2 Ushuaia in 1882 during the visit and shipwreck suffered by Bove in the Beagle Channel near Sloggett Bay. *Source* Engravings from the Giacomo Bove Expedition, after Bove (1883b)

a good mountaineer ascended the rugged main peaks, made the first descriptions of the slate schists and quartzose volcanic rocks of the island, assigning them to the Devonian and Carboniferous respectively. Today we know that they belong to the Beauvoir Formation of Neocomian age and the Lemaire Formation of Jurassic age (Caminos and Nullo 1979). In addition, they toured the Beagle Channel, visiting Ushuaia and reaching Punta Arenas and making some observations on the coast of Tierra del Fuego.

The merit of this expedition (Bove 1883a, b), despite not having been able to reach the Antarctic coasts, is that it was the first carried out by Argentina to these southern lands.

8.2 The French Expedition to Cape Horn

Scientific missions from France, Germany and Brazil at the end of 1882 in support of the International Polar Year of 1882–1883 proposed by the International Congress of Meteorology in 1879, headed to the Magellanic region to observe this unusual astronomical phenomenon known as the transit of Venus. This is characterized when Venus gets in the way between the orbit of the Earth and the Sun. The French government sent a scientific mission to Cape Horn between 1882 and 1883 (Figs. 8.3 and 8.4). The French navy assigned the transport La Romanche under the command of Captain Louis-Ferdinand Martial to transport scientists and their teams. He settled in Orange Bay on the Hardy Peninsula, to study the transit of Venus.

The scientific significance of this mission was due to its astronomical contributions and terrestrial magnetism, as well as the meteorological, ethnographic, botanical and zoological results presented in a complete seven-volume work initially edited by Louis-Ferdinand Martial. His stay in the region coincided in part with the Bove expedition that was undertaking its return.

Fig. 8.3 View of the mission in Orange Bay and the La Romanche transport (photographic album of the Cape Horn expedition, 1882–1883. *Source* Hyades (1887)

Fig. 8.4 Erratic block on the beaches of Orange Bay and granitic rocks along the coast. *Source* Hyades (1887)

The geological part was in charge of Dr. Paul Daniel Jules Hyades (1847–1919), a doctor at the Museum of Natural History in Paris, who stayed in Orange Bay for a year. He also toured Packsaddle, Tékinika and Ponsonby Sound, crossing on one occasion to Ushuaia. Based on the samples collected and others provided by Lovisato from those obtained by the Italian-Argentine expedition in the Beagle Channel and the Isla de los Unidos, he produced a detailed report on the geology and petrography of the Magellan and Fuegian region (Hyades 1887). This work begins with a comprehensive description of previous geological observations of the region beginning with Bougainville in 1767, Hall in 1820, King in 1827 and 1831, Darwin in 1832, Dumont d'Urville and de Jacquinot in 1838, Wilkes in 1839, culminating with those of Bove and Lovisato of 1882. The meticulous evaluation of the geological contributions of these previous investigations is striking. The rest of the work is a detailed petrographic description of all the available samples from various sectors of the Magellanic archipelago and localities in Tierra del Fuego. According to Kranck (1932), his main contribution is having described for the first time the diorites of the Beagle Channel together with schists that show evidence of contact metamorphism given the presence of andalusite and garnet.

8.3 Gustav Steinmann's First Campaigns to South America

In 1882 Gustav Steinmann (1856–1929) visited South America for the first time, when he was an assistant at the University of Strasbourg (Fig. 8.5). He was part as an "*astronomic assistant*" of a German mission, possibly linked to the International Polar Year, that came to the south of Chile to see the transit of Venus.

Steinmann was a graduate of the University of Munich and had his first indirect contact with South America with the topic of his habilitation thesis. He was a professor of Geology from 1885 in Jena, moving the following year to the University of Freiburg im Breisgau (Seibold and Seibold 2010). He not only stood out for the quality of his research but also for his concern in founding and building research centers such as the Geological-Paleontological Institute of Freiburg and Bonn, which became leading institutions in Germany. During this period, he selected several young German geologists who came under his recommendation to work in the nascent Direction of Mines and Geology.

In his first geological campaign to Chile, he spent eight months from 1883 to 1884 and made important observations along the coast, exploring the Southern Patagonian Mountain Range. He redefined Darwin's Navidad Formation (1846), establishing the Navidad Stage, which overlayed the Cretaceous-age Quiriquina Stage (Steinmann 1883; Steinmann et al. 1895). The fossils found in Quiriquina were the subject of Otto Rudolph Wilckens' doctoral thesis in 1904 under the supervision of Steinmann. In 1883 he toured the region between Puerto Hambre and Lago Argentino, visiting the Patagonian Mountains, finding frequent ammonites of Lower Cretaceous age in the highly deformed slate shales and sandstones with Inoceramus from the upper Cretaceous, and Tertiary deposits slightly tilted to horizontal without deformation

Fig. 8.5 Gustav Steinmann, professor of numerous geologists who came to Argentina. *Source* Biblioteca Digital F.C.E.N

with abundant fossils (Steinmann 1883, 1908). He carried out a second campaign between 1895 and 1897. The Tertiary marine invertebrates were studied by Wilckens for his *Habilitationsschrift* in Freiburg in 1905 (Fig. 8.6), as well as later in 1907 those collected by Hauthal between Lago Argentino and the Seno de Última Esperanza (Wanner 1943).

Steinmann in subsequent years basically studied the Andean regions of Peru, Bolivia and the extreme north of Argentina. His most important South American work is the *Geology of Peru*, where he establishes for the first time the successive orogenies that uplifted the Andes (Steinmann 1929). His legacy to the most universal Geology was his definition in 1927 of what is known today as the "*Steinmann Trinity*" formed by serpentine, pillow basalts and deep-water sediments, an association that years later would be linked to the oceanic crust and that it was essential to recognize obduction in collision zones. He further founded the German Geological Society (Geologische Vereinigung) in 1910 and its journal (Geologische Rundschau), of which he was director and editor until his death in 1929.

Fig. 8.6 Gustav Steinmann (1856–1929) with his disciple Otto R. Wilckens (1876–1943) who studied the invertebrate faunas of Argentina and Chile collected by the master. *Source* Geologische Rundschau, 1926, vol. 17a, and Beck (1943–45)

8.4 The Princeton University Expedition to Patagonia

The discoveries of the Ameghino brothers awakened the interest of the scientific world in the geological and paleontological problems that the writings of Dr. Florentino Ameghino had raised. John Bell Hatcher (1861–1904), a paleontologist at Princeton University (Fig. 8.7), decided to carry out a series of geological and paleontological studies in the territory of Santa Cruz between lakes Pueyrredón and San Martín (Hatcher 1897).

The scientific expedition to Patagonia, which Hatcher planned, resulted in notable success (Dingus 2018). Although the main objective of the expeditions was to make collections of fossil vertebrates and invertebrates from Patagonia, their stratigraphic studies allowed us to recognize for the first time the Cretaceous-Tertiary sequence of the Southern Patagonian Cordillera. His studies identified unknown geological formations and laid the foundations for current knowledge of the Patagonian Andes of Santa Cruz.

The observations were made in three expeditions: March 1896 to July 1897, November 1897 to November 1898 and December 1898 to September 1899. The preliminary results, together with the narrative of the expedition, were published in Hatcher (1897, 1900, 1903). The collections made and their importance are highlighted when examining the paleontological monographs of the invertebrates of the Cretaceous and Tertiary and the vertebrates of the Cenozoic, together with the geology, stratigraphy, zoology and botany that were brought together in eight large

Fig. 8.7 John Bell Hatcher in 1885 and the wagon he used on his three expeditions to Patagonia, during a winter crossing, currently preserved at the Killik Aike Norte Estancia. *Source* Courtesy of Yale Peabody Museum and Hatcher (1903)

volumes that present the results of this expedition (Scott 1903–1932). This set of contributions constitutes one of the most important works carried out on this region.

The authors of this work thank Florentino Ameghino, Carlos Berg and Perito Moreno for the hospitality provided in their museums and for having made available a wide variety of fossils from existing collections to be examined with absolute freedom. Unfortunately, Hatcher died in 1904 at the age of 42, a victim of typhoid fever. In their works Hatcher (1900, 1903) and Ortmann (1902) criticized Ameghinian ideas about the stratigraphy and age of the Tertiary terrains of Patagonia and, in response to these criticisms, F. Ameghino wrote *L'age des formations sédimentaires de la Patagonia (1901–1903).*

8.5 The Sixth International Congress of Geography in London

The First International Polar Year of 1882–1883 meant a wake-up call for the scientific world about the polar regions, still virgin to science. However, it was the 6th International Congress of Geography, held in London in 1895, that promoted scientific research in Antarctica. In those years the interior of this continent was completely unknown, as was a large part of its coast. At this congress, a motion was adopted declaring that "*the exploration of the Antarctic Regions*" was "*the greatest geographical exploration yet to be accomplished*". Over the next 25 years, no fewer than sixteen major expeditions were sent to Antarctica during an episode widely known as "*the heroic era of Antarctic exploration*".

This interest continued with the 7th International Congress of Geography in Berlin in 1899, which would give rise to the *Great International Antarctic Expedition.* These expeditions inaugurated a new stage in the history of Antarctic exploration and its geological knowledge.

8.6 The First Swedish Expedition to Patagonia and Tierra del Fuego

In the summer of 1895–1896, Otto Nordenskjöld (1869–1928) led a Swedish expedition to Tierra del Fuego and southern Patagonia. He had obtained a doctorate in Mineralogy and Geology at the Universität Greifswald in northern Germany in 1894 and carried out petrography studies in Paris that same year. Although his training was that of a geologist, his main interests, influenced by his uncle, Baron Nordenskiöld, were oriented towards the geographical exploration of unknown areas. This is how he undertook his journey to the Magellan region of the Patagonian Cordillera whose only antecedents were Steinmann's reconnaissance in 1883 and Hatcher's preliminary reports, without other modern scientific investigations.

The campaign lasted almost 2 years until April 1897 and, in addition to an extensive report published in 1907, produced a map of the foothills of the Cordillera and the extra-Andean region, whose main problem was the lack of a topographic base on which to record the observations (Fig. 8.8). He carried out studies from the 51° S latitude of the mountain range and the Sound of Última Esperanza to the Sound of the Admiralty and Beagle Channel (Nordenskjöld 1897, 1907). Its initial objectives were to know the extent of the ice in Patagonia, study the extent of the glaciers of the last glaciation, compare these processes with those of northern Europe and analyze the relationship of the existing glacial deposits with the "*great Patagonian gravels*" of the south. From Patagonia.

His research covered the Tertiary deposits whose marine fossils were studied by Steinmann, while the plants were described by Per Dusén (in Nordenskjöld 1907). Special attention is paid to the crystalline rocks, extending from the Admiralty Sound to the vicinity of the Beagle Channel. He describes the petrography of the granites of the axial region, as well as that of different metamorphic rocks and slates, which constitute the country rock. Although he is not certain of their age, he doubts that they are Paleozoic as interpreted by Lovisato on the States Island. He suggests for these rocks a Mesozoic age is most likely.

They studied the composition of the granites of the Patagonian batholith and doubted the existence of basic rocks described by other authors in the absence of a petrographic description. Steinmann produces a brief report on the marine fossils of Patagonia and Tierra del Fuego included in the final results of the Swedish expedition (Steinmann and Wilckens in Nordenskjöld 1907). The scientific results of this expedition were later published, covering geology along with pioneering botanical, zoological, and ethnographic contributions made by various authors (Nordenskjöld 1905, 1907). Since his expedition report was written after the Antarctic expedition, Nordenskjöld is struck by the geological similarity of the two mountain ranges. He

Fig. 8.8 Otto Nordenskjöld doctor in Mineralogy and Geology from the *Universität Greifswald* in 1894. *Source* Duse (1905)

concludes that the study of Graham Land in Antarctica can yield a lot of information about what the different stages of the glaciation were like in the Magellanic region.

8.7 The First Swedish Expedition to Antarctica

Nordenskjöld's first expedition to Patagonia was followed by a brief expedition with scientists and mining specialists to Alaska to the gold prospects newly discovered at Klondyke, paid for by Swedish financiers. In the summer of 1900, he accompanied the Danish Lieutenant Georg Carl Amdrup on an expedition to Greenland, where for two months he gained experience in the equipment and organization for his polar expedition. He had encountered difficulties in financing his Antarctic expedition since the Swedish Academy did not collaborate, receiving only private funds, and the German and English governments made substantial but insufficient contributions. The Argentine state provided generous contributions in kind. Nordenskjöld had to personally take out large loans to make the expedition possible. He hired the Norwegian captain Carl Anton Larsen (1860–1924) who had made an expedition from 1892 to 1894 to Antarctica. The experience and scientific interest shown by Larsen, who had found fossil plants on the Peninsula, stimulated him to take him on as captain of a whaling ship that he had purchased, which he named Antarctic. At the end of 1901 he left Gothenburg with an outstanding team of young scientists (Fig. 8.9). Another geologist, Johan Gunnar Andersson (1874–1960), recently graduated from the University of Uppsala, participated; the zoologist Karl A. Andersson (1875–1968); meteorologist Anders G. Bodman (1875–1960); the botanist Carl J. F.

Fig. 8.9 The Antarctic leaving Gothenburg in October 1901 and trapped in the Antarctic ice in 1902. *Source* Dirección Nacional del Antártico, DNA-IA archives and Duse (1905)

Fig. 8.10 José M. Sobral in Cerro Nevado doing measurements, and the shelter declared a national historical monument. *Source* Dirección Nacional del Antártico, DNA-IA archives

Skottsberg (1880–1963) and the cartographer Samuel A. Duse (1873–1933). While passing through Buenos Aires, Ensign José María Sobral (1880–1961), appointed by the Argentine Navy, was added to the expedition (Fig. 8.9).

In February 1902, Nordenskjöld settled with his companions on Seymour Island (now Marambio) and built the shelter on Snow Hill (Fig. 8.10). The Antarctic leaves the expedition members and Andersson with Skottsberg visit the Georgia Islands where they carry out topographical, geological, and botanical studies. They return to the Malvinas Islands where they spend the winter. They study the "*stone rivers*" (stone runs) and among the processes that form them, they propose solifluction, which in more extreme conditions than the current ones would cause these *stone rivers* (Andersson 1906a). They visit Cape Meredith (Cabo Belgrano) and describe the metamorphic and igneous rocks (Fig. 8.11). After two months in Tierra del Fuego Andersson and Skottsberg head to Antarctica to look for the scientists at Snow Hill.

Meanwhile, Nordenskjöld carries out geological observations that reveal a rich fossil fauna of ammonites, other marine mollusks, and plant remains. They traveled more than 600 km by sleigh the following summer, but when they observed that the sea was freezing again and they did not come to look for it, they prepared to spend another winter in the Cerro Nevado shelter.

Fig. 8.11 The two geologists of the Swedish expedition, Otto Nordenskjöld and Johan G. Andersson upon their return to Sweden, and Ensign Sobral, later the first Argentine geologist to graduate abroad. *Source* Library of Congress archives, Washington, and Duse (1905)

When the Antarctic is trapped in the ice, Andersson and two crew members head to the northern end of the peninsula to try to see a ship to ask for help (Fig. 8.12). They had to spend the southern winter of 1903 in Esperanza Bay, in a stone shelter for nine months feeding on penguins and seals, until they were rescued by the Argentine corvette ARA Uruguay, commanded by Julián Irizar. During his stay in Esperanza Bay, Andersson explored and mapped the northern tip of the Antarctic Peninsula. On Mount Flora, discovered by him, he collected Jurassic fossil plants and minerals. Upon his return in 1906, he published in Uppsala a work on the geology of Graham's Land (Andersson 1906b).

The corvette Uruguay under the command of Julián Irizar managed to bring the scientists from Cerro Nevado and the shipwrecked people from the Antarctic to

Fig. 8.12 John G. Anderson, Toralf Grunden and Samuel A. Duse when they arrived at Snow Hill after seven months in Hope Bay. *Source* Duse (1905)

Buenos Aires in the first days of December 1903. Despite its eventful end and the great difficulties it endured, the expedition was considered a scientific success, as they had explored much of the east coast of Graham Land, including Cape Longing, James Ross Island, the Joinville Island Group, and the Palmer Archipelago. The expedition collected valuable geological samples and numerous fossils, and its results earned Nordenskjöld important recognition (Reguero and Moly 2007). However, the enormous cost of the expedition left Nordenskjöld deeply in debt for several years.

The results were widely disseminated, highlighting the publication of "*Antarctica: or two years amongst the ice of the South Pole*" by Nordenskjöld and Andersson (1905) and that of Sobral (1904) (see Marenssi 2007). For his part, Andersson made known his observations in Esperanza Bay, the Malvinas Islands and in Tierra del Fuego (Andersson 1906a, b, 1907a, b).

At the end of the expedition, José M. Sobral resigned from the navy and went to study geology at the University of Uppsala, where he graduated in 1912, becoming the first Argentine geologist.

8.8 The Scottish Antarctic Expedition

Among the various national expeditions organized after the International Geographical Congress, one was the Scottish one led by William Speirs Bruce (1867–1921), one of the ablest polar scientists of his generation (Fig. 8.13). He studied at the School of Medicine at the University of Edinburgh and from 1892 he participated as a surgeon and naturalist in a British Whaling Exploration campaign where he had the opportunity to explore Antarctica. He participated in several subsequent Arctic explorations until he decided to organize his own Antarctic expedition. After numerous financial problems, he was able to organize the first Scottish Antarctic expedition which made important observations in the Weddell Sea and discovered Coats Land.

The expedition set out in November 1902 on the Scotia, stopped at the Malvinas Islands in December, and continued south to conduct the first oceanographic exploration of the Weddell Sea. They carried out numerous surveys, finding during the dredging a calcareous block carrying archaeocyathids, which were not identified at the time. After touring the Weddell Sea, they take refuge on Laurie Island in South Orkney. On this island, they founded a meteorological observatory known as the Omond House and wintered in 1903 (Fig. 8.14). They carried out geological observations and the Scotia left part of the crew at the observatory and left for the Malvinas Islands to equip themselves and make repairs to the Scotia. Unable to carry them out, he heads to Buenos Aires where, through the help of the director of the La Plata Museum, Perito Moreno manages to interest the authorities (see Carrasquero 2021). The Scotia arrived in Buenos Aires in December 1903 a few days after the arrival of the corvette Uruguay with the rescue of the Swedish scientists. The observatory facilities are transferred to the Argentine government, with the Scotia returning to put Argentine scientists to work on Laurie Island. Starting in 1904, they took charge

Fig. 8.13 William Bruce and the Omond House Meteorological Observatory founded in March 1903. *Source* Wikimedia and Bruce (1907)

Fig. 8.14 The Scotia at Laurie Island. The Scotia Sea was named in honor of this glorious ship of the Scottish expedition. *Source* Wikimedia and Bruce (1907)

of the meteorological observatory, an activity that continues today. Bruce donates a series of samples to the La Plata Museum with which this institution began its Antarctic collection in 1904 (Carrasquero 2021).

The Scottish expedition between 1902 and 1904 carried out the first topographical survey and scientific investigation of Laurie Island and completed an extensive program of oceanographic research in the Scotia and Weddell seas. When the expedition returned to Scotland, Bruce published the scientific results of the expedition in a series of high-quality volumes, but when he reached the eighth volume on geology, his funds ran out and the series was abandoned. However, several contributions that had been planned for that volume were produced, while unpublished proofs and archival notes survived. From these James H. H. Pirie (1878–1965), the expedition's doctor, in charge of geological observations, was able to reconstruct the volume planned by Bruce. His geological observations were commendable, with the notable exception of a major misidentification of a graptolite that was inexplicably supported by eminent British experts. They mistook poorly preserved Triassic plant remains for a graptolite. The archival material was preserved (Stone 2018).

8.9 The Third Swedish Expedition to Patagonia

Between 1907 and 1909, the third Swedish expedition to the Magellanic region was carried out under the direction of Carl Skottsberg (1880–1963), who had participated in the first Antarctic expedition as a botanist between 1902 and 1904. The geologist Percy Quensel was part of this expedition. (1881–1966), who concentrated on the eruptive rocks of the Patagonian Mountains, and Thore Gustaf Halle (1884–1964), a geologist and paleobotanist who graduated from the University of Uppsala who mainly studied the Malvinas Islands.

Percy Quensel graduated as a geologist from the University of Uppsala in 1905. He did postgraduate studies at the universities of Graz and Heidelberg until 1907, and then participated for almost two years with Skottsberg in the third expedition to the Magellan region (Fig. 8.15). Although he was in the Malvinas Islands, South Georgia and Tierra del Fuego, his main work was to survey the geology of the Patagonian Mountain Range, especially its granitic and volcanic rocks. His *Geologische-petrographische Studien in der patagonischen Cordillera*, was his doctoral thesis defended at the University of Uppsala in 1911 (Mason 1968). He produced the first geological map of this sector of the Patagonian Andes and made a detailed study of the petrography and chemical characteristics of its igneous rocks.

Fig. 8.15 Carl Skottsberg, head of the expedition and botanist, and Percy Quensel, geologist who described the igneous rocks of the Southern Patagonian Mountains. *Source* Wikimedia

He was characterized by his precise mineralogical studies and his mastery of petrography that allowed him to characterize the Patagonian batholith from the Última Esperanza Sound to the San Martín and Belgrano lakes and their Tertiary stocks such as the Payne, Fitz Roy and San Lorenzo hills. He characterized the alkaline rocks of the sub-Andean sector as the Essexites of the Carbon River. He crossed the mountain range at the latitude of Aysén, joining his observations with those of Steffen and Krügger (Quensel 1911). He later described the Jurassic quartziferous porphyries, today known as the El Quemado Formation, from Santa Cruz to Tierra del Fuego. He draws his attention to the monotonous and constant characteristics of the volcanic series in a north–south direction and how in a transverse direction they quickly pass from undeformed volcanic rocks to metamorphic rocks in the western sector with high deformation (Quensel 1913). In July 1964, the Argentine Geological Association, in an "*act of just recognition of the distinguished researcher of our Patagonian Mountains*", at the initiative of Bernabé Quartino, unanimously appointed him as an honorary member, although according to its statutes, he should have been a Corresponding Member, given that he was never based in the country (Anónimo 1981).

Thore G. Halle completed his doctoral thesis "On the geological structure and history of the Malvinas Islands", defended at the University of Uppsala in 1911 (Fig. 8.16). His geological studies in the Malvinas made it possible to identify the occurrence of the typical *Glossopteris flora* and thus date the Lafonia tillites as Permian in age (Halle 1911). He established the areal extent of glacial deposits and recognized the Devonian sedimentary rocks previously described by Darwin. He also studied the Jurassic fossil plants found by Andersson in Graham Land (Halle 1913). His studies of the Lafonia Tillite allowed Hans Keidel to correlate them with those of Sierra de la Ventana and those of South Africa (Keidel 1913).

8.10 The Expeditions of the German Scientific Society

Deserve to be highlighted the expeditions of the German Scientific Society, which between 1910 and 1916, carried out various explorations in the Patagonian Mountain Range between 40° and 52° south, both on the Pacific and Atlantic sides.

Geographers, geologists, and other naturalists participated, following in the footsteps of Perito Moreno, studying the most unexplored regions by the boundary commission. Their results were presented in detail in a work where they describe the sectors studied (Kölliker et al. 1917). The contribution is poorly known, just a synthesis of the general geological history of the Patagonian Andes carried out by Dr. Lutz Witte (1885–1963) and the geographical-geological results of the expedition presented by Dr. Fritz Reichert (1878–1953).

The region studied ranges from Cerro Tronador in the north to the Lake Viedma region in the south. They made the first scientific expedition to the Continental Ice where they surveyed the main topographic features and made magnetic observations.

Fig. 8.16 Thore G. Halle, geologist and paleobotanist and his doctoral thesis defended in 1911 at the University of Uppsala. *Source* Svenskt biografiskt lexicon (1971) and Hall (1911)

In its tectonic schemes, which are detailed in the southern sector of the mountain range, a metamorphosed core with remains of ammonites is observed, which rides above intensely folded black slates also carrying ammonites. These slates in turn override Lower Cretaceous marine deposits and Upper Cretaceous continental sequences in the weakly folded frontal region (Fig. 8.17). This sequence is overlain by Cenozoic deposits and sub-horizontal basalts of the Plateau region (Witte 1917).

Fig. 8.17 Fault thrust that overrides metamorphosed sandstones with ammonites on top of strongly folded black slates in the headwaters of Lake Viedma. *Source* Witte (1917)

8.11 Hydrological Studies Commission of the Ministry of Public Works

The region of the upper basin of the Chubut River was analyzed between 1911 and 1914 by Bailey Willis (1857–1949), as part of his works by the Hydrological Studies Commission of the Ministry of Public Works for the construction of the railway from Nahuel Huapi to San Antonio Oeste (Fig. 8.18). Willis was a prominent Mining Engineer who graduated from the University of Columbia in New York, and acted as a geologist at the United States Geological Survey, years later being its director. In the 4 years that he was in the country, he made important contributions, where he evaluated the watershed between 39° and 44°S and its potential natural resources (Willis 1914b, c, 1916). These studies had the collaboration of two other geologists, doctors J.R. Pemberton of Stanford University and C.W. Washburne of the United States Geological Survey. While in these works, he frequently describes and illustrates geological features, it is in his contributions to the XIIth International Geological Congress in Toronto in 1913, where he presents his main geological results (Willis 1914d, e).

In the first of them on the Forty-first parallel Survey of Argentina, he concentrates on the general geological characteristics of the coastal belt and its lowlands, the Somuncurá plateau, the pre-Andean depressions and the structural characteristics of the Andes Mountains.

It describes the metamorphic basement, the granites, the Jurassic schists of the Patagonian Andes, the abundant rhyolites and the Cenozoic deposits (Willis 1914d). In the second contribution, he makes interesting analyses of the Cordillera between 39° and 44° S, placing emphasis on the interoceanic divide that is located at the eastern foot of the Andes at these latitudes (see Riccardi 2019). Although both works concentrate on the Rio Negro sector, when he describes the western sector he reaches the latitude of Corcovado, with detailed analyzes and mainly hydrological maps (Willis 1914e).

8.12 Final Considerations

The brief description presented shows us that in parallel to the programs of Argentine institutions, the southern regions were a source of academic interest in various European countries. They organized scientific expeditions to Patagonia, Tierra del Fuego and Antarctica, motivated mainly by the geographical-geological curiosity of discovering unexplored lands. These explorations had a strong adventure bias that complicated the plans for their execution. The vicissitudes experienced have been related by our first geologist José María Sobral, who after spending two winters in Antarctica at the invitation of Nordenksjöld decided to study geology at the University of Uppsala in Sweden.

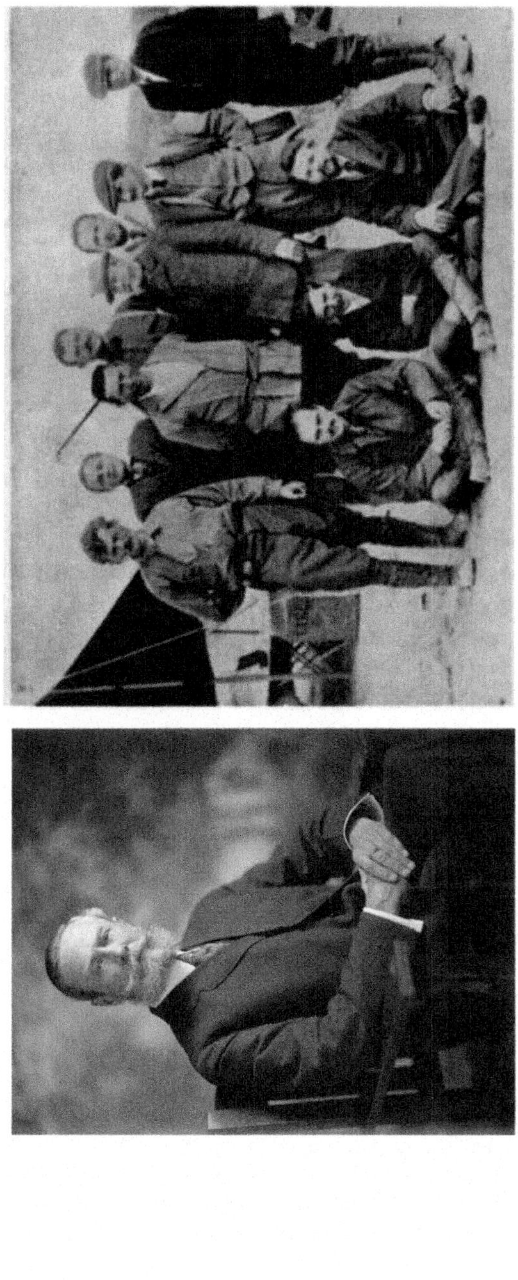

Fig. 8.18 Bayley Willis at the USGS and the Patagonian survey commission with Willis in the middle of the second row. The tallest standing is geologist J.R. Pemberton from Stanford University. *Source* Willis (1914a, 1943)

However, the knowledge of stratigraphy and paleontological content that allowed ages to be assigned to the different sequences, previously unknown, was due to the studies of Steinmann and Wilckens. The latter studied the collections made by Steinmann for his doctoral thesis, as well as those from Southern Patagonia made by Rodolfo Hauthal.

Among the different expeditions, the Swedish ones stand out from a geological point of view, especially the third, due to the results obtained. The geologists Quensel and Halle were motivated by a clear academic interest in carrying out research whose results allowed them to present their respective doctoral theses. This modality will increase in the following decades where various universities chose our region to have original topics that would allow them to develop a thesis. The vicissitudes of the second Swedish expedition, the first on the Antarctic continent, show the tenacity and spirit of sacrifice of those explorers who were not daunted by any problem.

Other expeditions were of a more general nature, where geological studies were not the main objective of the investigations, such as the Italo-Argentine expedition, the French one, the Scottish one and even that of the German Society of Buenos Aires. In them, geology was a complement to other observations such as botany, zoology and ethnography.

Some expeditions were guided by paleontological and stratigraphic problems trying to follow in the footsteps of the Ameghino brothers, among which that of Princeton University stands out for its results and the excellence of its final publications. There were other paleontological explorations, but on a smaller scale, carried out by other institutions.

When these expeditions are analyzed together, it is realized that they provided an excellent overview of the general geological conditions of the Patagonian and Fuegian region, and that they aroused the interest of our institutions in achieving a more systematic knowledge of these regions.

References

Ameghino F (1901–1903) L'age des formations sédimentaires de Patagonie. Anales de la Sociedad Científica Argentina 54:283–342, Buenos Aires

Andersson JG (1906a) Solifluction, a component of subaërial denudation. J Geol 14:91–112

Andersson JG (1906b) On the geology of Graham Land. University of Uppsala, Geological Institutionen, Bulletin 7:19–71, 6 láms., Stockholm

Andersson JG (1907a) Geological fragments from Tierra del Fuego. University of Uppsala, Geological Institutionen, Bulletin 8:168–183, 4 láms., Stockholm

Andersson JG (1907b) Contributions to the geology of the Falkland Islands. Wissenschaftliche Ergebnise der Schwedischen Südpolar Expedition 1901–1903, 3(2):1–38, Stockholm

Anónimo (1981) Homenaje al doctor Percy Dudgeon Quensel Miembro Honorario de la A.G.A. Revista de la Asociación Geológica Argentina 36(1):1–3

Beck H (1943–45) Otto Rudolf Wilckens, In Memoriam. Mitteilungen der Geologischen Gesellschaft Band 36–38:285–294, Wien

Bove G (1881) Proyecto de una Expedición Antártica Argentina. Boletín del Instituto Geográfico Argentino 2:81–90

Bove G (1883a) Expedición Austral Argentina, Informes preliminares presentados a S.S.E.E. los ministros del Interior y de Guerra y Marina de la República Argentina. Publicación del Instituto Geográfico Argentino, 217 p, Buenos Aires

Bove G (1883b) Patagonia-Terra del Fuoco, Mari Australi, Rapporto del Tenente Giacomo Bove capo della spedizione al comitato centrale per le Esplorazioni Antartiche. Parte I, 186 p, Genova

Bruce WS (1907) Report on the scientific results of the voyage of S.Y. "Scotia" during the years 1902, 1903 and 1904, under the leadership of William S. Bruce. Volume 2, Oliver & Boyd, 394 p, Edinburgh and London

Caminos RL, Nullo FE (1979) Descripción geológica de la Hoja 67e, Isla de los Estados, Territorio Nacional de la Tierra del Fuego, Antártida e Islas del Atlántico Sur. Escala 1:200.000. Boletín del Servicio Geológico Nacional 175:1–52, Buenos Aires

Carrasquero SI (2021) The geological collection from the Scottish National Antarctic Expedition (1902–04) in the Museo de La Plata, Argentina. Scott J Geol 57:1–7

Darwin C (1846) Geological observations on South America. Being the third part of the geology of the voyage of the Beagle, under the command of Capt. Fitzroy, R. N. during the years 1832 to 1836. Smith, Elder and Co., 279 p, London

Dingus L (2018) King of the dinosaur hunters: the life of John B. Hatcher and the discoveries that shaped paleontology. Pegasus Books, 544 p

Duse FA (1905) Unter Pinguinen und Seehunden. Verlag Wilhelm Baenfch, 262 p, Berlín

Fabre S, Console F, Pantaloni M (2017) 1876–1881: domenico Lovisato and the geology of Calabria (southern Italy). Ital J Geosci 136(3):471–484, Roma

Giodino IB (2013) Tras las huellas de Giacomo Bove en el fin del mundo. Editora Cultural Tierra del Fuego, 161 p, Ushuaia

Halle TG (1911) On the geological structure and history of the Falkland Islands. Academic dissertation, Bulletin Geological Institute 11:115–129, Upsala

Halle TG (1913) Some Mesozoic plant-bearing deposits in Patagonia and Tierra del Fuego and their Floras. Kungel Svenska Vetenskapasa Akademiens Handlingar 51:3–58, Stockholm

Hatcher JB (1897) On the geology of southern Patagonia. Am J Sci (4) 4(23):327–354, New Haven

Hatcher JB (1900) Sedimentary rocks of Southern Patagonia. Am J Sci (4) 9(50):85–108, New Haven

Hatcher JB (1903) Narrative and geography. Report Princeton University Expedition to Patagonia 1:1–314, New Haven

Hyades P (1887) Mission Scientifique du Cap Horn (1882–1883). IV Geologie, 249 p, Paris

Keidel J (1913) Über das Alter, die Verbreitung und die gegenseitigen Beziehungen der verschiedenen tektonischen strukturen in den argentinischen Gebirgen. XII° Session du Congrés Géologique International, Compte Rendu 1, 671–687, Toronto

Kölliker A, Kühn F, Reichert F, Tomsen A, Witte L (1917) Patagonia: Resultados de las expediciones realizadas en 1910 a 1916. Editado por la Sociedad Científica Alemana, 431 p, Buenos Aires

Kranck EH (1932) Geological investigation in the Cordillera of Tierra del Fuego. Acta Geogr 4(2):1–231, Helsingfors

Lista R (1881) La Espedicion de Bove. Boletín del Instituto Geográfico Argentino 2:55–57

Lovisato D (1883a) Apuntes geológicos sobre la Isla de los Estados. Expedición Austral Argentina. Informes preliminares presentados a S.S.E.E. los ministros del Interior y de Guerra y Marina de la República Argentina, Publicación del Instituto Geográfico Argentino, pp 47–53, Buenos Aires

Lovisato D (1883b) Una escursione geologica nella Patagonia e nella Terra del Fuoco. Bolletino della Società Geografica Italiana, Serie II 8(5):333–347 y (6):420–443

Marenssi S (2007) Doctor José María Sobral (1880–1961): de los hielos antárticos al olvido argentino. Revista de la Asociación Geológica Argentina 62(4):495–497

Mason B (1968) Memorial of Percy Dudgeon Quensel. Am Mineral 53(3–4):590–596

Nordenskjöld O (1897) A journey in South Western Patagonia. Geogr J 10:401–410

Nordenskjöld O (1905) Die Krystallinen Gesteine der Magellans-länder. Wissenschaftliche Ergebnise der Schwedischen Expedition nach den Magellansländer. Stockholm 1(6):175–240

Nordenskjöld O (1907) Wissenschaftliche Ergebnise der Schwedischen Expedition nach den Magellansländer 1895 a 1897. Band 1 (Geologie, Geographie und Anthropologie): 257 p; Band 2 (Zoologie): 170 p, Stockholm
Nordenskjöld O, Andersson JG (1905) Antarctica: or two years amongst the ice of the South Pole. Hurst and Blackett Ltd., 608 p, London
Ortmann A (1902) Tertiary invertebrates. Reports of the Princeton University Expedition to Patagonia 1896–99, vol IV:47–247, Princeton
Pérez M (1950) La primera expedición Antártica Argentina con propósitos científicos, se proyectó en el año 1881. Universidad Nacional de Cuyo, Boletín de Estudios Geográficos 7:93–125, Mendoza
Quensel PD (1911) Geologisch-petrographische Studien in der patagonischen Cordillera. University of Uppsala, Geological Institutionen, Bulletin 9:1–113
Quensel PD (1913) Die Qaurzporphyr und Porphyroidformation in Südpatagonien und Feuerland. University of Uppsala, Geological Institutionen, Bulletin 12:9–40
Riccardi AC (2015) Panorama histórico integral de las instituciones dedicadas a la geología en la Argentina, ca. 1880–1950. Manuscrito inédito preparado para el Diccionario histórico de las Ciencias de la Tierra en la Argentina (Archivo Histórico del Museo de La Plata) Prohistoria Ediciones
Reguero MA, Moly JJ (2007) Otto Nordenskjöld, su visión profética de la Antártida. Revista del Museo de La Plata, Museo 3(21):79–86
Riccardi AC (2019) Bayley Willis: un geólogo yanqui y el desarrollo del norte de la Patagonia. Revista de la Facultad de Ciencias Exactas, Físicas y Naturales 7 (Suplem 1):63–83, Córdoba
Scott WB (1903–1932) Reports of the Princeton University Expedition to Patagonia 1896–99, vols I–VIII, Princeton
Seibold E, Seibold I (2010) Gustav Steinmann (1856–1929): Ein deutscher Ordinarius der Kaiserzeit. Int J Earth Sci (Geologische Rundschau) 99(Suppl 1):S3–S15
Sobral JM (1904) Dos años entre los hielos—1901–1904. Imprenta de J. Tragan, 364 p, Buenos Aires
Steinmann G (1883) Reisenotizen aus Patagonien. Neues Jahrbuch für Mineralogie, Geologie und Paläontologie, Band 2:255–258. Stuttgart
Steinmann G (1908) Das Alter der Schieferformation im Feuerlande. Centralblatt für Mineralogie, Geologie und Paläontologie 1908:193–194
Steinmann G (1929) Geologie von Perú. Karl Winter, 448 p, Heidelberg
Steinmann G, Deecke W, Möricke W (1895) Das Alter und die Fauna der Quiriquina-Schichten in Chile. Beitrage zur Geologie un Palaeontologie von Südamerika, Neues Jahrbuch Mineralogie und Palaeontologie, 1–118
Stone P (2018) The Scottish National Antarctic Expedition, 1902–1904: reconstructing the missing geological report. Arch Nat Hist 45:350–362
Wanner J (1943) Otto Wilckens. Geol Rundsch 33.499–506
Willis B (1914a) El norte de la Patagonia: Naturaleza y riquezas. Scribner Press, Nueva York, vol 1, 500 p, vol 2, 11 mapas
Willis B (1914b) Northern Patagonia: character and resources. Scribner Press, Nueva York, vol 1, 464 p, vol 2, 11 maps
Willis B (1914c) The physical basis of the Argentine Nation. J Race Dev 4:443–460
Willis B (1914d) Forty-first parallel Survey of Argentina, un mapa. Congrès Géologique International, Compte Rendu de la XIIe Session, Canada (1913), 713–731, Ottawa
Willis B (1914e) Physiography of the Cordillera de los Andes between latitudes 39° and 44° South. Congrès Géologique International, Compte Rendu de la XIIe Session, Canada (1913), 733–756, Ottawa
Willis B (1916) The awakening of Argentina and Chile. Natl Geogr Mag 30(2):121–142
Witte L (1917) La geología de la región del Lago Viedma. In: Kölliker et al (eds) Patagonia: resultados de las expediciones realizadas en 1910 a 1916, Parte III, Editado por la Sociedad Científica Alemana, pp 275–327, Buenos Aires

Willis B (1943) Historia de la Comisión de Estudios Hidrológicos del Ministerio de Obras Públicas—1911–1914. Ministerio de Agricultura, Dirección de Parques Nacionales y Turismo, 170 p, Buenos Aires

Chapter 9
The Beginnings of the Direction of Mines and Geology

Abstract The beginning of the twentieth century marked the need to create an Argentine Geological Survey to study the country's potential in resources such as minerals, water, coal, and oil. Eng. Enrique M. Hermitte was put in charge of the new Division of Mines, Geology and Hydrology in 1904. Hermitte, in collaboration with Bodenbender, hired young German graduates with excellent recommendations to create a good Geology department. This was put in charge of Hans Keidel, a prestigious young geologist with a good track record who, during his twenty years as head of geology, managed to form a team of almost 15 German geologists who traveled the country from Jujuy to Tierra del Fuego. Important regional studies were carried out, with the survey of precise maps in the Puna, the Sierras Pampeanas, the Cordillera Oriental, and the Sierras Subandinas. These maps and their corresponding descriptions changed the geological knowledge of the country and contributed to the search for underground water. Numerous perforations were carried out to supply drinking water to localities in the interior producing basic and indispensable geological knowledge of the country. The new data allowed the evaluation of numerous manifestations of coal and other minerals. Various areas were incorporated into the geological knowledge, especially those of the new hydrocarbon basins of central Patagonia and Neuquén. The studies were successful and allowed a new vision of the regional geology of large sectors of the national territory in a few years.

Keywords Enrique Hermitte · Young Germans · Regional surveys · Hans Keidel · Hydrogeology

One of the first measures of Manuel Quintana's government when he assumed the presidency (1904–1906) was to restructure the Ministry of Agriculture. The designated minister was Wenceslao Escalante, who was in charge of organizing the new structures. This is how in 1904 he brought together a new Division of Mines, Geology and Hydrology, the previous Division of Mines and Geology, and the Commission for Studies of Water Tables and Coal Deposits. He appointed engineer Enrique Hermitte (1871–1955) as head of the new division, who was in charge of the Commission of Water and Coal Deposits (Fig. 9.1).

V. A. Ramos, *History of the Geological Sciences in Argentina*, The Latin American Studies Book Series, https://doi.org/10.1007/978-3-031-83139-3_9

Fig. 9.1 Engineer Enrique M. Hermitte in charge of the new Division of Mines, Geology and Hydrology. *Source* Segemar archives

The engineer Hermitte has studied at the *École Nationale Superieure des Mines* in Paris between 1891 and 1894 where he had graduated as a Civil Mining Engineer, a title that he revalidated upon his return to the Faculty of Exact, Physical and Natural Sciences of the University of Buenos Aires. Aires. Hermitte's management was associated with extensive knowledge of our mineral resources as evidenced by his exhibition on Coal, petroleum and water in Argentine, a work presented at the Saint Louis Exposition in the United States in 1904. In this work, he consistently reviewed the potential of coal, oil and groundwater in Argentina. He analyzed the oil from Cacheuta, Jujuy and Neuquén, presenting chemical analyzes carried out by Dr. Enrique Herrero Ducloux (Hermitte 1904). His management at the head of this institution was full of achievements and he allowed it to be consolidated throughout his 18 years of practice, hiring a plethora of young geologists who allowed him to implement important regional and specific studies. In the incorporation conference to the National Academy of Exact, Physical and Natural Sciences he makes an excellent summary of the creation of the institution and its management (Hermitte 1928).

9.1 Background of the Mines and Geology Division

The director of the previous Department of Mines and Geology was the engineer Henry Davis Hóskold (1829–1904), who was also Inspector General of Mines. Hóskold was a Mining and Civil Engineer, like Hermitte, very prestigious given that he was also a Member of the North England Institute of Mining and Mechnical Engineers and the American Institute of Mining Engineers of New York, Corresponding Member of the New York Academy of Sciences, Fellow of the Royal Geographical Society and the Geological Society of London, among other famous institutions. In addition, he was well known for having developed the so-called *Hóskold Formula* in 1877, an algorithm that was used for decades to calculate the present value of a mine. He represented us with his works at the *Exposition Universelle* in Paris in

1889 and at the Chicago Exhibition in the United States in 1893, where he received one of the most important prizes (Hóskold 1889, 1893). This synthesis of Argentine mining is one of the first and most complete inventories of our mineral resources, illustrated with geological maps and sections of the main mining districts (Fig. 9.2). He published for the San Luis Exhibition in the United States an Official report upon the mines, mining, metallurgy, and mining laws of the Argentine Republic (Hóskold 1904), which he could not present due to his unexpected death.

The position of Inspector of Mines was created by Justo J. Urquiza during his presidency in 1857 to appoint the following year the mining engineer August Bravard, a graduate of the *École des Mines of Saint-Étienne*, in France. Unfortunately, he died in the line of duty in the Mendoza earthquake of 1861 (see Chap. 4). As governor of the province of San Juan, Domingo F. Sarmiento hired the military mining engineer Francis Ignacio Rickard in 1862 as Inspector General of Mines of the Nation with the agreement of the central government (Catalano 2004). Major Rickard made important observations in the Tontal mines and made a very complete report on provincial mining. Bartolomé Mitre supports Major Rickard in his position and when Sarmiento becomes president, he receives a complete report on the state of mining in the country (Rickard 1869a, b). Julio Argentino Roca created the Mines Section during his first presidency and appointed engineer Luis Albertini in 1885 to carry out geological and mining studies throughout the country. However, it was Miguel Juárez Celman who raised the rank of Section to the General Department of Mines and Geology in 1887, appointing the engineer Hóskold.

Engineer Hóskold was General Inspector of Mines for almost 20 years, having an outstanding work in promoting mining, and as director of the Department of Geology and Mines created in 1887. In those years he fought for the need to hire geologists, given which had few technical personnel and did not count on them. In addition to

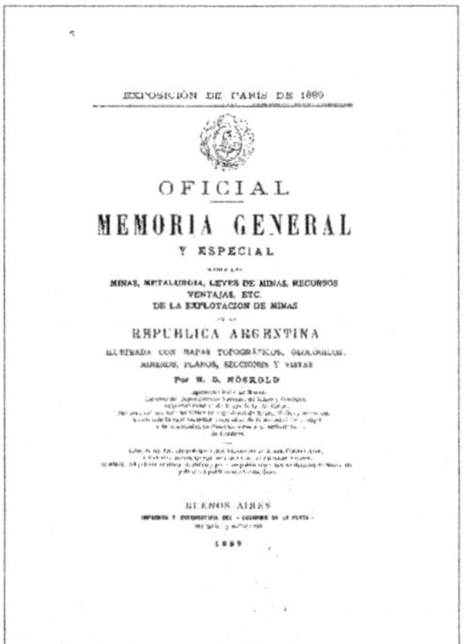
OFICIAL
MEMORIA GENERAL
Y ESPECIAL
MINAS, METALURGIA, LEYES DE MINAS, RECURSOS
VENTAJAS, ETC.
DE LA EXPLOTACION DE MINAS
REPUBLICA ARGENTINA
BUENOS AIRES
1889

Fig. 9.2 Synthesis of the mineral resources of Argentina by Davis Hóskold presented at the Paris Exhibition of 1889 and illustration of the La Peregrina Mine (after Hóskold 1889)

mining studies, he was the one who promoted the need to survey the topography and geology of the country on a scale of 1: 200,000 (Reverberi 2004), a project that became a reality many years later.

9.2 The Beginnings of the Mines, Geology and Hydrology Division

The year 1904 is marked by great institutional progress and an advance of modernity with the new building of the National Congress, the inauguration of several hospitals, new avenues and an important extension of the railway system. It was at that time that the Division of Mines, Geology and Hydrology was restructured, and engineer Hermitte was placed in charge as director. This new organization was divided into three sections: mines, geology and hydrology, which oversaw the systematic work of drilling, geology and exploration of mineral resources, water, coal and hydrocarbons (Zappettini 2004).

The engineer Hermitte invites Dr. Guillermo Bodenbender to organize the Geology Section (Camacho 2008). Bodenbender had arrived in the country in 1884 and was a member of the National Academy of Sciences in Córdoba, which in those first years of the century was going through critical budget problems and its relationship with the university was in crisis.

Bodenbender took charge of the organization of the section, while he carried out various studies on coal deposits in various areas of the country and produced the first 1: 1,000,000 scale map of the province of Córdoba in 1905, published by the division. (see details in Chap. 6). Although he returned to Córdoba for health reasons, he continued collaborating with the division as an *ad-honorem* geologist until 1925.

At the end of 1902, Dr. Rodolfo Hauthal had visited the University of Strasbourg, who had been commissioned by the authorities to hire a chemist with knowledge of mining and geology. There he met a young chemist Federico (Fritz) Reichert (1878–1953), an excellent professional and good mountaineer who was enthusiastic about the possibility of coming to work in the Andes. He had obtained a doctorate from the University of Strasbourg in 1902, when Alsace was under German rule. "*I clung to the skirts of that 'old man's' frock coat and begged him in every tone for him to intercede on my behalf*," writes Reichert. Hauthal recommended his hiring, which was accepted, he married, and Reichert arrived in the country in May 1904 to join the Division's laboratories. In addition to the laboratory work, he was entrusted with exploring the borates of the Puna de Atacama, a mission that excited him, not only from a professional point of view but also as a fervent mountaineer. A few weeks after arriving in Buenos Aires, he undertook his task in the southern Puna, studying the borate deposits and touring the Puna desert on muleback between June and September 1904. On that occasion, he ascended the Antofalla Volcano and the Nevado de Chañi, not being able to reach the summit at Nevado de Cachi (Fig. 9.3). At the top of Chañi he found the testimonies left by the Swedish geologist Nordenskjöld of a previous

Fig. 9.3 Federico Reichert in his camp at the foot of Nevado de Cachi in 1904. *Source* Segemar archives

climb. The result of this work is his pioneering study on the deposits of borates and other exploitable mineral products in the *Territory of the Andes*, which describes for the first time the ulexite deposits of the borate mines of the Puna (Reichert 1907). The chemist Pedro Arata offers him to set up the laboratories and become a professor of Agricultural and Analytical Chemistry at the recently founded Higher Institute of Agronomy and Veterinary Medicine, a position he accepts, basically because he had long summer vacations to make his Andean expeditions. This institute became the Faculty of Agronomy of the University of Buenos Aires where he taught until his retirement in 1936.

From these laboratories came the first analyzes of the oil found in Comodoro Rivadavia in 1907, a study carried out with Walther Schiller in 1909, a young comrade of his ascents in the Alps. With Schiller, they toured and climbed several peaks in the region south of Aconcagua. He was a founding member of the Argentine Society of Geographic Studies GAEA and was linked by his scientific interest and his vocation as a mountaineer to the German expeditions from 1910 to 1917 in Patagonia. He is considered one of the fathers of Argentine Andinism (Arko 1966).

Recommended by Professor Gustav Steinmann, Juan (Hans) Keidel arrived in the country in 1906 (Fig. 9.4), and took charge as head of the Geology Section, replacing Bodenbender.

Juan (*Hans*) Keidel received his degree in mining engineering from the University of Berlin in 1899 and a doctorate from the University of Freiburg in 1904 under the direction of Gustav Steinmann with a thesis on the Carboniferous of Tien-Shan, where he had participated in a pioneering exploration of that remote region of Asia. As soon as he arrived, he began to plan the program of topographic-geological surveys at a scale of 1:200,000 that was implemented in 1910 and to carry out the first explorations in the Precordillera in the north of Mendoza, especially of the Rhaetian coals and their structure.

Shortly after Keidel arrived in Buenos Aires, Eduard Suess (Fig. 9.5), father of modern tectonics and the term Gondwana, wrote to Steinmann asking if the Sierras de la Ventana were a virgation of the Andes. Steinmann referred him to Juan Keidel,

Fig. 9.4 Guillermo Bodenbender (1857–1941), who was in charge of the recently created Geology Section; Juan Keidel (1877–1954) and Ricardo Stappenbeck (1880–1963), the first two young geologists who joined the section graduated from the universities of Freiburg and Berlin. *Source* A.N.C. and Segemar archives

his young disciple who had just joined the Geological Survey. When he received the letter, he consulted with Rodolfo Hauthal, who worked at the La Plata Museum, and with other geologists who knew the area and sent Suess an extensive response, where he stated "*it is a Paleozoic mountain range developed on the margin of Gondwana... and it has no relations with the Andes*". Suess was so pleasantly impressed by the response that he had it published directly in the annals of the Vienna Academy of Sciences, becoming an extensive 25-page article that was Keidel's first work on the Ventania System (Keidel 1907). This event marked the beginning of Keidel's interest in the geology of Sierra de la Ventana.

The occurrence of Paleozoic conglomerates in the Sierra de la Ventana was known since the work of Hauthal (1892). Halle (1911) had described for those years the late Paleozoic glacial deposits of Hill Cove in the Malvinas Islands, which he correlated with the Dwyka Tillite of the Cape system. Keidel found great similarity in the descriptions of the Lafonia Tillites of Malvinas with Hauthal's (1892) descriptions of the Sauce Grande conglomerates and decided to visit the Sierra de Pillahuincó for the first time in 1912. On this visit he found strong evidence of its glacial origin, such as clasts with pentagonal shapes and strong striations (Fig. 9.6). The following year he presented his observations at the 12th International Geological Congress in Toronto where he presented "*About the age, distribution and mutual relations of the different tectonic structures in the Argentine mountains*" (Keidel 1913a, b). In that work, he describes for the first time the glacial origin of the Sauce Grande Tillites and describes the structure of the Ventania system, where the intense deformation and its northeast vergence stand out (Fig. 9.7). Around 1916, and as a result of new study trips, he presented in a very complete contribution his observations and theories about the genesis of continents and oceans that, at that time, occupied the high scientific centers of Europe. Keidel (1916) exposes the similarity in composition and structure between the Sierras de la Ventana and the Cape Mountains in South Africa. This

Fig. 9.5 Gustav Steinamn (1856–1929) thesis director of Keidel and Eduard Suess (1831–1914) a referent in the tectonics of those times. *Source* Wilckens (1930) and Wikimedia

Fig. 9.6 Keidel's first visit in 1912 to the Sierra de la Ventana. He is seen sitting on the Sauce Grande Tillite of upper Carboniferous age (Keidel 1916)

work had important consequences in the work of Wegener (1915) and the foundation of the Continental Drift (see Ramos 2013). Based on the new observations, Wegener changed the discussion on the Geology Chapter of his first edition, describing the correlation between the Ventania and Cape System proposed by Keidel (Wegener 1915 and successive editions).

The results of these investigations were anticipated in a conference given at the Argentine Scientific Society in 1913, where Keidel presented his new developments

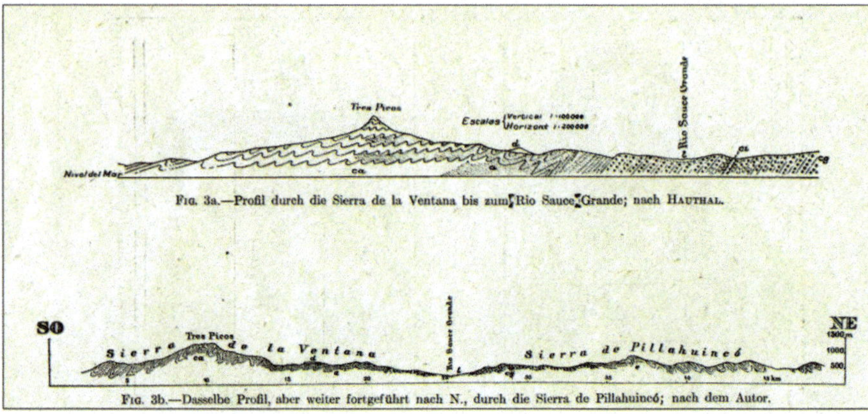

Fig. 9.7 Section of Cerro Tres Picos surveyed by Hauthal in 1892, and a new one presented by Keidel (1913a, b) at the 12th International Geological Congress of Toronto in 1913

on the "*Geology of the mountains of Buenos Aires and their relations with the mountains of South Africa and the Andes*".

Although his interest was in elucidating the nature of Sierra de la Ventana, upon his arrival in Argentina, in addition to a brief reconnaissance of the Precordillera north of Mendoza, he had to carry out a series of studies in northern Argentina. These investigations allowed him to recognize for the first time the presence of Precambrian deposits in unconformity below the Paleozoic sequences and the complex thrust structures produced by the Andean movements in the Eastern Cordillera of Salta and Jujuy (Borrello 1952). A preliminary report published on the structure of the Quebrada de Humahuaca (Keidel 1908–1910) is followed by his proposal of the Gondwanides, as a late Palaeozoic belt, together with the description of other Permian glacial sequences of northwest Argentina (Keidel 1921a). However, his masterpiece on the Eastern Andes of Salta and Jujuy was published years later, already retired, based on his initial observations (Keidel 1943).

In those first years between 1915 and 1920 he studied the stratigraphy of the Precordillera of San Juan and Mendoza, presenting his classic detailed profiles of the early Paleozoic sequences (Keidel 1921b) (Fig. 9.8).

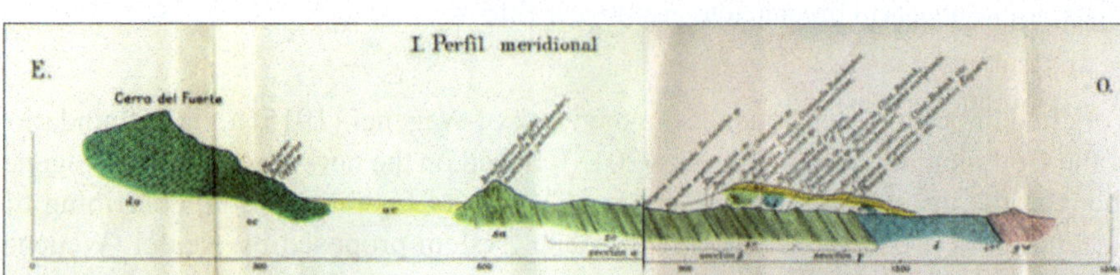

Fig. 9.8 Classic profiles of the lower Paleozoic in Cerro del Fuerte, Precorillera sanjuanina (after Keidel 1921b)

Keidel, in addition to his geological investigations, produced an important change in his management as Head of the Geology Section. In his 1908 report, he presented a new approach to planning future geological work, which until that date was related to the presence of mineral deposits. He maintained the need to also study other regions and carry out strictly scientific investigations, establishing which regions would be appropriate to carry them out, so that the country had for the first time an organic plan for geological activities (Camacho 2001).

The second geologist to be hired by the Mines Division who arrived in the country in 1906 was Ricardo (*Friedrich August Richard*) Stappenbeck. He received his doctorate from the University of Berlin in 1904, under the direction of the famous paleontologist Josef Felix Pompeckj. His thesis was purely paleontological. He remained in the country for 12 years, becoming head of the Hydrogeology Section until 1927, when he returned to Germany. He has produced a fundamental work, the *Geologie und Grundwasserkunde der Pampa* (Geology and Hydrogeology of the Pampa) in 1926. Details of his hydrogeological work were analyzed by Tineo (2008) and his scientific career with a complete bibliography was presented by Olsacher (1963).

The discovery of oil in the San Jorge Gulf basin in 1907 motivated a series of studies that were led by Stappenbeck (see Chap. 5), who was sent to Comodoro Rivadavia to supervise drilling in 1908. On that occasion he made an excellent study of the groundwater potential of the region, established the geology of the different sequences and analyzed the oil horizon, being one of the first works related to the hydrocarbon potential of this basin (Stappenbeck 1908, 1909). These works are presented together with one of the first geological maps of the basin between Lake Muster and the Atlantic coast at a scale of 1: 500,000.

One of the first regional surveys that Stappenbeck was commissioned to do was that of the Precordillera of San Juan and Mendoza (Fig. 9.9).

As Bruno Baldis said in 1990 on the occasion of inaugurating Stappenbeck Plaza in the center of the town of Zonda during the Argentine Geological Congress in San Juan "*this young geologist, at only 25 years old, arrives in San Juan and performs an extraordinary task, leaving their mule trains from this plaza to cover a large part of the San Juan Precordillera*". Without adequate maps, based on the railroad layout in the foothills, he created the first geological map of the San Juan and Mendoza Precordillera at a scale of 1: 500,000 where the main characteristics of the region are already outlined (Stappenbeck 1910). The fossils found by Stappenbeck were classified by Emanuel Kayser (1845–1927) who had already studied those collected by Bodenbender and Stelzner (Kayser 1876). These fossils were later the subject of a complete review by Clarke (1912). The absence of petrographic laboratories in Buenos Aires made it necessary to resort to the Petrographic Institute of the University of Munich, which studied the 400 samples obtained by Stappenbeck in the Precordillera and others from a brief excursion in 1907 to the Sierra de Pie de Palo. A complete petrographic report was published by Otto Stieglitz shortly afterwards (Stieglitz 1914).

Between 1909 and 1910 he also raised the eastern slope of the Cordón del Plata, establishing the stratigraphy and structure of the Frontal Mountain Range of Mendoza

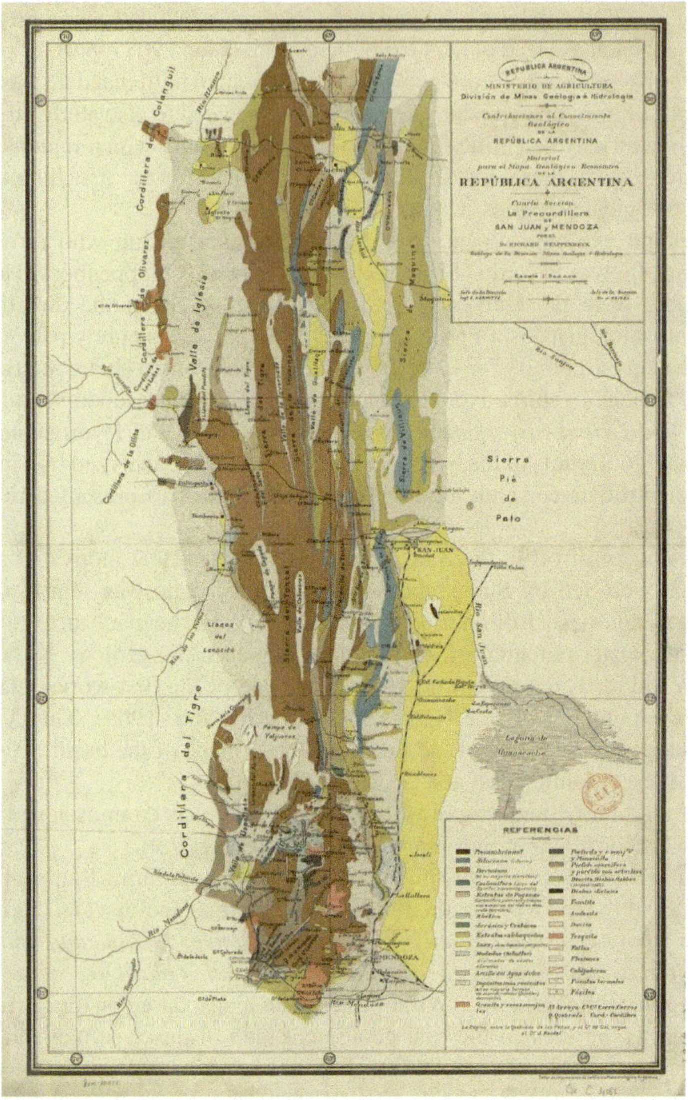

Fig. 9.9 First geological map of the San Juan and Mendoza Precordillera published by Stappenbeck (after Keidel 1910). Note the precision of the geological data. At the same time, it indicates that the Sierra de Las Peñas and Cerro La Cal were previously mapped by Juan Keidel

Between 1909 and 1910 he also raised the eastern slope of the Cordón del Plata, establishing the stratigraphy and structure of the Frontal Mountain Range of Mendoza, which was published years later and evaluated the oil potential of the Cacheuta basin (Stappenbeck 1917).

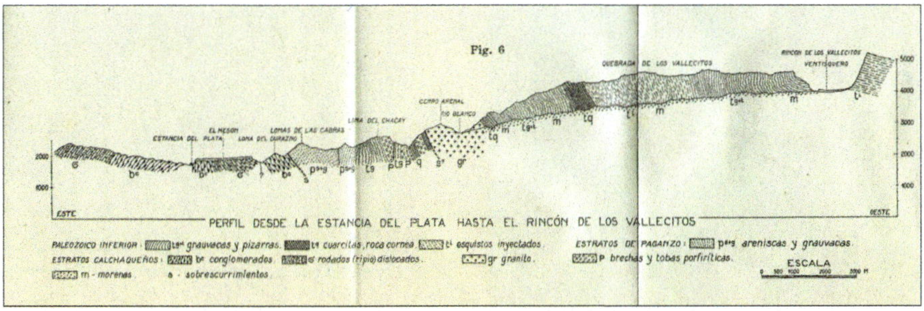

Fig. 9.10 Classic structural section of the Frontal Cordillera in the eastern foothills of the Cordón del Plata in Mendoza (Stappenbeck 1917)

Between 1909 and 1910 he also raised the eastern slope of the Cordón del Plata, establishing the stratigraphy and structure of the Frontal Mountain Range of Mendoza (Fig. 9.10), which was published years later and evaluated the oil potential of the Cacheuta basin (Stappenbeck 1917).

He also made a complete inventory of mineral deposits and application rocks in Argentina (Stappenbeck 1918). In 1909, the Hydrogeology Section was created and was entrusted with the execution of the *Hydrogeological Map* and the study of all the drilling cores carried out by the Hydrology Section, under the direction of Julio Krause, in addition to the preparation of the corresponding profiles. Stappenbeck was appointed to head the new Section. Thus, starting in 1910, he began an extensive geological and hydrogeological study of the Pampas region in the center of the country, presented in various reports that he summarized in his magnum opus of 1926. He died in Munich in 1963.

Windhausen comments in his book that all underground drilling took place between 1910 and 1920, and that after 1920 no further deep drilling was carried out in search of water (Windhausen 1931, p. 502). In almost all the provinces, aquifers of different importance and nature were recognized and in some, extensive artesian basins were found, such as those located to the east of the cities of San Juan and Mendoza and Bahía Blanca, about which Hermitte (1910) and Stappenbeck (1926).

With the arrival of new German geologists, the Mines Division plant was expanded, highlighting at this stage the contributions of Anselmo Windhausen, Enrique Gerth and Pablo Groeber.

Anselmo Windhausen (1882–1932) arrived in the country in April 1909 (Fig. 9.11), with good geological experience and his doctorate obtained in Goettingen in 1907, under the supervision of Adolf von Koenen (1837–1915). His thesis topic was Die geologischen Verhältnisse der Bergzüge westlich und südwestlich von Hildesheim. His first experiences were in the Unicorn Cavern in the Harz Mountains, where he worked on several occasions to survey the cavern and its geology, making novel contributions to Quaternary geology, observations that continued to be valid after more than 70 years (Vladi 1981).

Fig. 9.11 Anselmo Windhausen, Enrique Gerth, and Pablo Groeber joined the Geology Section between 1909 and 1911. *Source* Segemar archives

When Rodolfo Hauthal returned to Germany to take over the direction of the Roemer-Museum in Hildesheim, Lower Saxony, he met a student who worked at the museum and was doing his thesis in this region. It was the opportunity for Windhausen to encounter the characteristics of Argentine geology (Hünicken 1986). He accepted an offer from the Argentine government and came to the country to work in the Geology Section.

Windhausen's first commission was in the southern sector of the province of Mendoza, and shortly after he was in charge of organizing a museum for the Division of Mines, Geology, and Hydrology, which prepared the materials that were to be exhibited in 1910 at the Exhibition of the Centennial in Buenos Aires; and that, a year later, they would be part of the sample that would be presented at the Industries and Work Exhibition in Turin and Roubaix where the Grand Prize was won (Argüello and Sachi 2020).

Another geologist who arrived at that time was Enrique Gerth (1884–1971) who arrived in Buenos Aires in 1910 to join the Geology Section. He came with the recommendation of Gustav Steinmann, who was his professor at Freiburg, but he presented his doctorate at the University of Bonn in 1908, and his habilitation in 1910. Shortly after his arrival, his research covered part of the Sierras Pampeanas. He was commissioned to carry out a study of the geological conditions for the provision of groundwater in the province of San Luis. During this period, he made several study trips to the mountain region, especially to the province of Mendoza.

After completing this task, Dr. Gerth took charge of the reconnaissance of the Mendoza Mountains in the section between the Diamante and the Atuel rivers. The results of this research revealed his great work capacity and his excellent academic training since his geological observations gave a strong boost to the knowledge of the stratigraphy and paleontology of these regions. He made a precise topographic-geological map on a scale of 1: 200,000, covering a wide area in which, with special detail, he concentrated on the characterization of the Jurassic and Lower Cretaceous marine deposits that have exceptional development in the investigated area

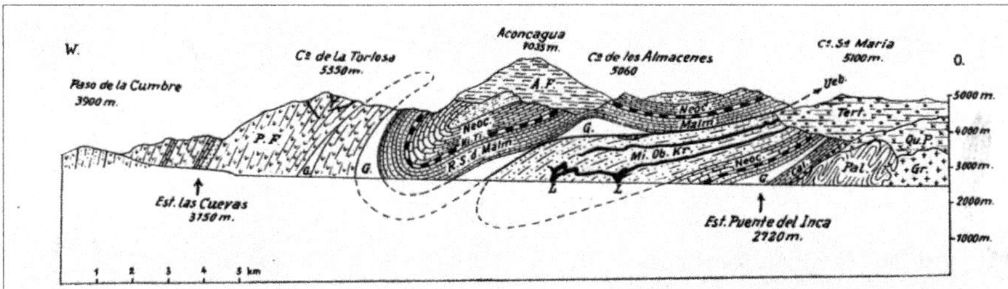

Abb. 12. Profilentwurf durch den Gebirgsstock des Aconcagua nach der Profilansicht von SCHILLER (nach GERTH, 1926).
Pal = metamorphes Paläozoikum, gr = Granitintrusion in diesem, QuP = triadische Quarzporphyre, Cal = Callovien, G = Gips des unteren Kimmeridge, PF = Porphyritformation mit Einlagerungen von Kreidekalk, AF = alttertiäre Andesitserie

Fig. 9.12 Structural section through Cerro Aconcagua where Gerth interprets the thrusts identified by Walther Schiller (Gerth 1926)

(Gerth 1913). Gerth completed his work with the study of the Neocomian ammonites collected by him, offering a paleontological zonation of the collected fossils.

At the end of his contract in Argentina in late 1913, he made a long tour of the Andes of Bolivia and southern Peru. Between 1916 and 1918 he participated in the First World War as a military geologist for the German army. In 1918 he was appointed full professor at the University of Bonn (Mon 2021). He completed his prewar investigations which were published in Gerth (1921, 1926) and translated by Bodenbender in Gerth (1925a, b, 1931). In one of his last works, he revalues the structural section of the Aconcagua fold and thrust belt, made by Walther Schiller years ago (Fig. 9.12), being one of the few accredited in the thrusts identified by this author (Gerth 1926).

Walther Schiller (1879–1944) had been hired by Perito Moreno to take charge of the Mineralogy and Geology section at the La Plata Museum, to replace Rodolfo Hauthal, who had returned to Germany. Schiller had completed his doctorate in 1903 at the *Albert Ludwigs Universität Freiburg*, a prestigious old German university where Gustav Steinmann taught and was his director. He did military service during part of the years 1903–1905, in the battalions of Alpine hunters, in Alsace. In 1904 he published a work that involved a detailed geological survey of a harsh region of the Alps, disturbed by tectonic dislocations of a hitherto unknown type. This young researcher discovered out-of-sequence thrusts, which put younger rocks on top of older rocks, unlike traditional thrusts, which had a strong impact on his director and the specialists of the epoch. He was recognized as a shrewd and meticulous researcher and, at the same time, an excellent mountaineer with solid training and exceptional physical resistance (Fossa Mancini 1944).

As soon as he arrived in Argentina in 1905, he expressed his desire to closely examine the colossus of the Andes and reach its long-awaited summit (Fig. 9.13). Already in the summer of 1906, together with Roberto Lehmann Nitsche (1872–1938), anthropologist and researcher at the La Plata Museum, also a graduate of Freiburg, they had made an excursion to Puente del Inca, reaching 6,000 m high in Aconcagua (Cuerda 1997). The preliminary results of this recognition were made

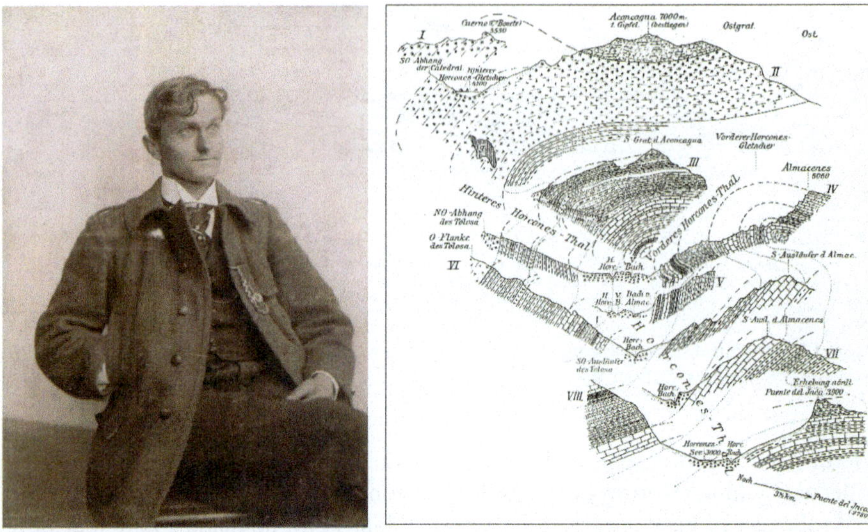

Fig. 9.13 The young geologist Walther Schiller, and the first structural sketch of the Horcones River and Aconcagua in the High Cordillera of Mendoza. *Source* Segemar archives and Schiller (1907)

known by Schiller (1907) with the first sketches that showed the complexity of the structure.

Shortly after his arrival, he was invited to work on behalf of the Geology Section in the *Alta Cordillera.* Since 1907, he was in charge of a geological commission whose objective was the survey of the Main Cordillera of San Juan and Mendoza, making important contributions to the structure and stratigraphy of this region (Schiller 1908, 1912). Among his contributions, the identification of overthrusts such as that of Cerro Penitentes stands out, "*a phenomenon that was not known in such a great way in all of South America*" (Schiller 1912: 51). In turn, his ascents of Aconcagua, although he was not able to reach the summit, allowed him to recognize that it was not a volcano, since its summit was formed by tuffs and deformed volcanic rocks.

Between 1910 and 1911 he carried out stratigraphic and structural studies of the Comodoro Rivadavia sector related to the discovery of oil, partly commissioned by the General Direction of Petroleum Exploitation of the Ministry of Agriculture, which he published years later with a complete detailed map of the surroundings of Comodoro Rivadavia (Schiller 1925).

Upon returning from a campaign in Comodoro Rivadavia in 1915, he left the country traveling incognito to his homeland. He left everything he had, including his wife, to join as a volunteer in the Alpine Hunter Corps since Germany had entered the First World War. He was sent as soon as he arrived at the battlefront (Hernández 2020) and returned to Argentina in 1919.

9.3 The Creation of the General Direction of Mines, Geology and Hydrology

A new hierarchy of the old division in 1911, always under the supervision of engineer Hermitte, allowed the regional work of topography and geology to continue to be intensified, launching the geological map at a scale of 1:200,000. These activities, together with the creation of a seismological station in Mendoza in 1907 (Camacho 2001), allowed the division's organization to be consolidated after the creation of the Topography Section and the Mineral Waters Section. These works were complemented by important drilling plans in search of water in different sectors of the country (Hermitte 1914).

The imminent international conflict would affect the oil supply, so Keidel, Windhausen, and Groeber began an extensive mapping of the territory of Neuquén. The existence of oil springs in the Cerro Lotena region led to Windhausen being sent to study this region between 1912 and 1913. Upon his return, he gave a lecture at the Argentine Scientific Society on the Geology of the Neuquén Oil Province in September 1913. where he indicated the three most favorable areas; one of them was the discovered oil in Plaza Huincul years later. This gave rise to a controversy that continues today about who first identified the Plaza Huincul sector where the oil discovery well was made in the Neuquén basin in 1918 (Windhausen 2007).

It has always attracted attention that Juan Keidel, head of the Geology Section, has ordered his work on the Geological Report on the Challacó Oil Field to be printed in the Peuser publishing house as printing by the author (Keidel 1913a, b). In this work, he indicates the locations of the main anticline and the proposed well, which was the discovery of oil years later. However, in this publication, he mentions that in the Zapala area, the axes have a west–east direction according to Dr. Windhausen (unpublished observation) and indicates the date of the article as June 23, 1913, months before the Windhausen conference. Our interpretation is that the Challacó area was known to Keidel previously and when he saw Windhausen's report, he rushed to publish his previous observations.

Finally, Windhausen's observations, very important both in the biostratigraphy of the Neocomian and in the "aberrant structures" of the east–west direction, to highlight that they did not follow the northern direction of the Andean structures, were published later (Windhausen 1914). These aberrant structures are those that De Ferraris (1947) later identified as the Huincul High. This work is also "*...a study of the oil region of the central part of Neuquén (Lotena and Covuncó Hills)*", which partially coincides with that described by Keidel (1913a, b).

Several years later Keidel carried out an important synthesis of Neuquén geology based on the studies carried out in those years and analyzed Windhausen's contributions with a critical perspective (Keidel 1925).

Windhausen was subsequently sent to Patagonia to support the Comodoro Rivadavia oil field studies (Figs. 9.14 and 9.15), which he carried out between 1918 and 1923 (Windhausen 1921, 1923). Windhausen had anticipated in the *American Journal of Science* the distribution of marine intrusions from the Atlantic Ocean with

Fig. 9.14 Anselm Windhausen west of the Gulf of San Jorge c. 1919, with the baquiano and his loads, where for almost four years he toured the central region of the basin, making important contributions and studies on the control of the deposits (Windhausen 1924)

quite a precision that contrasts with the low level of later paleogeographic interpretations such as those of Mercerat (1925, his figure p. 251). This comparison only highlights the solid academic training of Windhausen, who made important contributions to the paleogeography of the Meso-Cenozoic marine ingressions on the eastern coast of Patagonia related to the breakup of the Gondwana (Borrello 1964).

Shortly afterward Windhausen resigned from the Direction of Mines and accepted a teaching position at the University of Córdoba (Storni 1947). In those years he produced a gigantic work such as his Argentine Geology in two volumes published by Editorial Peuser, which constituted not only the first complete treatise on regional geology of the time but also an effort by a single author that has never been surpassed (Windhausen 1931).

At the request of Juan Keidel, head of the Geology Section, Pablo (*Paul Friedrich Karl*) Groeber (1885–1964) arrived in Argentina in 1911 (Fig. 9.16). There, his impressive work began, which has led him to be considered one of the great masters of Argentina's geology and a generator of a school of disciples who, both in academia and industry, laid the basis of our current knowledge.

Pablo Groeber received his doctorate from the Ludwig-Maximilians-Universität of Munich in 1906, when he was only 22 years old, under the direction of Ernst Wilhelm Benecke, famous paleontologist at the *Geologisch-Paläontologisches Institut* in Strasbourg. He had made two large expeditions to the Tien Shan and came with a very good background (Leanza and Ramos 2011).

Upon his arrival, he was assigned to the province of Neuquén to work on the 1: 200,000 scale topographic-geological survey program (Fig. 9.17) that Keidel had implemented in 1911, following a previous proposal by Hóskold.

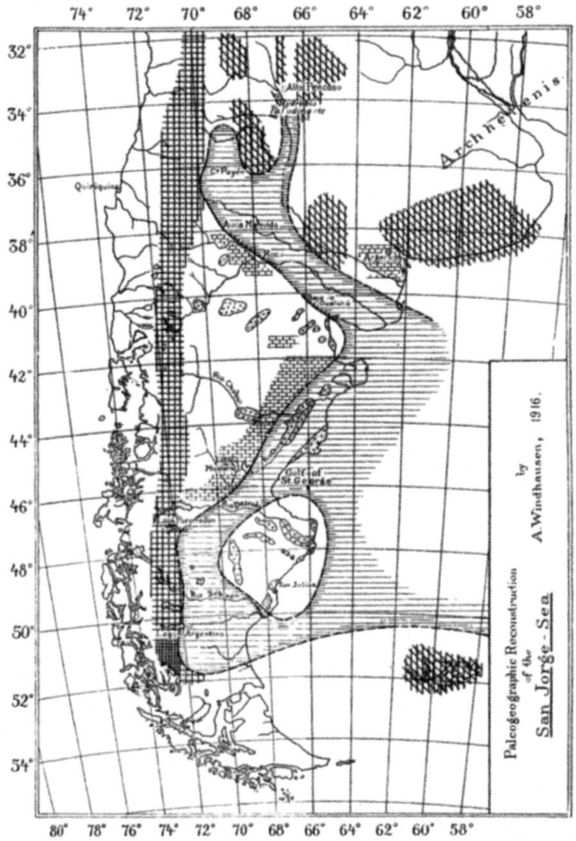

Fig. 9.15 Paleogeographic sketch of Patagonia published by Windhausen (1918) in the American Journal of Science. Note his proposal for the "*Isla del Deseado*" as a positive element in Atlantic ingressions (after Windhausen 1918).

Fig. 9.16 Pablo Groeber in his youth and as an old man in 1959, with his disciples Christian Petersen, Armando Leanza, and Oscar Ruiz Huidobro (Leanza and Ramos 2011)

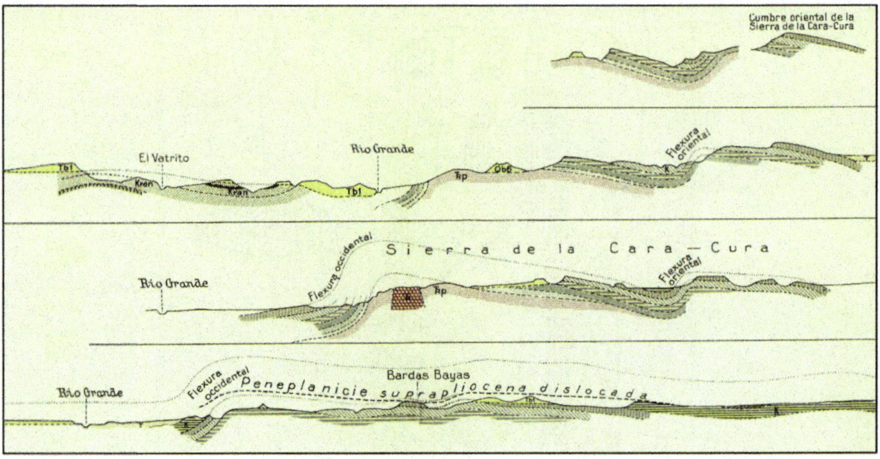

Fig. 9.17 Structural section of Hoja 31 c, Confluence of the Grande and Barrancas rivers, one of the first maps surveyed at a scale of 1: 200,000, where you can see the west-vergence of the deformation in the Sierra de Cara Cura (Groeber 1933)

He had begun the survey of Hoja 31c, Confluence of the Grande and Barrancas rivers, where he had to carry out both the topography and the geology survey between 1914 and 1917, which was published years later (Groeber 1933). Engineer Hermitte interrupted this survey to study the disasters on the Colorado River that occurred on December 29, 1914, since he knew the region before the flood. It was his third campaign to the mountain range and his studies immediately after the flood were submitted to the Ministries of the Interior and Public Works "*to prevent new misfortunes.*" This report was published the following year and became his first contribution to our country (Groeber 1916). This study and the interpretation of the natural processes that led to the outbreak and the subsequent flooding are still valid, as well as the estimated volume, which has made it one of the three largest natural alluviums recorded worldwide and one of the first being studied (see details in Ramos 2017).

Groeber during those years worked intensely in Neuquén producing fundamental works such as *Stratigraphy of the Dogger in the Argentine Republic* and *Age and extension of the structures of the Cordillera between San Juan and Nahuel Huapi* (Groeber 1918a, b a and b), which began a series of contributions to knowledge of the Andes Mountains that became classic works of obligatory consultation (Ramos and Leanza 2011).

Guido Bonarelli (1871–1951), a native of Ancona, Italy, arrived in the country in 1911, hired by Eng. Hermitte, as an expert in oil exploration (Fig. 9.18). He had graduated from the University of Turin, where he obtained his doctorate in 1894 under the direction of Carlos Fabrizio Parona, one of the most important Italian geologists and paleontologists of the time and a great expert on the Alps. Dr. Pierina Pasotti also had him as director of her thesis in 1927 and she was for years an admirer of his wisdom. After graduating, Bonarelli was hired by Royal Dutch, Shell's predecessor

Fig. 9.18 Guido Bonarelli when he arrived in our country and already matured in Italy (Bonarelli 2002)

oil company, to explore Borneo, southern Spain, Algeria, and Tunisia between 1901 and 1907 (Concheyro and Montenegro 2011).

His most important work was the survey and evaluation of the oil potential of the Subandean Sierras, so-called by Bonarelli, where he made important contributions. His first results were presented in a public conference given at the Argentine Scientific Society in 1913 on the "*Oil Formation of Salta and Jujuy*", where he disseminated the important oil potential of northern Argentina, pioneeringly described by Brackebusch (1883). The results were published in three geological contributions that presented the first maps and structural sections (Fig. 9.19) and an analysis of the petroleum system of the Subandean Sierras (Bonarelli 1913, 1914, 1921). In the latter, he carries out a complete analysis of the stratigraphy and development of the oil formation that opens important perspectives for the region's hydrocarbons.

Interestingly, the first well proposed by Bonarelli in the Capiazuti anticline in Salta had to compete in 1917 with the first drilling in Plaza Huincul located by Keidel. Given the lack of funds that the Direction of Mines had, Hermitte summoned both geologists and after hearing the potential of both regions in a heated discussion, they could not reach an agreement, so they had to make the decision by flipping a coin. Plaza Huincul won, whose first well was successful and Capiazuti was temporarily suspended, according to what Engineer Enrique Cánepa (1967) tells us. Sometime later, when this last location was drilled, it turned out to be dry. However, during Bonarelli's lifetime, he did not learn of the success of the drilling carried out later in the structures he selected in the Subandean Sierras, given that the first ones did not reach the depth necessary to find the oil formation.

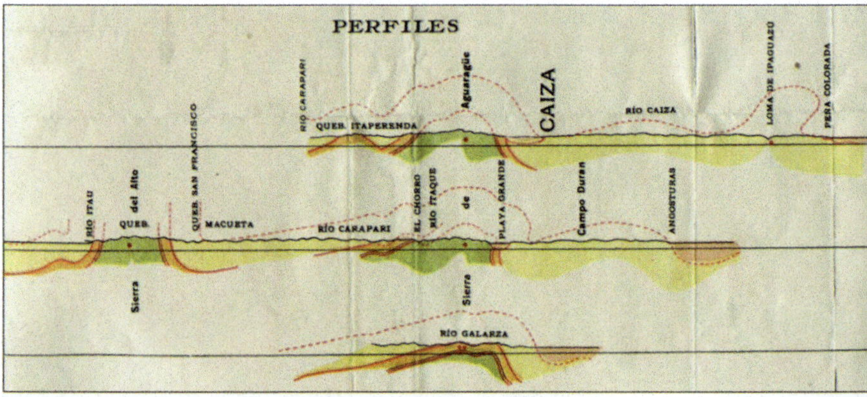

Fig. 9.19 The first structural sections of the Subandean Sierras were surveyed by Guido Bonarelli in 1913 (Bonarelli 1921)

During his first stay in Argentina, Bonarelli worked between 1911 and 1918 in the Direction of Mines. In addition to studies in northern Argentina, he worked in Mendoza, Santa Cruz, and Tierra del Fuego, among other sectors (Bonarelli 1917; Bonarelli and Nágera 1921).

The year after the arrival of Groeber and Bonarelli, Walther Penck (1888–1923) arrived in Buenos Aires in 1912 (Figs. 9.20 and 9.21). This young German was the only son of the famous geographer Albrecht Penck who was the first to propose a world map on a scale of 1: 1,000,000. He received his doctorate in 1910 from the Ruprecht-Karls-Universität in Heidelberg, the oldest university in Germany. He had previously been to the continental United States and to Hawaii, where he was impressed with the lava lakes of Kilauea. His thesis on the Predazzo mountains in the Trentino foothills, currently Italy, caused a strong impact due to the novelty of its explanations of the magmatic dynamics associated with tectonic development (Backlund 1923).

The Direction of Mines entrusted him with the survey of the southern edge of the Puna de Atacama. Between 1912 and 1914 he explored topographically and geologically an area of about 14,000 square km with great difficulty, with inaccessible heights, where he tested his ability as a mountaineer. A high hill in the Catamarca Puna of more than 6,500 m is named after him. The raised region comprised Hojas 11c and 12c, and a large part of Hojas 11b, 12b, 13b and 13c. He presented a preliminary report (Penck 1914b), and years later published his new findings on the relationships between volcanism and orogeny (Fig. 9.22), although his work transcended due to his analysis of geomorphological processes (Penck 1920, 1924). He spent 2 years in our country and upon returning to Germany in 1914 he participated as a volunteer in the First World War. In 1915 he accepted a teaching position at the University of Istanbul, where he founded the Geological Institute. He stayed there until 1918 when he fell ill and returned to Leipzig where he died very young in 1923 (Amferer 1924).

Fig. 9.20 Walther Penck in the Puna; as director of the Geological Institute of the İstanbul University, and his last years in Leipzig. *Source* Yalçın and Penck (2019) and Yalçın (2015)

Fig. 9.21 During his work between 1912 and 1914 in the southern Puna (Photographic archive of Gerhard Penck in Centro Cultural Argentino de Montaña). *Source* Segemar archives

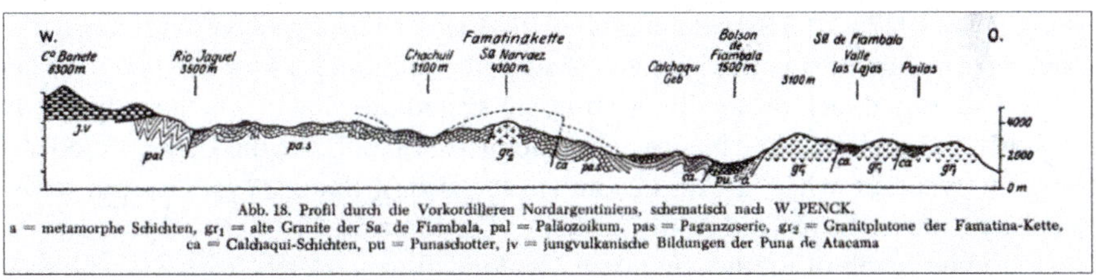

Fig. 9.22 Structure of the Sierras Traspampeanas between Cerro Bonete and the Famatina and Fiambalá mountain ranges (after Penck 1914a, 1920)

Ricardo (*Richard*) Wichmann (1881–1930) also arrived in the country in 1912 to join the Direction of Mines (Fig. 9.23). A native of Hamburg, he had completed his doctorate at the University of Goettingen under the direction of Adolf von Koenen.

His first work was the control of the Argerich drilling in 1913 in the surroundings of Bahía Blanca, which culminated in the survey of the first geological map at a

Fig. 9.23 Ricardo Wichmann was a great expert on Extra-Andean Patagonia, where he surveyed extensive areas through numerous and extensive fieldworks, especially the Somun Cura Massif in Río Negro and Chubut. *Source* Biblioteca Digital F.C.E.N

scale of 1: 200,000 published, where he studied in detail the Cenozoic deposits, to evaluate the groundwater of the region. The map was released in 1916 and the text in 1918 (Wichmann 1916, 1918).

After 1917 and 1918, when he was controlling the drilling of the Comodoro Rivadavia oil field (Wichmann 1921), he dedicated himself to studying the extra-Andean regions of Neuquén, Río Negro, and Chubut, making numerous contributions to their geology. Among his contributions are the identification of the lacustrine deposits of the strata with dinosaurs, the discovery of *Baculites* in the Rocanense that allowed him to recognize the beginning of the transgression in the Maastrichtian and establishing the stratigraphic sequence of the Cenozoic deposits of those regions (Wichmann 1927a, b). His campaigns of 1922 and 1923 were epic with more than six continuous months in the field and thousands of kilometers traveled (Wichmann 1934). Due to the lack of a more appropriate scale base (Fig. 9.24), he mapped the extra-Andean region of Río Negro at a scale of 1: 500,000. He died unexpectedly in 1930, hit by a tram when he was heading to the port of Buenos Aires to wait for his brand-new wife who arrived from Germany (Groeber 1931).

On the occasion of the International Geological Congress held in Stockholm in 1910, Juan Keidel met the Swedish geologist Helge B. Backlund (1878–1958) and invited him to work in Argentina. He accepts the invitation and arrives in the country in 1912. Backlund was born in Dorbat, currently Estonia, which was under Swedish rule. Backlund studied geology at the *Université Impériale de Saint-Pétersbourg*, graduating in 1902, and completed his doctorate at the *Universität Wien* in 1908. He worked at the *Musée Geologique Pierre le Grand prés l'Academie Imperiale des Sciences de Saint Pétersbourg*, where he published his thesis on the olivine group

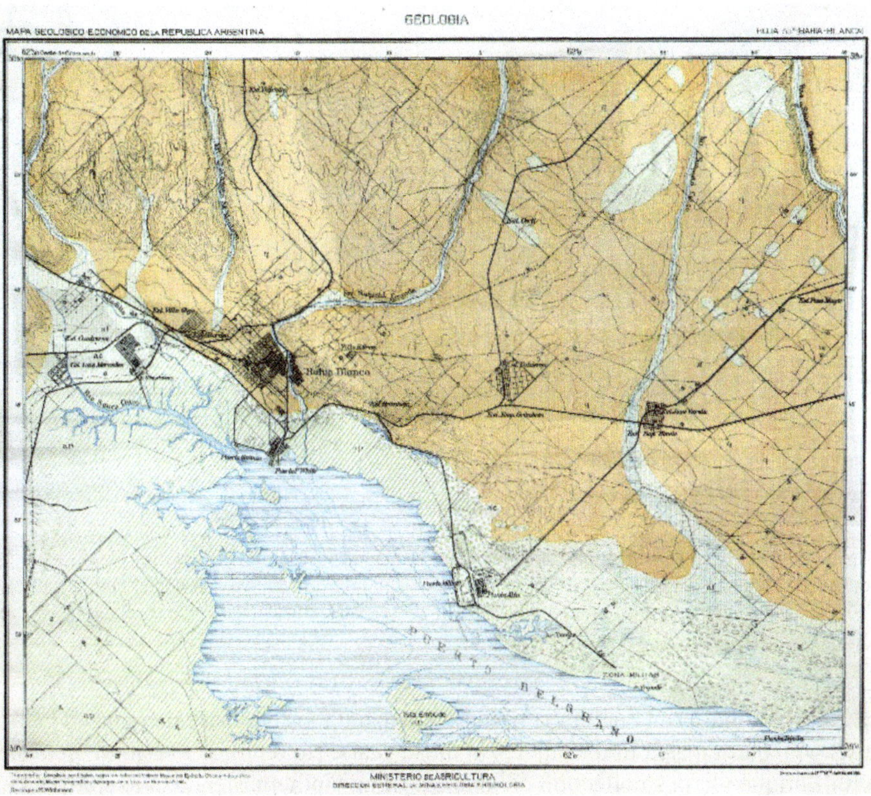

Fig. 9.24 Hoja 35 m Bahía Blanca, first sheet published in 1916 in the Directorate of Mines by Ricardo Wichmann (after Wichmann 1916 y 1918)

in 1909. He made several expeditions to Mongolia and Siberia before coming to Argentina.

One of his first observations in our country was at Cerro Negro of Olavarría where he first recognized the occurrence of mylonites (Backlund 1913). Subsequently, he worked for two years in the Cordillera Principal in southern Mendoza and the Cordillera del Viento in Neuquén (Fig. 9.25). He carried out four campaigns between the Barrancas and Neuquén rivers, and paid special attention to eruptive rocks, being one of the first to identify the migration of igneous rocks towards the foreland. His main contribution was to recognize the existence of two cycles, one pre-Andean and the other Andean of similar compositional variations, represented by effusive and intrusive rocks. Both cycles form a clan with similar chemical characteristics in their temporal evolution, but he interpreted the Permo-Triassic rocks as subaerial in contrast to the Jurassic-Cretaceous rocks, which, due to the presence of spilites, he associated with a marine environment (Backlund 1926).

Fig. 9.25 The Swedish petrologist Helge B. Backlund made important observations in the eruptive rocks of the Cordillera Principal of southern Mendoza and northern Neuquén (Segemar archives). Proposed an orogenic development of magmatism and its migration towards the foreland (Backlund 1921–1923, 1926)

Backlund has been one of the best prepared and most globally recognized geologists who participated in the studies in those years. He returned to Saint Petersburg with Enrique Gerth's collection of samples from southern Mendoza to study them, along with his own, but upon his return from one of his expeditions to Mongolia, the Russian Revolution broke out in 1917. Although he had everything ready to publish, they did not allow him to remove the material studied and he had to flee to Sweden, where he settled at the University of Abo, in Turku. There he began the study of Gerth's samples, who collaborated again for their processing and subsequent publication (Backlund 1921–1923).

Backlund's theoretical contributions were widely recognized. He postulated the formation of granites from metasomatism of sedimentary rocks, the origin of Rapakivi granites, and reworked the stratigraphy of the basement of the Baltic Shield. In 1943, he received the Gustav Steinmann medal for his contributions to the formation of granites by granitization, a currently discarded hypothesis that had important followers in the first half of the twentieth century. In addition, he has received numerous distinctions for his contributions, such as his incorporation as a member of the Swedish and German academies and the Royal Society of Edinburgh (Lundegardh 1958).

In 1912, the geologist Roberto Beder (1888–1930), a prestigious Swiss mineralogist, was hired. He was born in Zürich, received his doctorate in 1909 at the *Universität Zürich*, and later completed his habilitation with Viktor Goldschmidt in Heidelberg. With his arrival in Buenos Aires, the preparation of collections of minerals and rocks from the main Argentine deposits began for distribution to primary, secondary, and higher education establishments (Figs. 9.26, 9.27 and 9.28), a task that between 1914 and 1924 was in charge of Beder (Fig. 9.29).

Beder was sent to carry out studies of different mineral deposits starting in 1913, in the north of Córdoba and south of Santiago del Estero, San Luis, and later Catamarca, where he studied the Capillitas mining district. He carried out excellent mineralogical studies on the wolframites of the Sierra del Morro, the diopside in the crystalline limestones of the Sierras de Córdoba, the topaz of San Salvador in Catamarca, among many other contributions (see Sureda 2008). In 1922 he was appointed member of the National Academy of Sciences and in 1926 professor of Mineralogy and

Fig. 9.26 Building where the Mines and Geology Division operated at Maipú 1241, a few meters from the current Plaza San Martín in Retiro (Segemar Archives). The building was demolished in 1930 as part of the plaza expansion plan. *Source* Segemar archives

Fig. 9.27 The laboratories of the Mines and Geology Division and the institution's first museum of rocks and minerals. In the background on the right, you can see the Argentine Exhibition Pavilion, designed and built in France for the Paris Exposition of 1889 and brought to Buenos Aires (Segment Archives). The three Ford Ts parked on the street correspond to models from 1923 to 1925. *Source* Segemar archives

Fig. 9.28 Interior of the museum designed in 1911 whose furniture was brought from Dresden, Germany, and manufactured by Casa Kuhnscherf e Hijos, along with volume I of the Catalog of Minerals, Rocks, and Fossils started in 1905 that covers the first years. *Source* Segemar Archives and Courtesy of Luis A. Galindo Castro, 2023

Fig. 9.29 Roberto Beder and Hans Rassmuss worked in the Direction of Mines from 1912 to 1924 and 1926, respectively. *Source* Segemar archives

Petrography at the National University of Córdoba, from where he continued to make fundamental contributions to the crystalline limestones of Córdoba and their metamorphism (Beder 1922), one of his most important works on the Sierras of Córdoba, together with studies of different mineral districts (Beder 1928a, b, 1929). He died in 1930 at the young age of 42 because of dental malpractice, followed by an even more unfortunate tracheotomy (Sureda 2008).

Juan (Hans) Rassmuss (1886–1971) arrived in 1912 hired by the Direction of Mines and remained working in our country until 1926. Rassmuss was born in Berlin, where he completed his studies, obtaining a doctorate at the University of Berlin, shortly before coming to Argentina.

One of his first works was on the basement of the Sierra de Aconquija during the summer of 1913–1914, which included a large part of Hojas 12e and 13e of Tucumán and Catamarca. Based on these surveys he published his classic study on *General geological features of the Sierras Pampeanas* in 1916, where he established the differences between two different types of gneisses, the classic gneisses of Suncho and that of Piscuyacu, and recognized the existence of compressional faults that uplifted the basement blocks. He draws attention to the different vergence of the Sierras Pampeanas to the east and west of Capillitas with faults tilting in different directions (Rassmuss 1916). Later he made a more complete report of the Andalgalá region (Rassmuss 1922a, b, c), where his observations were linked to Walther Penck's previous studies located further west.

In addition to these studies, he carried out regional geology research where he evaluated coal finds, and analyzed lignite areas where he highlighted their geology and structure. He evaluated the possibilities of coal deposits in Metán basins in Salta, Marayes in La Rioja, Ñirehuao south of Bariloche, those in the southern Mendoza, and the Rafaelite veins in Neuquén and other regions (Rassmuss 1920, 1921, 1922a, b, c, 1923). He studied the geology, hydrogeology, and morphology of the Sierra de Velasco as a consequence of a strong decrease in the flow of the La Rioja River. He studied and described an oil exploration well in Tartagal, where he interpreted the stratigraphy and structure of the Subandean Sierras. He also analyzed the geology of the region, structure, and origin of the hot springs of Rosario de la Frontera.

He worked in the Direction of Mines until 1923, when he resigned and began working at the Tornquist Bank where he carried out oil exploration studies until 1926. From 1927 he worked in Peru for a few years, returned to Germany, and worked in a Romanian oil company, where he was captured by the Russians and deported to that country. In 1946, through the Vatican, he was able to reunite with his family in France and return to Peru (see González 2008). He died in Santiago de Chile in 1971.

One of the last to join the Direction of Mines was a Finnish geologist, Hans Magnus Hausen (1884–1979), hired as a petrographer (Fig. 9.30). Born in Helsinki, Hausen received his doctorate from the *Helsing Ülikoolis* (University of Helsinki) in 1914, arriving in the country that same year.

One of his first tasks was a reconnaissance of the Sierra de Umango in La Rioja to study its crystalline rocks previously interpreted as archaic and which according to his observations would be of Paleozoic age. He was at the institution until 1917 and made two important contributions. The first was a geological profile of the edge of the Puna of Atacama and in the Eastern Cordilleras of Salta and Jujuy, with a description of some mining deposits in the Puna, which was later published in the Bulletin of the National Academy of Sciences (Hausen 1925). His other contribution has been the first study of the petrography of Misiones where he characterized the basalts. This began with the analysis of samples collected by others and from a drilling in 1914,

Fig. 9.30 Hans Hausen on his visit to Argentina in 1914–1917 and working in the field at a mature age (Uppslags-verket Finland 1968, and Gaigalas 2008)

but in 1916 he made the first field study to establish the stratigraphic and petrographic characteristics of those rocks in the field.

His precise descriptions of the basalts and associated sedimentary rocks were later published (Hausen 1919). He also made a study trip to the sub-Andean region of La Rioja. Upon his return, he worked at the Geological Survey of Finland and finally at the University of Abo, in Turku, Sweden, until 1951. After retiring he served for many years as a state geologist in the Canary Islands and wrote popular and scientific works on the history of the geology and mineralogy of Finland. He died in Helsinki in 1979 (Raukas 1984).

9.4 Hydrogeological Studies

The extension of the railroad to the Pacific revealed the need in 1883 to find water in the environs of San Luis. A report by Eduardo Aguirre, then professor of Geology at the University of Buenos Aires (see Ramos 2018), suggested the favorable conditions that Balde, a town located a few kilometers west of San Luis, presented. The drilling was a success when water was found at a depth of 600 m in 1891 and became the first artesian well in the country. This well marks the beginning of hydrogeological research of our subsoil (Fig. 9.31).

When the Division of Mines, Geology, and Hydrology was created in 1904, among its objectives was the realization of deep exploration drillings. A few years later, in 1909, Law 6816 was passed, which specifically assigned the Division the responsibility for studies and drilling for the exploration of underground water. That same year, the Hydrogeology Section was created and placed in the hands of Stappenbeck,

Fig. 9.31 El Balde near San Luis, Argentina's first water artesian well in production since 1891, and Eduardo Aguirre who made the hydrogeological study (Pérez 1949, and Ramos 2018)

who, at the beginning of his studies, laid the foundations for a series of drilling operations throughout the country. In 1912, the Argerich drilling allowed the discovery of the deep thermal aquifer of the Bahia Blanca artesian basin in the province of Buenos Aires. At that time, deep basins were also explored in Mendoza, Salta, Santiago del Estero, and Tucumán, among others. In fact, for example, in 1913, three deep drilling operations were started in Tunuyán, Mendoza; between 1906 and 1917, more than 20 deep wells were drilled in various localities in the province of Tucumán. Since its creation, the institution has carried out commendable hydrogeological work and also social action, by supplying water to railway stations and to hundreds of cities and towns in the intermountain valleys (Zappettini 2004),

9.5 Final Considerations

The boost that Eng. Hermitte gave to geological investigations between 1904 and 1922 through his management, together with the success achieved with the oil discoveries in Comodoro Rivadavia in 1907 and in Plaza Huincul in 1918, have been extraordinary. Much of the success is due to the initial recommendations of Guillermo Bodenbender and Juan Keidel, which allowed us to bring together a group of young German geologists, who together with those from Italy, Finland, and Sweden formed a team of excellence. These young people were attracted by the possibilities they found in Argentina, both because of the vast unexplored areas and because of the pay offered. The Argentine government had been offering tempting work contracts to young scientists since 1904, with salaries more than three times higher than what was usual in Germany at the time (Vladi 1981). The average age of these young people ranged between 27 and 28 years, most of them almost recent PhDs, who came to a completely different country to try their luck. Half of them stayed to live in Argentina.

The pioneering push produced by the National Academy of Sciences gave way to the systematic work of the Direction of Mines and Geology (Leanza 2008). The energy and preparation of these young people changed the geological knowledge of northern Argentina with the studies of Penck, Keidel, and Bonarelli; the central mountain region with Stappenbeck, Schiller, Groeber, Keidel, Backlund and Gerth; the Sierras Pampeanas with Rassmuss and Beder, and the Patagonian region with Wichmann, Windhausen and Schiller. In a few years, a large part of the country was covered with maps and geological studies.

This progress in knowledge was also linked to Argentina's economic development and incipient wealth at the end of the nineteenth century, which extended to the first two decades of the twentieth century. However, we have seen that the available budget began to decrease, even for tasks as important as oil and mineral exploration. After the First World War, the crisis slowly took hold. Those salaries could no longer be offered to these newly graduated young people, who became more than fourteen foreign geologists, and in the 1920s problems began. Engineer Hermitte retired in 1922, Juan Keidel resigned that same year, and great changes began to occur in the Direction of Mines, which will be analyzed in the next chapters.

However, those 20 years of economic wealth, and one could also say geological prosperity, remain a fundamental milestone in the progress of the country and the knowledge of its subsoil. The rapid advance of geological studies is slowing down, but what has been done remains a solid foundation to continue growing in future decades.

References

Amferer O (1924) Walther Penck. Verhandlungen der Geologischen Bundesanstalt, Seite 81:81–82, Viena

Argüello GL, Sacchi GA (2020) Nuevas aristas del perfil de Anselmo Windhausen. Revista Facultad de Ciencias Exactas, Físicas y Naturales, 7 (suplem. 1):25–29, Córdoba

Arko V (1966) El Dr. Federico Reichert, padre del Andinismo de Exploración de la Argentina. Centro Cultural Argentino de Montaña, en línea, Bariloche

Backlund HG (1913) Algunas observaciones sobre rocas notables procedentes de Olavarría (Provincia de Buenos Aires). Boletín de la Dirección General de Minería, Geología e Hidrología, Serie B (2):1–25, Buenos Aires

Backlund HG (1921–1923). Der magmatische Anteil der Cordillera von süd-Mendoza. Acta Akademiae Aboensis, Mathematica et Physica Institut 2(1) y 3(1): 298 p, Åbo

Backlund HG (1923) Walther Penck. Geologiska Föreningens i Stockholm förhandlingar 45(5):445–447, Stockholm
Backlund HG (1926) Magmatic activity and mountain folding in the Andes of south Mendoza. Geol Mag 63(9):410–422
Beder R (1922) Estudios geológicos de la sierra de Córdoba, especialmente de las calizas cristalino-granulosas y su fenómenos de metamorfismo. Boletín de la Dirección General de Minas, Geología e Hidrología, (Serie B, Geología) 33:1–86, Buenos Aires
Beder R (1928a) La sierra de Guasayán y sus alrededores. Una contribución a la geología e hidrología de la provincia de Santiago del Estero. Ministerio de Agricultura, Dirección General de Minas y Geología, Publicación 39:1–173, Buenos Aires
Beder R (1928b) Los yacimientos de mineral de plomo en el departamento de Yaví de la provincia de Jujuy, con algunas consideraciones geológico-económicas sobre el norte de las provincias de Jujuy y Salta. Ministerio de Agricultura, Dirección General de Minas y Geología, Publicación 38:1–60, Buenos Aires
Beder R (1929) Los yacimientos minerales de manganeso en el norte de la provincia de Córdoba y sur de Santiago del Estero. Anales del Museo Nacional de Historia Natural Bernardino Rivadavia 36:431–504
Bonarelli G (1913) Las Sierras Subandinas del Alto y Aguaragüe y los yacimientos petrolíferos del distrito minero Tartagal, departamento de Orán, provincia de Salta. Anales del Ministerio de Agricultura, Sección Geología, Mineralogía y Minería 7(4):1–50
Bonarelli G (1914) La estructura geológica y los yacimientos petrolíferos del distrito minero de Orán, provincia de Salta. Boletín de la Dirección General de Minas, Geología e Hidrología, serie B(9):1–43
Bonarelli G (1917) Tierra del Fuego y sus turberas. Anales Ministerio de Agricultura. Sección Geología 12(3):1–66, Buenos Aires
Bonarelli G (1921) Tercera contribución al conocimiento geológico de las regiones petrolíferas subandinas del norte (provincias de Salta y Jujuy). Anales del ministerio de Agricultura de la Nación, Sección Geología, Mineralogía y Minería 15(1):1–96
Bonarelli G, Nágera JJ (1921) Observaciones geológicas en las inmediaciones del Lago San Martín (Territorio de Santa Cruz). Boletín de la Dirección General de Minas 27 B:1–39, Buenos Aires
Bonarelli L (2002) Guido Bonarelli (1871–1951), vida y obra científica. Ediciones del 15° Congreso Geológico Argentino, 136 p, Buenos Aires
Borrello A (1952) Profesor doctor Juan Keidel, homenaje en su septuagésimo quinto aniversario. Revista de la Asociación Geológica Argentina 7(3):145–156
Borrello AV (1964) La contribución del Dr. Anselmo Windhausen al conocimiento paleogeográfico de la Patagonia. Comisión de Investigación Científica, Provincia de Buenos Aires, 10 p, La Plata
Brackebusch L (1883) Estudios sobre la formación petrolífera de Jujuy: Informe oficial. Boletín de la Academia Nacional de Ciencias 5:137–184, Córdoba
Cánepa EP (1967) Un recuerdo académico Enrique M. Hermitte (24-IV-1871-28-I-1955). Anales de la Academia Nacional de Ciencias Exactas, Físicas y Naturales 22:55–66, Buenos Aires
Camacho HH (2001) Las Ciencias Geológicas en Argentina hasta 1939. Saber y Tiempo 12:177–210
Camacho HH (2008) La contribución de la Dirección General de Minas, Geología e Hidrología de la Nación a la formación de la primera generación de geólogos argentinos, y la actuación del Ing. Enrique M. Hermitte. INSUGEO, Serie Correlación Geológica 34:103–108
Catalano E (2004) Antecedentes y estructura histórica de la minería argentina. In: Lavandaio E, Catalano E (eds) Historia de la Minería Argentina, Anales del Instituto de Geología y Recursos Minerales 40(1):1–176
Clarke JM (1912) El Devoniano de la Argentina. Anales del Ministerio de Agricultura de La Nación, Sección Geología, Mineralogía y Minería 8(2):1–353, Buenos Aires
Concheyro A, Montenegro T (2011) Guido Bonarelli, explorador y geólogo incansable: pionero en la prospección de hidrocarburos en la argentina. Revista de la Asociación Geológica Argentina 68(3):337–345

Cuerda A (1997) Walther Schiller (1879–1844). Prohombres del Museo, Revista Museo 1997, 25–30 p, La Plata

De Ferraris C (1947) Edad del Arco o Dorsal Antigua del Neuquén Oriental de acuerdo con la estratigrafía de la zona inmediata. Revista de la Sociedad Geológica Argentina 2(3):256–283

Fossa Mancini E (1944) Walther Schiller. Revista del Museo de la Plata, Nueva Serie, Sección Oficial, 193–213, La Plata

Gaigalas, A (2008) Quaternary research in the Baltic countries. In: Grapes RH, Oldroyd D, Grigelis A (eds) History of Geomorphology and quaternary geology, geological society. Special Publications 301, pp 129–140, London

Gerth H (1913) Stratigraphie und Bau der argentinischen Kordillere zwiscren demr Rio Grande und Río Diamante. Zeitschrift der Deutsche Geologische Gessellschaft 65(1):568–575, Hannover

Gerth H (1925[a]) Contribuciones a la estratigrafía y paleontología de los Andes argentinos con la colaboración de los paleontólogos. Actas de la Academia de Ciencias de la República Argentina 9:10–55, Córdoba

Gerth H (1925b) La fauna Neocomiana de la Cordillera Argentina en la parte meridional de la Provincia de Mendoza. Actas de la Academia de Ciencias de la República Argentina 9:57–132, Córdoba

Gerth H (1921) Die geologische Geshichte der argentinischen Pampas und ihrer Gebirge. Vortrag 3/12/ 1921, Vergaering, 33–35, Den Haag

Gerth E (1926) Orogenese und Magma in der argentinischen Kordillere. Geologische Rundschau 17a:62–107, Berlín

Gerth E (1931) La estructura geológica de la Cordillera Argentina entre el río Grande y el río Diamante en el sud de la provincia de Mendoza. Actas de la Academia Nacional de Ciencias 10(2):125–172, Córdoba

González OE (2008) Juan Rassmuss (1886–1971): su contribución a la geología Argentina. Instituto Superior de Correlación Geológica, Serie Correlación Geológica 24:137–149, Tucumán

Groeber P (1916) Informe sobre las causas que han producido las crecientes del río Colorado (territorios del Neuquén y La Pampa) en 1914. Boletín de la Dirección General de Minas, Geología e Hidrología,11, Serie B (Geología):1–29, Buenos Aires

Groeber P (1918a) Estratigrafía del Dogger en la República Argentina. Estudio sintético comparativo. Boletín de la Dirección General de Minas, Geología e Hidrogeología 18, Serie B (Geología):1–81, Buenos Aires

Groeber P (1918b) Edad y extensión de las estructuras de la Cordillera entre San Juan y Nahuel Huapi. Physis 4:208–240, Buenos Aires

Groeber P (1931) El Dr. Richard Wichmann y su obra científica. Physis, Revista Sociedad Argentina de Ciencias Naturales 10:267–278

Groeber P (1933) Confluencia de los ríos Grande y Barrancas (Mendoza y Neuquén). Boletín de la Dirección Nacional de Geología y Minería 38:1–72, Buenos Aires

Halle TG (1911) On the geological structure and history of the Falkland Islands. University of Uppsala, Geological Institutionen, Bulletin 11:626–629

Hausen J (1919) Contribución al estudio de la petrografía del territorio nacional de Misiones. Boletín de la Dirección Nacional de Minas, Geología e Hidrología 21:1–39 p, 5 lám., Buenos Aires

Hausen J (1925) Sobre un perfil geológico del borde de la Puna de Atacama, con una descripción de algunos yacimientos mineros en la Puna de Salta y Jujuy. Boletín de la Academia Nacional de Ciencias 28:1–95, Córdoba

Hauthal R (1892) La Sierra de la Ventana (Provincia de Buenos Aires). Revista del Museo de La Plata 3:3–11

Hermitte E (1904) Coal, petroleum and water in Argentine, paper presented in the Saint Louis Exposition. Ministry of Agriculture, 85 p, Buenos Aires

Hermitte E (1910) Las investigaciones geológicas, mineralógicas e hidrológicas en la República Argentina. Necesidad de fomentarlas. Memoria presentada al Congreso Científico Internacional Americano, Dirección de Minas, Geología e Hidrología, Buenos Aires

Hermitte E (1914) Memoria de la Dirección General de Minas, Geología e Hidrología. Anales del Ministerio de Agricultura, Sección Geología, Mineralogía y Minería 11(1):3–14, Buenos Aires
Hermitte E (1928) El mapa geológico-económico de la República Argentina. Anales de la Sociedad Científica Argentina 106:118–138
Hernández JH (2020) Biografía del geólogo Walter Schiller. Centro Cultural Argentino de Montaña. http://www.culturademontania.org.ar/Historia/biografia-walter-schiller.html
Hóskold HD (1889) Memoria General y Especial sobre las Minas, Metalurgia, Leyes De Minas, Recursos, Ventajas, etc. de la explotación de Minas en la República Argentina. Informe preparado para la Exposición de París. Departamento Nacional de Minas y Geología, Imprenta de Obras, 600 p, Buenos Aires
Hóskold HD (1893) Informe general sobre las condiciones de minerales, de productos metalúrgicos y maquinaria minera, para la exposición de Chicago, Departamento Nacional de Minas y Geología, Imprenta de Obras, Buenos Aires
Hóskold HD (1904) Official report upon the mines, mining, metallurgy and mining laws of the Argentine Republic. South American Bank Note Company, 475 p, Buenos Aires
Hünicken MA (1986) Anselmo Windhausen y su contribución al conocimiento de la geología argentina. Boletín de la Academia Nacional de Ciencias 57(1–2):159–168, Córdoba
Kayser E (1876) Ueber primordiale und untersilurische Fossilien aus der Argentinischen Republik. Paleontologischer Theil en Paleontographica, Cassel Supp. 3(2):1–33
Keidel H (1907) Ueber den Bau der argentinischen Anden. Sitzungsberichte der kaiserlichen Akademie der Wissenschaft, Mathematisch-Naturwissenschaftliche Classe 114(1):649–674, Wien
Keidel J (1908–10) Estudios geológicos en la Quebrada de Humahuaca, en la de Iruya y algunos de sus valles laterales. Memoria de la División de Minas, Geología e Hidrología correspondiente al año 1908. Anales del Ministerio de Agricultura de la Nación 5(11):2, Buenos Aires
Keidel J (1913) Über das Alter, die Verbreitung und die gegenseitigen Beziehungen der verschiedenen tektonischen strukturen in den argentinischen Gebirgen. XII° Session du Congrés Géologique International, Compte Rendu 1, 671–687, Toronto
Keidel J (1913) Informe geológico sobre el yacimiento petrolífero de Challacó (Territorio del Neuquén). Ediciones Peuser, 46 p, Buenos Aires
Keidel J (1916) La geología de las Sierras de la Provincia de Buenos Aires y sus relaciones con las montañas de Sudáfrica y Los Andes. Anales del Ministerio de Agricultura de La Nación, Sección Geología, Mineralogía y Minería 11(3):1–78, Buenos Aires
Keidel J (1921a) Sobre la distribución de los depósitos glaciares del Pérmico conocidos en la Argentina y su significación para la estratigrafía de la serie del Gondwana y la paleogeografía del hemisferio austral. Boletín de la Academia Nacional de Ciencias 25: 239 368, Córdoba.
Keidel J (1921b) Observaciones geológicas en la Precordillera de San Juan Mendoza. Anales del Ministerio de Agricultura de La Nación, Sección Geología, Mineralogía y Minería 15(2):1–103, Buenos Aires
Keidel J (1925) Sobre la estructura tectónica de las capas petrolíferas en el Oriente del Territorio del Neuquén. Ministerio de Agricultura, Dirección General de Minas, Geología e Hidrología, Publicaciones 8:1–67, Buenos Aires
Keidel J (1943) El Ordovícico inferior en los Andes del norte argentino y sus depósitos marino-glaciales. Boletín de la Academia Nacional de Ciencias 36:140–229, Córdoba
Leanza HA (2008) Los aportes de algunos ilustres geocientíficos del Segemar al conocimiento geológico del Territorio Nacional. In: Aceñolaza FG (ed) Los geólogos y la geología en la historia argentina. InsuGeo, Serie Correlación Geológica 34:165–178
Leanza H, Ramos VA (2011) En memoria al Dr. Pablo Groeber (1885–1964), maestro de la geología neuquina. In: Leanza H et al (eds) Geología y Recursos Naturales de la Provincia del Neuquén, 17ª Congreso Geológico Argentino, Relatorio: 3–12, Neuquén
Lundegardh H (1958) Helge G. Backlund, 1878–1958. Geologiska Föreningen i Stockholm Förhandlingar 80(4):488–491, Stockholm

Mercerat A (1925) Rasgos que caracterizan los problemas de la geología argentina. Revista del Museo de La Plata 28:243–328

Mon R (2021) Heinrich Gerth (1884–1971). In: Ramos VA (ed) Los miembros correspondientes de la Asociación Geológica Argentina, Publicación Especial 15:25–28

Olsacher J (1963) Dr. Ricardo F. A. Stappenbeck. Revista de la Asociación Geológica Argentina 18(3–4):113–115

Penck W (1914a) Principales resultados provenientes de los estudios hidrográficos y geológicos en el valle de Fiambalá (Provincia de Catamarca). Anales de la Sociedad Científica Argentina 77:224

Penck W (1914b) La estructura geológica del valle de Fiambalá y las cordillera limítrofes al Norte de Tinogasta. Boletín del Ministerio de Agricultura 17:270–277

Penck W (1920) Der Südrand der Puna de Atacama. Abhandlungen Mathematisch Physik Klasse der Sächisischen Akademic der Wissenschaften. Band 37, 420 p, Leipzig

Penck W (1924) Die morphologische Analyse: ein Kapitel der physikalischen Geologie. Geographische Abhandlungen; Reihe 2, Heft 2, 283 p, Stuttgart

Pérez M (1949) Balde, provincia de San Luis, una instalación humana en la región árida argentina. Boletín de Estudios Geográficos 3:11–40

Ramos VA (2013) Keidel (1877–1954) y Du Toit (1878–1948), dos pioneros de la deriva continental. 3º Congreso Argentino de la Historia, Actas: 159–168, Salta

Ramos VA (2017) El aluvión del río Colorado de 1914: Primera contribución geológica de Groeber desde su llegada a la Argentina. Revista de la Asociación Geológica Argentina 74(1):9–18

Ramos VA (2018) El desarrollo de la Geología en la Universidad de Buenos Aires. Revista Facultad de Ciencias Exactas, Físicas y Naturales, Nueva Serie 5(Suplem. 1):23–32, Còrdoba

Ramos VA, Leanza H (2011) Historia de la evolución del conocimiento geológico. In: Leanza H et al (eds) Geología y Recursos Naturales de la Provincia del Neuquén, 17ª Congreso Geológico Argentino, Relatorio: 13–27, Neuquén

Rassmuss J (1916) Rasgos geológicos generales de las Sierras Pampeanas. Boletín de la Dirección General Minas, Geología e Hidrogeología, Serie B (Geología) 13:3–18, Buenos Aires

Rassmuss J (1920) Geología de los Yacimientos de Carbón en la República Argentina. Boletín de la Dirección General de Minas, Geología e Hidrogeología, Serie B (Geología) 23:10–27, Buenos Aires

Rassmuss J (1921) Observaciones geológicas en Salta: La cuenca de Metán. Boletín de la Dirección General de Minas, Geología e Hidrogeología, Serie F (Informe Prelim. y Comun.), 4:25–37 and 1 pl., Buenos Aires

Rassmuss J (1922a) Apuntes geológicos sobre el hallazgo de carbón al sur del lago Nahuel Huapí. Boletín de la Dirección General de Minas, Geología e Hidrogeología, Serie B (Geología), 28:5–22 and 1 pl., Buenos Aires

Rassmuss J (1922b) La cuenca de Marayes. Boletín de la Dirección General de Minas, Geología e Hidrogeología. Serie B (Geología) 32:7–21, 2 maps, 5 sections and 3 pls., Buenos Aires

Rassmuss J (1922c) El perfil de la perforación Capiazuti en la región petrolífera de Tartagal (Provincia de Salta). Boletín de la Dirección General de Minas, Geología e Hidrogeología, Serie F (Informes Preliminares y Comunicaciones) 5:43–46, Buenos Aires

Rassmuss J (1923) Breves apuntes geológicos sobre la parte del territorio del Neuquén entre Auca Mahuida y El Tromen. Boletín de la Dirección General de Minas, Geología e Hidrogeología, Serie F (Informes Preliminares y Comunicaciones) 6(1):15–20, Buenos Aires

Raukas A (1984) Hans Hauseni mälestuseks. Loodusajakiri MTU, Eesti Loodus (Naturaleza de Estonia) 9:613–615

Reichert F (1907) Los Yacimientos de Boratos y otros Productos Minerales Explotables del Territorio de los Andes (Puna de Atacama). Anales del Ministerio de Agricultura. Sección de Geología, Mineralogía y Minería 2(2):1–105, Buenos Aires

Reverberi O (2004) Historia de la Minería Argentina. Capitulo Organismos Nacionales de Minería. Cronología de su organización y evolución. Anales Segemar 40(1):177–225, Buenos Aires

Rickard FI (1869a) Informe sobre los Distritos Minerales. Minas y establecimientos de la República Argentina 1868–1869. Ministerio del Interior, 179 p, Buenos Aires
Rickard FI (1869b) The mineral and other resources of the Argentina Republic in 1869. Longmans, Green and Co., 330 p, London
Schiller W (1907) Geologische Unterschungen bei Puente del Inca (Aconcagua). Vorlaufige Mitteilung. Neues Jahrbuchfür Mineralogie. Paläontologie und Geologie, Beilage Band XXIV:716–736, Stuttgart
Schiller W (1908) Cordillera alta a ambos lados del Río Cuevas (Río Mendoza). Anales del Ministerio de Agricultura de la Nación. Sección Geología, Mineralogía y Minería 5(2):65–71, Buenos Aires
Schiller W (1912) La Alta Cordillera de San Juan y Mendoza y parte de la provincia de San Juan. Anales del Ministerio de Agricultura de la Nación. Sección Geología, Mineralogía y Minería 7(5):1–68, Buenos Aires
Schiller W (1925) Estratigrafía, tectónica y petróleo de Comodoro Rivadavia (Chubut). Anales del Museo de La Plata 2(1):1–56
Stappenbeck R (1908) Geologische Beschreibung der Umgebung des Sees Musters in Patagonie. Akademie der Wissenschften 117:1243–1249, Wien
Stappenbeck R (1909) Informe preliminar relativo a la parte Sudeste del Territorio del Chubut. Anales del Ministerio de Agricultura de La Nación, Sección Geología, Mineralogía y Minería 4(1):1–20, Buenos Aires
Stappenbeck R (1910) La Precordillera de Sanjuán y Mendoza. Anales del Ministerio de Agricultura de La Nación, Sección Geología, Mineralogía y Minería 4(3):1–187, Buenos Aires
Stappenbeck R (1917) Geología de la falda oriental de la Cordillera del Plata, Provincia de Mendoza. Anales del Ministerio de Agricultura de La Nación, Sección Geología, Mineralogía y Minería 12(1):1–49, Buenos Aires
Stappenbeck R (1918) Los yacimientos de minerales y rocas de aplicación de la República Argentina. Boletín de la Dirección General de Minas, Geología e Hidrología 19:1–107, Buenos Aires
Stappenbeck R (1926) Geologíe und Grundwasserkunde der Pampa. Pub. Schweizerbart, 409 p, Stuttgart
Stieglitz O (1914) Contribución a la petrografía de la Precordillera y del Pie de Palo. Boletín de la Dirección de Minas, Geología e Hidrología 10:1–97, Buenos Aires
Storni CD (1947) Doctor Anselmo Windhausen, en el 15° Aniversario de su muerte. Revista de la Sociedad Geológica Argentina 2(2):89–94, Buenos Aires
Sureda RJ (2008) Historia de la Mineralogía. InsuGeo, Serie Correlación Geológica 23:1–190, Tucumán
Tineo A (2008) Ricardo Stappenbeck: el primer hidrogeólogo en Argentina. InsuGeo, Serie Coirrelación Geológica 24:127–136, Tucumán
Vladi F (1981) Anselm Windhausen - Zum 100. Geburtstag eines Harz-Höhlenforschers. Unser Harz 29(10):187–190, Clausthal
Wegener A (1915) *Die Entstehung der Kontinente und Ozeane*. Friedrich Vieweg & Sohn, Braunschweig, 1st edition, 1915; 2nd edition, 1920, 3rd edition, 1922; 4th edition, 1924; 4th revised edition, 1929
Wichmann A (1916) Hoja 35 m, Bahía Blanca. Mapa Geológico-económico de la República Argentina, escala 1:200.000
Wichmann A (1918) Geología e Hidrología de Bahía Blanca y sus alrededores (provincia de Buenos Aires). Anales del Ministerio de Agricultura de la Nación, Sección Geología, Mineralogía y Minería 13(1):1–67, Buenos Aires
Wichmann R (1921) Estudio geológico de la zona de reserva de la explotación nacional de petróleo en Comodoro Rivadavia (Territorio Nacional del Chubut). Boletín de la Dirección General de Minas, Geología e Hidrología (Serie B Geología) 25:1–18, Buenos Aires
Wichmann R (1927a) Sobre la Facies Lacustre Senoniana de los Estratos con Dinosaurios y su fauna. Boletín de la Academia Nacional de Ciencias 30:383–405, Córdoba

Wichmann R (1927b) Resultados de un viaje de estudios geológicos en los territorios del Río Negro y del Chubut efectuado durante los meses de enero hasta junio del año 1923. Dirección General de Minas, Geología e Hidrologla, Publicación 33:1–59, Buenos Aires

Wichmann R (1934) Contribución al conocimiento geológico de los territorios del Neuquén y del Río Negro. Boletín de la Dirección de Minas y Geología 39:1–27, 24 pls., Buenos Aires

Wilckens O (1930) Gustav Steinmann, sein Leben und Wirken. Geol Rundsch 21:389–415

Windhausen A (1914) Contribución al conocimiento geológico de los territorios de Río Negro y Neuquén, con un estudio de la región petrolífera de la parte central del Neuquén (Cerros Lotena y Covuncó). Anales del Ministerio de Agricultura de la Nación. Sección Geología, Mineralogía y Minería 10(1):1–69, Buenos Aires

Windhausen A (1918) The problem of the Cretaceous-Tertiary boundary in South America and stratigraphic position of the San Jorge Formation in Patagonia. Am J Sci, Series 4, 45:1–53, New Haven

Windhausen A (1921) Ein Blick auf Schichtenfolge und Gebirgsbau im südlichen Patagonien. Geol Rundsch 12(3–5):109–137

Windhausen A (1923) Cambios en el concepto de las condiciones geológicas del yacimiento petrolífero de Comodoro Rivadavia. Boletín de la Academia Nacional de Ciencias 27:1–8, Córdoba

Windhausen A (1924) Líneas generales de la constitución geológica de la región situada al oeste del golfo de San Jorge. Informe sobre reconocimientos efectuados en los años 1919–1923. Boletín de la Academia Nacional de Ciencias 27:167–320, Córdoba

Windhausen A (1931) Geología Argentina, Geología Histórica y Regional del Territorio Argentino. J. Peuser, 2:645 p, Buenos Aires

Windhausen RA (2007) Anselmo Windhausen y su polémica con el Coronel Enrique Mosconi, Reseña Histórica. Petrotecnia (agosto, 2007), 130–132

Yalçın MN (2015) 100 Yılın Hikâyesi-İstanbul Darülfünunu Arziyat Dar-ül Mesaisi'nden İstanbul Üniversitesi Jeoloji Mühendisliği Bölümü'ne (1915–2015). M. Namık Yalçın (Editör), İstanbul Üniversitesinde Jeoloji-Bir Asırlık Geçmiş. Gürsoy Grup Kültür Yayınları 4:11–31, Istanbul

Yalçın MN, Penck GW (2019) Auf den Fussspuren eines deutschen Geowissenschaftlers in Istanbul (1915–1918): Prof. Dr. Walther Penck. Studies in Ottoman Science, Osmanlı Bilimi Araştırmaları 20 (2): 1–58, Istanbul

Zappettini E (2004) Servicio Geológico Minero Argentino, 1904–2004, 100 Años al servicio del desarrollo nacional. Secretaría de Minería, Segemar, 106 p, Buenos Aires

Chapter 10
End of the German Hegemony in the Direction of Mines and Geology

Abstract Argentine geologists began to work at the Geological Survey, among whom the first two stand out: Franco Pastore and Juan José Nágera, who graduated from the University of Buenos Aires in 1914 and 1915. However, the first Argentine geologist was José María Sobral, who after participating in the Swedish expedition to Antarctica, went to Sweden with Otto Nordensjöld to study at the University of Uppsala, graduating in 1912 as a geologist. At the beginning of the 20s, the economic crisis reduced the budget of the Direction of Mines and Geology. Due to this, most of the German geologists left the institution, and Hermitte and Keidel retired as director and head of the Geology division. Jose M. Sobral was appointed as the new director of Mines and Geology. These were very hard years with a low budget, and a reduced number of geologists in the survey. Among them, Augusto Tapia stood out for his Antarctic research and Luciano Catalano for his mining studies. The most notable geological contributions were those carried out by Sobral's group of Swedish geologists based on his relations with his former colleagues at the University of Uppsala. Among them, the one who made the most important contribution was Carl Caldenius for his glaciological studies in Patagonia. From this group of Swedish geologists, Tor Hagerman also stood out in sedimentology, and Eric Ljunger in his surveys of the Bariloche region. In 1930, when Sobral left the institution, dark years came for the Directorate of Mines and Geology, with no budget and almost no geologists.

Keywords Sobral director · First Argentine geologists · Antarctic glaciology · Swedish geologists · Glaciations

In 1922, important changes occurred in the General Direction of Mines and Geology. Engineer Enrique Hermitte leaves the management after almost 20 years of successful management. Dr. Juan Keidel resigns as Head of Geology and begins his career as a professor at the University of Buenos Aires and later simultaneously at the National University of La Plata. Many of the first German geologists have left this region or the country, leaving only a few such as Pablo Groeber and Ricardo Wichmann.

V. A. Ramos, *History of the Geological Sciences in Argentina*, The Latin American Studies Book Series, https://doi.org/10.1007/978-3-031-83139-3_10

Fig. 10.1 The young José María Sobral in 1905, when he was studying Geology at the University of Uppsala on the recommendation of Otto Nordenskjöld, with whom he shared two years in Antarctica and the diploma where he was named Knight of the Order of the King on May 25, 1904, in recognition his merits.[1] *Source* Wikimedia and after Marenssi (2007)

The reasons for this change may be associated with a drastic decrease in the budget, which is now half of what it had been years before (Camacho 2001, p. 188).

Dr. José María Sobral (1880–1861), the first Argentine geologist, carried out successful geological studies at the University of Upsala, Sweden, where he received his doctorate in 1913 (Fig. 10.1). He was going to return to the country in 1914 to join the General Directorate of Mines and Geology, but he asked Eng. Hermitte, then director, to delay his entry until next year. Before his return, he had advised him on the installation of a petrographic laboratory for the institution (unpublished letter from the family collection of Mr. Hermitte).

10.1 The First Argentine Geologists

The importance acquired by geological investigations in the country had decided the Division of Mines, Geology and Hydrology to collaborate in the training of Argentine geologists. He hired students from the University of Buenos Aires with the position of Second-Grade Geologists. The first to be hired were Elías Pelosi and Franco

[1] *Undertecknad Ordenskansler gor harmed veterligt. Att det behagat H. M. Konungen under 25 Maj 1904 i nåder utnamna till Riddars af. Kungl Svarads Orden, andra Klassan, Lojtnanten i argenntinska flottan Jose M. Sobral* (The Chancellor of the undersigned Order hereby acknowledges: That it pleased His Majesty the King on May 25, 1904 to appoint Knight of the King's Order, Second Class, Lieutenant of the Argentine Navy José M. Sobral).

Fig. 10.2 Franco Pastore and Juan José Nágera, the first two geologists to graduate from the University of Buenos Aires, went to work at the General Direction of Mines and Geology (Groeber 1938; Barrio 1966)

Pastore (Camacho 2001). This is how Franco Pastore (1885–1958) entered as Second-Grade Geologist and later Juan J. Nágera (1887–1966) in 1912 (see Malumián 2009) (Fig. 10.2). Engineer Hermitte had invited Pastore, still a student, to enter in 1910 and gave him support to make two campaign trips to do his thesis in the Sierra del Morro de San Luis, which he defended in 1914 at the University of Buenos Aires (Pastore 1914, 1915). He became the first geologist to graduate in Argentina. Pastore was appointed Geologist of the Direction of Mines and Geology in 1914, replacing Walther Penck.

Juan José Nágera carried out studies in the Sierra Baya of the province of Buenos Aires (Fig. 10.3), which constituted his doctoral thesis, defended in 1915 and later published by the General Directorate (Nágera 1919). His thesis received the "Carlos Berg" prize, half of which he donated to create a natural park in the surroundings of Buenos Aires. As a teacher, he took his students to the tomb of General Belgrano and took an oath that they would never abandon the study of Natural Sciences (Barrio 1966). He was head of the Geology Division of the General Directorate of Mines and Geology, and later honorary deputy director during Tapia's management as director of the department. He made one of the first surveys in 1923 of the Hoyos del Campo del Cielo between Chaco and Santiago del Estero, known for meteorite impacts. Despite finding fragments, which he had analyzed and which confirmed that they were meteorites, he wrongly concluded that the depressions were not related to potential impacts (Nágera 1926a).

From that time, his most transcendental contribution was his "*Mar Libre*" doctrine published in 1926, where he postulated that the territorial, jurisdictional or coastal sea extends to the continental edge, which he called the "*Argentine epicontinental sea*". On this basis, years later, Decree 1386 of 1944 extended our sovereignty over the epicontinental sea and our submarine platform, our country being the first to

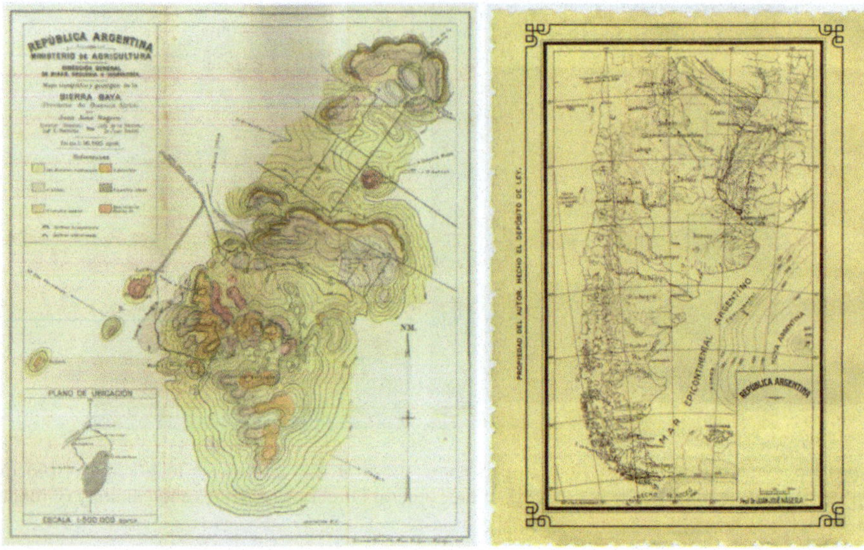

Fig. 10.3 Map of the thesis published in 1919 and one of the geographical postcards printed by Nágera in 1934 to disseminate the concept of the "*Argentine epicontinental sea*" (after Nágera 1919 and Malumián 2011). On the back, he says: "*Young student: The Republic needs one hundred geologists, study Geology, the Homeland will thank you.*"

extend its sovereignty to the limit of the continental shelf, followed sometime later by the United States and Mexico.

In 1926, he was commissioned to write a new program for the teaching of Geography in secondary schools and published his *Geographic Atlas of the Argentine Republic*, one of the first with a notable representation of the mountain and mountain systems of the country (Nágera 1926b), which was not accepted for official use in schools due to the omission of the Malvinas Islands. This atlas was dedicated to "*The future hundred geologists that the republic needs*". It is interesting to note that the financing for printing this beautiful atlas and subsequent geographical postcards comes from a lottery ticket won by Nágera. Juan Carlos Riggi relates that "*getting out of a taxi, a disabled man offered Nágera lottery tickets for sale, which were acquired as help, and were awarded*" (in Malumián 2009, p. 26). These funds were used for printing and subsequent free distribution.

His idealistic vision of Geology is expressed by Augusto Tapia in a letter that he addresses to Nágera upon leaving his position as director. Although he carried out important geological works that led him to be considered registration number one in the *Consejo Profesional de Geología* (Nágera 1940), his greatest significance has been for his contributions to geography, represented in his book Argentine Physical Geography (Nágera and Manito 1938) and by the *Nágera Doctrine* that founded the sovereignty of Argentine resources on our continental shelf (Nágera 1926a, b, c).

Fig. 10.4 Augusto Tapia in the South Orkney Islands and view of the Laurie Island observatory in 1920. *Source* Segemar archives

His concern for history led Nágera to study and publish his classic work on "*Puntas de Santa María del Buen Aire, founding place of Don Pedro de Mendoza*" (Nágera 1936 and subsequent editions).

In 1915, Augusto Tapia (1893–1966) joined the General Directorate of Mines and Geology (Fig. 10.4). He graduated as an expert from the Santa Catalina Agronomic School in 1912, dependent on the University of La Plata in 1912. In 1913, he traveled to Europe and the following year he took part in a measurement campaign in the territory of Formosa.

In the first years of the Direction of Mines, he carried out numerous works and reports on Applied Geology, especially for the siting of hydraulic works (Groeber and Tapia 1926), some of them co-authored with Bonarelli.

In 1920 he was invited to be part of the scientific expedition to the South Orkney Islands (Fig. 10.5), at the Laurie Island meteorological observatory, which since 1904 has been the oldest permanent base in the entire Antarctic territory. He carried out the first Antarctic glaciological studies in Argentina and unfortunately had an accident in which eight of his frozen fingers had to be amputated in precarious conditions (Capdevila 2013). However, he continued his work during those long months, wintering at the base and making new observations (Tapia 1925a). Upon his return to the continent due to fatality in the line of duty, he could have retired in 1921. However, he took the opportunity to train his hands and use them to return to writing and drawing. This is how the following year he carried out important studies on the Aconquija glaciers of Catamarca and Tucumán (Alonso 2018), work that would be published in 1925 by the Argentine Society of Geographic Studies (Tapia 1925b). He is rightly considered the first Argentine Antarctic glaciologist.

His later works were dedicated to the Pampas Plain, reflected in countless publications and reports that allowed him to acquire extensive knowledge about this sector of our country (Tapia 1935, 1937). Tapia was the continuator of Stappenbeck's work in the field of Hydrogeology. In addition, he was an excellent teacher, serving between 1924 and 1948 as a professor of Argentine Physical Geography at the National Military College and simultaneously, between 1931 and 1933 at the Army Higher Technical School. He has also been a professor of Political Economic Geography

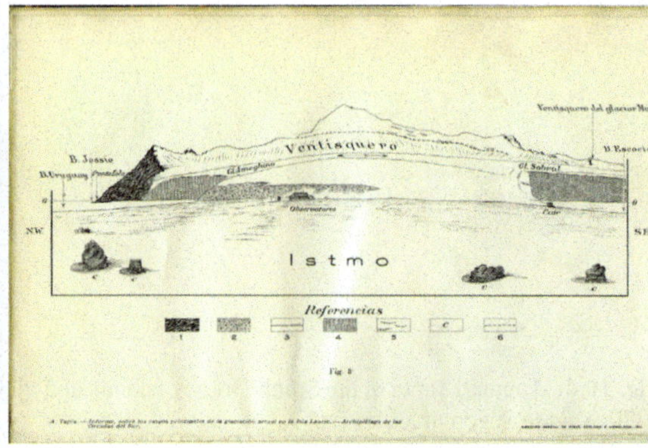

Fig. 10.5 Augusto Tapia at a mature age and his 1925 work of Laurie Island (after Groeber 1938 and Tapia 1925a, b)

and Argentine Physical Geography at the Faculty of Humanities and Educational Sciences of the University of La Plata, where he was Director of the Institute of Geography. He was also a professor at the National University of the South in Bahía Blanca. He was the National Director of Mines, Geology, and Hydrogeology in 1944 (Fig. 10.6).

José María Sobral finally joined the Direction of Mines in 1915 and was in charge of petrographic studies of samples from different sectors of the country (Marenssi 1907). Eng. Hermitte, who had a high opinion of him as a petrographer, when he presented one of the first studies carried out by Sobral for publication, clarified to the minister that "*it is the first work of that category done by an Argentinian*", which is endorsed by Stappenbeck (Hermitte in Sobral 1918). In those years he had continued the petrographic and field studies in the province of Misiones initiated by Hausen, along with reconnaissance in mineral districts of Mendoza and other places.

10.2 Sobral as New Director in 1922

When Eng. Hermitte resigned, he proposed that Sobral succeed him, who took over the management in a period characterized by a sharp decrease in resources. Prior to his entry, Sobral communicated from Uppsala in May 1915 to Eng. Hermitte: "*Studying European magazines, I have had the opportunity to observe that foreign geologists employed in our country publish their work in some of them (in Austrian and German magazines). I believe that if a state hires a person to perform a job, giving them a remuneration, a salary, that job belongs to the institution with its current value and its consequences. That person has not only received his salary but also probably many facilities of all kinds, trips, payments, etc. That is why it seems natural to me that the state takes full advantage of the expenses incurred, not only in*

Fig. 10.6 First hydrogeological map of the Argentine Republic published by the Direction of Mines and Geology made by Augusto Tapia in 1941. *Source* Segemar archives

the practical consequences of the study but also by requiring that the publications be made in the national language and in the bulletin of the interested institution or in another national magazine (the latter with due authorization from the head of the institution). All nations in the world apply what I have just explained because these publications enrich and give value not only to the nation but also to the language that thus becomes more important every day... You. Surely you have thought about these things and I bring them to your attention as a compatriot and with the hope that the work of our future petrographic laboratory will be printed in Spanish and in the institution's bulletin" (unpublished letter from Eng. Hermitte's family collection).

Sobral took over the leadership in 1922, which he held until the military coup of 1930 when he ominously left office (see details in Malumián 2009). During his administration, he made frequent complaints about the need for more resources that were very scarce, "*as well as the number of geologists... It is no longer disputed that research is the fundamental basis of mining exploitation, which is why all of it must be allocated to this Direction the means it needs... It currently has a very small body of geologists, because very little has been done to ensure that Argentine youth dedicate their energies to the study of a science that in this country requires sacrifices and deprivations*" complaint made in 1926 (Zappettini 2004).

The lack of resources led to an interruption in the publication of the bulletins. One exception was the first Argentine geological publication of the Antarctic region, which was published, corresponding to studies on the glaciation of Laurie Island in the South Orkney archipelago. One of the most notable works of that time was the compilation of the Geological Map of the Argentine Republic at a scale of 1:5,000,000 in 1923, which was never published. In 1925 he compiled and edited the Hydrological Map of Argentina at a scale of 1:2,000,000. Sobral himself, during his time as director, wrote various contributions that were published in local and foreign magazines, not being able to fulfill his desires expressed in the opportunity to Hermitte (Zappettini 2004).

During that time the staff of foreign geologists in the Geology Division had been reduced to Groeber and Wichmann. Beder had gone to the National University of Córdoba in 1926; Windhausen had been assigned to the brand-new state oil company like Bonarelli who, on his second visit to the country, was working at YPF. The Argentines were Juan J. Nágera, Franco Pastore, Augusto Tapia, Luciano Catalano and Remigio Rigal.

Dr. Luciano Roque Catalano (1890–1970) studied at the Faculty of Exact, Physical and Natural Sciences of the University of Buenos Aires, where he defended his doctoral thesis in Chemistry in 1917, under the supervision of Eng. Enrique Hermitte. The chemical analyses for the preparation of his thesis were carried out at the General Direction of Mines and Geology, where he entered at the invitation of Eng. Hermitte. His doctoral thesis consisted of the analysis of a complex mineral from the Sierra de Pie de Palo in San Juan that contained rare metals and rare earths (Catalano 1917). Dr. Enrique Herrero Ducloux advised him on the analyses and had a very demanding jury, in which Eduardo Holmberg participated. He specialized in geology, mineralogy, and physics, mastering these disciplines as evidenced by the numerous books and articles he published. His strength was geological chemistry (or chemical geology), becoming a referent in these topics for his time. He devoted himself to economic geology and industrial mobilization planning, making several contributions to these disciplines.

He was a pioneer in the exploration of our mineral riches, he traveled through the Puna in all directions, in long campaigns that lasted up to six months (Fig. 10.7). His research was largely focused on the study of borates in the salt flats of the Puna (Catalano 1927). This is how he discovered, among others, the Tincalayu deposit, one of the most important borax mines in the world (Alonso 2007). He completed the study of ulexite deposits in salt flats that Federico Reichert had begun in 1907. He wrote accurately about the genesis of borates (Catalano 1927). He described the main borate mines of the Puna, such as those of Salar de Cauchari, Coyahuaima, Antuco and Salar de Hombre Muerto. He has the merit of having been the first to warn about the presence of lithium in the salt flats, a metal whose exploitation began successfully in the brines of the Hombre Muerto salt flat in Catamarca (Catalano 1964a, b), being today exploited in various salt flats between Jujuy and Catamarca. Although his work in the Puna was carried out for the most part in the 1920s, when he became Undersecretary of Mining, he had his unpublished reports published between 1964 and 1965. On that occasion, we created a special collection, the Argentine Series, which consists of eight volumes comprising his important unpublished contributions. His comprehensive vision of economic development led him to propose that the economic organization of the Puna, around mining activity, needed railway branches that were its backbone. He carried out a complete and pioneering geological and geographical review of the Puna de Atacama in the Territory of the Andes where he presented a synthesis of his fieldwork carried out between 1923 and 1927 that he published at the Universidad del Litoral (Catalano 1930).

The volcanic phenomena and the consequences of the Quizapu-Descabezado eruptions, whose ashes fell from April 11 to 13, 1932, over a large sector of the provinces of Mendoza, La Pampa, and Buenos Aires, generated concern about their effects. The National Museum of Natural History entrusted Erwin Kittl, who was

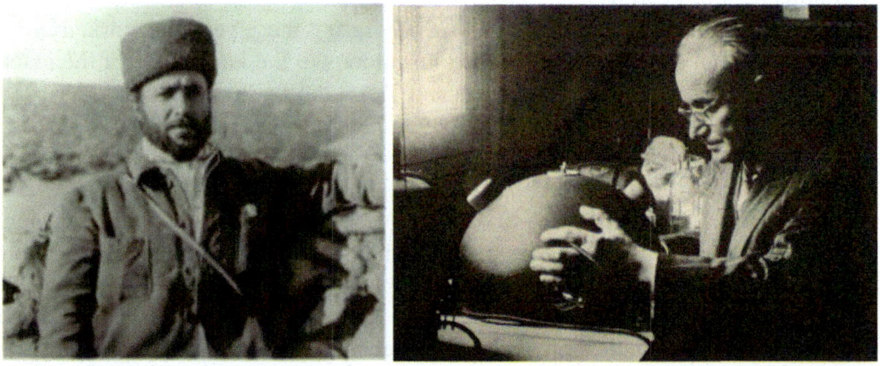

Fig. 10.7 Luciano Catalano young in the Puna, and at a mature age in the office (Alonso 2020a, b; Abad de Santillán 1971)

working in the Direction of Mines, with the evaluation of the affected sectors and the chemical and mineralogical composition of the ash deposits, with the collaboration of Luciano Catalano on behalf of the Geology Section (Zappettini 2004).

At the end of the 1930s, Catalano was president of the Board of Directors of the Direction of Mines, Geology, and Industries of Córdoba during the administration of Governor Sabattini. His name is associated not only with boron and lithium but also with iron since he prospected the province of Jujuy to corroborate its occurrence in the region of Palpalá, at the request of General Manuel Savio in 1939. Upon his return, Catalano warned that the discovery of Zapla iron was as important for the country as the chance discovery of oil in Comodoro Rivadavia, pointing out that "*the deposit is a sedimentary basin of hematite whose visible power ensures a reserve of at least 50 million, perhaps 100, and maybe I will fall short*". The first blast furnace in Argentina would be located there, with the creation of the state company Altos Hornos Zapla in charge of the Fabricaciones Militares, in which Catalano also served as Geological Advisor.

Likewise, he was a technical advisory advisor to the Commerce and Industry Commission of the Chamber of Deputies between 1941 and 1943, becoming Undersecretary of Mining of the Nation during the government of Arturo Illia between 1963 and 1966.

10.3 Carl Caldenius and the Swedish Geologists

Sobral as General Director, taking advantage of his good relations with the country where he had studied, developed a strong Swedish-Argentine cooperation program for which in 1925 he invited four Swedish geologists. These were Seth Rosén in stratigraphy; Nils Albert Lannefors, for mining studies who between 1926 and 1930 produced about twenty reports on different mines in the country; Sven A. Wassmann as metallurgist and Carl Caldenius to study Quaternary glacial deposits in southern Argentina (Camacho 2001).

The works of Carl Caldenius (1887–1961) during the 5 years he worked in our country stand out for his fundamental contributions to the knowledge of the extent and duration of glaciations in Patagonia and Tierra del Fuego (Figs. 10.8 and 10.9). Caldenius was recommended to Sobral by Gerhard de Geer (1858–1943), professor at Stockholm University and director of the Institute of Geochronology, shortly after he had completed his thesis in 1924 on stratigraphy and geochronology of varve sequences of glacial origin in a Sweden lake. De Geer wanted to correlate through varve studies the climatic changes observed in northern Europe with those of the southern hemisphere (Lundqvist 2006).

He carried out three extensive campaigns in Patagonia and Tierra del Fuego establishing the limits of the different glaciations, recognizing four episodes and the fluctuations along the foothills of the Andes where he produced precise geological maps. He finished his work in 1930, but due to the financial crisis, the publication of the

Fig. 10.8 Carl Caldenius in 1952 and details of his geochronological studies analyzing varve sequences carried out in Epuyén. *Source* Lundquist (1983) and Caldenius (1932)

Fig. 10.9 Caldenius in his camp in Patagonia analyzing a box with varves and trying to remove a cart stuck in the peat of Tierra del Fuego (photos by Román Guiñazú in 1927)

Direction of Mines and Geology was printed in Sweden (Caldenius 1932). It is interesting to note that he spent five years in the country and had the collaboration of two geological assistants, Isaías R. Cordini (1902–1966) and José Román Guiñazú (1897–1991), who had joined the division in 1925.

In his fieldwork, Caldenius studied the "*Rodados Patagónicos*" in various places in Patagonia, carrying out a complete analysis of their origin years later (Caldenius 1940).

Cordini worked for more than forty years at the institution, being a prominent limnologist and head of the Sedimentology Division. He received his doctorate from the University of Buenos Aires in 1942 and in 1946 he received the Grand National Science Prize for his work "*The Pilcomayo Rivers in the Patiño region*", which served as the basis for resolving the boundary dispute with Paraguay. He carried out numerous studies, among which his inventories of the Salinas of Argentina and his complete study of the diatoms of our country stand out.

Román Guiñazú, who had studied until the third year of medicine at the University of Buenos Aires, was a self-taught geologist. However, he did important work in his native province, from Hoja 24 f San Luis to the Sierra de Las Quijadas region, which due to its geological and paleontological riches he promoted as a natural reserve, today converted into a national park. Among his contributions he discovered the Tertiary flora of the Pichileufú stream in Patagonia, interpreting its tropical character. The collected fossils were sent to the University of Baltimore, where they were studied by Berry (1938) in a fundamental monograph.

Another participant in the Swedish-Argentine collaboration of those years was Eric Ljunger (1892–1954) (Fig. 10.10) who also graduated from the University of Uppsala in 1927 and that same year he came to work for the Directorate of Mines and Geology (Rolleri 2006). He carried out two extensive campaigns in 1928 and 1930 to the region between Cerro Tronador and Lake Nahuel Huapi. The geological survey and the maps produced of those little-known areas were published soon after (Ljungner 1931). The crisis of the 1930s and the departure of Sobral from the direction interrupted these studies. Less known is a third campaign carried out in the Patagonian Andes between the end of 1932 and mid-1934, sponsored exclusively by Swedish contributions. The results of this expedition were published post-mortem by his widow (Ljungner 1959).

Completing the Swedish cooperation came in 1927 Tor Henrik Hagerman (1899–1986), who had graduated from the University of Uppsala in 1925. The three years he worked in the Direction of Mines and Geology had as his main task the survey of the Sub-Andean Sierras of Santa Bárbara, Cachipunco, Centinela, Ronda and Maíz Gordo in the provinces of Salta and Jujuy. In these mountain ranges that are currently included in the Santa Bárbara System, he carried out precise sedimentological studies analyzing their relationships with the oil formation. During his stay at the institution, he presented two reports, a complete geological analysis of those mountain ranges and a report on the Poma earthquake presented in 1931, with a solid description of the effects and causes of the earthquake. This report, although unpublished, is almost the only existing work that describes this event a week after it occurred.

Fig. 10.10 Eric Ljunger and Tor Hagerman, two young Swedish geologists from the University of Uppsala came to collaborate with Sobral at the Direction of Mines and Geology. *Source* Segemar archives

When his contract with the Directorate of Mines and Geology ceased due to the severe budget crisis, Hagerman began working at YPF in 1932 (Fig. 10.11). Given his sedimentological formation, he was asked to study various Cretaceous-Tertiary units of northwest Argentina, which he presented the following year (Hagerman 1933). He continued working in the north and in 1936 he presented his doctoral thesis in Upsala on the granulometric studies of the Cretaceous-Tertiary sequences studied. His years of work at YPF marked a milestone in sedimentological studies of clastic sediments. A synthesis of his thesis is published in the *Boletín de Informaciones Petroleras* (Hagerman 1936) where he presents an interesting statistical analysis of the different fractions for its characterization (see details in Turner 1970). In 1940 he returned definitively to his country where he had a great academic performance, dying in 1986.

Among the other Swedish geologists who came for this agreement signed by Sobral, we must mention Nils Albert Lannefors, who carried out important economic evaluation studies of the mineral manifestations in those years, and Seth Rosen, a stratigrapher who carried out precise studies in the Marayes coal basin in San Juan (see Chap. 19).

10.4 Final Considerations

The management of José M. Sobral in charge of the General Direction of Mines and Geology ended ominously in 1930, coinciding with the interruption of the democratic system produced by the military revolution of Uriburu.

Despite Sobral's excellent academic preparation and the participation of a plethora of young Argentine geologists, the results of his management were not successful. In part when one compares Hermitte's management, with his important oil discoveries, with an extensive number of surveys undertaken by a respectable number of well-paid foreign geologists, the contrast is not surprising. Sobral had to deal with half the previous budget, with few resources to support foreign geologists, no budget to publish the surveys carried out, and a government that was facing the great economic crisis of 1930.

An important contribution was the agreement that Sobral made with the Stockholm Institute of Geochronology and the University of Uppsala. Caldenius' studies on the glaciation of the Patagonian Mountains and Tierra del Fuego marked the basis for later studies. The studies carried out by other Swedish geologists have been good contributions but unfortunately did not find a budget for their publication.

An experienced geologist like Pablo Groeber has in this period numerous publications in the Argentine Society of Geographic Studies and in the National Academy of Sciences and only a few brief notes edited by the Direction of Mines. Perhaps the only exception is his "*Fundamental Lines of the Geology of Neuquén, southern Mendoza and adjacent regions*", an extensive work and an essential contribution to the knowledge of the Neuquén basin and the adjacent mountain range (Groeber 1929). The first map surveyed a few decades ago must wait a few more years to be

Fig. 10.11 Tor Hagerman, standing first on the left, and Paulina Mühlmann, sitting on the right in 1934, reviewing the latter's thesis on San Bernardo Hill, Salta (after Tanuz 2021). Camp in Chorro, northern Argentina, standing with a beard in the center (after Lundqvist 2006)

published (Groeber 1933). The works carried out by Ricardo Wichmann must wait the next decade to be published and some of them were published postmortem.

Geologists such as Franco Pastore continued to carry out petrographic studies and surveys in the Sierras de Córdoba and San Luis, but the geological maps surveyed were published several years later. The other geologists did not contribute to maintaining the vigorous survey plan of the Geological Chart of the Argentine Republic that had had such a good start last decade, mainly due to the lack of resources and personnel to meet these objectives.

References

Abad de Santillán D (1971) Historia Argentina. Tipográfica Editora Argentina, Tomo 5, 476 p, Buenos Aires

Alonso RN (2007) Los Trabajos Geológicos de Luciano R. Catalano en la Puna Argentina (1923–1927). InsuGeo, Miscelánea 16:36, Tucumán

Alonso RN (2018) Augusto Tapia, el sabio mutilado. Diario El Tribuno, 9 de abril de 2018, Salta

Alonso RN (2020a) Historia del litio en la Puna. Revista Facultad de Ciencias Exactas, Físicas y Naturales, 7 (Sup 1):151–162, Córdoba

Alonso RN (2020b) Tor Hagerman y el petróleo salteño. Diario El Tribuno 1 de enero de 2020, Salta

Barrio A (1966) Juan José Nágera Ezcurra. Physis 26(171):203–205

Caldenius C (1932) Las glaciaciones cuaternarias de la Patagonia y Tierra del Fuego. Dirección General de Minas y Geología, Publicación 95:1–152, Buenos Aires

Caldenius C (1940) The Tehuelche or Patagonian Shingle formation, A contribution to the study of its origin. Geografiska Annaler 22(3–4):160–181, Stockholm

Camacho HH (2001) Las ciencias geológicas en la Argentina, hasta 1939. Saber y Tiempo 12:177–220

Capdevila R (2013) Augusto Tapia: primer glaciólogo antártico argentino. In: León Wöpke C, Jara Fernández M (eds) Esbozando la Historia Antártica Latinoamericana, Encuentros de historiadores antárticos latinoamericanos 1999–2011, Editorial L.W., 93–102, Santiago de Chile

Catalano LR (1917) Los metales raros: datos sobre los metales raros en la República Argentina y en particular del mineral complejo de las Sierras del Pie de Palo (San Juan). Facultad de Ciencias Exactas y Naturales, Universidad de Buenos Aires, unpublished PhD Thesis, 89 p, Buenos Aires

Catalano LR (1927) Geología química de los boratos. Formación de las cuencas y características generales de la superficie de los salares. Punta de Atacama, Territorio Nacional de Los Andes. Ministerio de Agricultura. Dirección General de Minas, Geología e Hidrología, Publicación 28:1–101, Buenos Aires

Catalano LR (1930) Puna de Atacama (Territorio de los Andes) Reseña geológica y geográfica. Universidad Nacional del Litoral, Publicación 8:1–106, Rosario

Catalano LR (1964a) Boro - Berilio - Litio, Una Nueva Fuente Natural de Energía. Ministerio de Economía de la Nación, Secretaría de Industria y Minería, Serie Argentina 3:1–99, Buenos Aires

Catalano LR (1964b) Estudio Geológico-Económico del Salar de Hombre Muerto (Puna de Atacama) (Territorio Nacional de Los Andes. Ministerio de Economía de la Nación, Secretaría de Industria y Minería, Serie Argentina 4:1–211, Buenos Aires

Groeber P (1929) Líneas fundamentales de la geología del Neuquén, sur de Mendoza y regiones adyacentes. Dirección Nacional de Geología y Minería, Publicación 58:1–109, Buenos Aires

Groeber P (1933) Confluencia de los ríos Grande y Barrancas (Mendoza y Neuquén). Boletín de la Dirección Nacional de Geología y Minería 38:1–72, Buenos Aires

Groeber P (1938) Mineralogía y Geología. Espasa-Calpe Argentina, 492 p, Buenos Aires

Groeber P, Tapia A (1926) Condiciones geológicas de la quebrada de Ullúm en relación con un dique de embalse, provincia de San Juan. Ministerio de Agricultura, Dirección General de Minas y Geología, Publicación 26:1–34, Buenos Aires
Hagerman T (1933) Informe preliminar sobre el levantamiento geológico del Departamento Santa Bárbara en la provincia de Jujuy. Boletín de Informaciones Petroleras 107:451–496
Hagerman T (1936) Investigaciones sobre el material clástico en formaciones del Norte Argentino. Boletín de Informaciones Petroleras 139:75–120
Ljungner E (1931) Geologischen Aufnahmen in der Patagonischen Kordillera: Vorläufige Mitteilung über staatliche Arbeiten am See Nahuel Huapi in dem Argentinischen Nationalpark unter dem 41. Breitengrad. University of Uppsala, Geological Institutionen, Bulletin 23:204–242
Ljungner E (1959) Nahuel Huapi. Ein geographischer Querschnitt durch die Anden in Patagonien. Dahlia Books Handelsbolag, 360 p, Uppsala
Lundqvist J (1983) Carl C:zon Caldenius and the Swedish work on Quaternary geology in Argentina. In Rabassa, J. (ed.) Quaternary of South America & Antarctic Peninsula 1. A.A. Balkema Publishers, 1–4, Rotterdam
Lundqvist J (2006) Carl Caldenius and other links between the Nordenskjöld expedition and recent Argentine–Swedish cooperation in Quaternary geology. In: Rabassa J, Borla ML (eds) Antarctic Peninsula and Tierra del Fuego: 100 years of Swedish-Argentine scientific cooperation at the end of the world. Taylor & Francis, 25–36, Leiden
Malumián N (2009) Juan José Nágera (22 de Mayo 1887- 15 de Mayo 1966), una rememoración del primer matriculado. Revista del Consejo Profesional de Geología, Geotimes 19:20–29
Malumián N (2011) Historia de la concepción de un espacio geográfico denominado plataforma continental. Revista de la Asociación Geológica Argentina 68(3):407–414
Marenssi S (2007) Doctor José María Sobral (1880–1961): de los hielos antárticos al olvido argentino. Revista de la Asociación Geológica Argentina 62(4):495–497
Nágera JJ (1919) La Sierra Baya, Estudio geológico y económico. Anales del Ministerio de Agricultura de la Nación, Dirección General de Minas, Geología e Hidrología, Sección Geología, Mineralogía y Minería 14(1):1–61, Buenos Aires
Nágera JJ (1926a) Los Hoyos del Campo del Cielo y el Meteorito. Ministerio de Agricultura, Dirección General de Minas, Geología e Hidrología, Publicación 19:1–30, Buenos Aires
Nágera JJ (1926b) Atlas de la República Argentina. Editor Talleres Gráficos Colombatti. 38 p, Buenos Aires
Nágera JJ (1926c) Mar Libre: Doctrina. Facultad de Humanidades y Ciencias de la Educación, Revista Humanidades 14:95–102, Buenos Aires
Nágera JJ (1936) Puntas de Santa María del Buen Aire, lugar de fundación de Don Pedro de Mendoza", Meditaciones geológicas, Buenos Aires, Imp. Francisco A. Colombo, 40 p, Buenos Aires
Nágera JJ (1940) Tandilia. Historia física de la provincia de Buenos Aires. Humanidades I:1–272, Buenos Aires
Nágera JJ, Manito O (1938) Geografía física de las Américas y de la República Argentina. Editorial Kapelusz, 2ª edición, 467 p, Buenos Aires
Pastore F (1914) Estudio geológico y petrográfico de la Sierra del Morro, San Luis. Tesis presentada para optar al grado de doctor en Ciencias Naturales, Facultad de Ciencias Exactas. Físicas y Naturales, 46 p, Buenos Aires
Pastore F (1915) Estudio geológico y petrográfico de la Sierra del Morro (Prov. de San Luis). Anales del Ministerio de Agricultura de la Nación, Sección Geología, Mineralogía y Minería 11(2):1–45
Rolleri E (2006) The work of Nordic geologist in Argentina. In: Rabassa J, Borla ML (eds) Antarctic Peninsula and Tierra del Fuego: 100 years of Swedish-Argentine scientific cooperation at the end of the world. Taylor & Francis, 3–24, Leiden
Sobral JM (1918) Estudio petrográfico de algunas rocas argentinas. Anales del Ministerio de Agricultura de la Nación, Sección Geología, Mineralogía y Minería 2:1–54, Buenos Aires

Tanuz MA (2021) Paulina Mühlmann, una pionera de la geología de campo. In: Guereschi A, Martino R, Ramos VA (eds) La Mujer en la Geología, Asociación Geológica Argentina, Publicación Especial 14:78–86

Tapia A (1925a) Sobre los rasgos principales de la glaciación actual en la Isla Laurie (Archipiélago de las Orcadas del Sur). Dirección General de Minas, Geología e Hidrología, Publicación 7:1–40, Buenos Aires

Tapia A (1925b) Apuntes sobre glaciarismo pleistocénico del Nevado del Aconquija. Anales de la Sociedad Argentina de Estudios Geográficos (GAEA) 4:1–365, Buenos Aires

Tapia, A (1935) Pilcomayo. Contribución al conocimiento de las llanuras argentinas, provincias del Chaco y Formosa. Boletín de la Dirección Nacional de Geología y Minería 40:1–143, Buenos Aires

Tapia, A (1937) Las cavernas de Ojo de Agua y Las Hachas. Historia geológica de la región de La Brava, Provincia de Buenos Aires. En relación con la existencia de hombre prehistórico. Boletín de la Dirección de Minas y Geología 43:1–212, Buenos Aires

Tapia A (1941) Mapa hidrogeológico general de la República Argentina. 1: 5.000.000. Dirección de Minas y Geología, Buenos Aires

Turner JCM (1970) Historia de la geología del noroeste argentino. En Primer Congreso Argentino de Historia de la Ciencia (1969), Boletín de la Academia Nacional de Ciencias 48 (primera parte):495–512, Córdoba

Zappettini E (2004) Servicio Geológico Minero Argentino, 1904–2004, 100 Años al servicio del desarrollo nacional. Secretaría de Minería, Segemar, 106 p, Buenos Aires

Chapter 11
The Foundation of *Yacimientos Petrolíferos Fiscales*

Abstract A series of small ventures attempted to exploit some oil seeps in northern Argentina and the Cacheuta region. The latter was achieved by the Polish geologist Rudolf Zuber, who managed to produce oil at the end of the nineteenth century for local use and to supply the railway fuel in Mendoza. The first important discovery was made in Comodoro Rivadavia in 1907, which revealed oil in one of the basins that years later became one of the most important in the country. The discovery was made through a systematic plan of drilling in search of water and eventually oil conducted by Eng. Hermitte. This discovery led years later to the discovery of oil in Plaza Huincul through careful geological studies carried out by Keidel, which constituted the pioneering drilling in the Neuquén basin in 1918. These discoveries led to the creation of *Yacimientos Petrolíferos Fiscales*, the first state oil company. This company was characterized in its beginnings by the participation of Italian geologists, firstly Guido Bonarelli and later Enrico Fossa Mancini, who developed an Italian exploration school. These geologists trained the first Argentine oil geologists who graduated from the universities of Buenos Aires, La Plata, and Córdoba. The exploration and geological studies carried out by several generations of Argentine geologists had important achievements in discovering oil and gas in various basins of the national territory.

Keywords Oil exploration · State oil company · Patagonian oil · Neuquén basin · Argentine pioneers

The creation of *Yacimientos Petrolíferos Fiscales* (Fiscal Oil Fields) was preceded by a series of events that marked the beginnings of modern oil exploration in our country. Its foundation allowed the creation of a solid school of earth sciences that played a leading role in Argentine geology's advancement throughout the last hundred years.

V. A. Ramos, *History of the Geological Sciences in Argentina*, The Latin American Studies Book Series, https://doi.org/10.1007/978-3-031-83139-3_11

11.1 The First Occurrences

The discovery of bitumen or pitch in our territory dates back to the year 1787 when a Franciscan friar, from the Missions of Tarija, reported on the existence of a pitch spring at the eastern foot of the Sierra de Aguaragüe in Salta. This appointment was followed by unsuccessful exploration developments in the provinces of Salta and Jujuy during much of the nineteenth century, either personally or through pioneering ventures by small mining companies that could not have sustainable development. Equivalent experiences were developed in the provinces of Mendoza and Neuquén on similar dates, with the drilling of superficial wells up to 360 m deep that turned out to be uneconomical. An excellent account of these initial activities is found in the classic work of Yrigoyen (1983), complemented by the observations of Alonso (2016, 2020) in northwest Argentina and the synthesis of Riccardi (2018).

The first modern geological study of these manifestations is due to the reconnaissance work of Luis Brackebush in the Laguna de la Brea in Salta, which allowed the definition of the Petroleum Formation (Brackebush 1883), today recognized as the Yacoraite Formation of Maastrichtian-Danian age.

A few years later, in 1886, the Mendoza Oil Company directed by Carlos Fader hired a famous Polish geologist, Dr. Rodolfo Zuber (Fig. 11.1) in Europe, along with other technicians and engineers. Zuber had studied chemistry and mineralogy at the University of Lviv, now in western Ukraine. He did complementary studies between 1881 and 1882 at the University of Vienna with Edward Suess among his professors and returned to Lviv where he defended his doctoral thesis in 1883. Since 1881 he had been interested in the geology of petroleum in the Carpathians, a topic that was his main interest until his last days (Miecznik 2021; Alonso 2015).

Zuber arrived in Mendoza in 1886 and worked for 6 years in Argentina. He studies the oil springs of Cacheuta, being accompanied in the beginning by Brackebush. He conducted the first geological study of the Cacheuta oil manifestations in Mendoza (Zuber 1890a, 1892). He assigned the oil formation to the Rhaetian and made a detailed map of the area. He successfully directed the Mendocina Company's exploration campaign, which drilled about 30 wells and produced about 8,000 cubic meters of oil (Riccardi 2018).

The success obtained led him to examine the coals described by Burmeister as carboniferous between Mendoza and San Juan, finding that they were associated with Triassic fossil plants. He also made a six-week expedition to the south of the Diamante River, where he studied the manifestations of San Rafael and crossed the Andes to the latitude of the Tinguiririca volcano, an experience full of vicissitudes (Zuber 1895).

The success achieved led the company to send him to the Strait of Magellan in 1887, where he studied the lignite-bearing banks on the Brunswick Peninsula and those of the Marta mine in the Skyring Sound, publishing his results in Zuber (1887).

In 1888 he undertook an expedition to Jujuy where he again studied the La Brea lagoon in the northern extension of the Sierra de Santa Bárbara. His geological results led to the formation of a new mining company to attempt its exploitation again (Zuber

Fig. 11.1 Rudolf Zuber (1858–1920) in Argentina and already an established academic in Poland. Note that he was armed with a rifle, and he wore large spurs on his riding boots (Miecznik 2021)

1890b, c). On this occasion, he takes advantage of an invitation and examines the silver mines of Cerro Rico in Potosí, Bolivia.

His contributions to petroleum geology during his six years in Argentina were recognized by the National Academy of Sciences, which named him the corresponding academic in 1890. In turn, Joaquín Frenguelli dedicated a new genus and species to him in 1943, *Zuberia zuberi*. a Triassic plant. Upon his return to Poland, he has an intense international activity studying oil fields in Venezuela, Spain, Africa, and in different parts of Europe and Asia. He was undoubtedly the most prominent petroleum geologist who worked in our country before the discovery of oil in Comodoro Rivadavia in 1907.

The San Rafael region had been visited on a brief excursion in 1892 by Zuber who described the carbonaceous manifestations (Zuber 1892). Rodolfo Hauthal, head of the Geological Section of the Museo de la Plata, made a couple of years after a detailed survey between the Diamante and Atuel rivers in the south of Mendoza to characterize these manifestations. Given the characteristics of these coals, he called them "rafaelite", which is a variety of asphaltite, pyrobitumen with high energy value (Hauthal 1895).

11.2 The Discovery of Oil in Comodoro Rivadavia

The discovery made in Comodoro Rivadavia in 1907 (Fig. 11.2) has two different and partly conflicting stories about its origins. The official history that has been documented in an extensive work carried out by YPF on the occasion of the 25th anniversary of this discovery called "*Development of the Fiscal Petroleum Industry (1907–1932)*" advocates a fortuitous event whose merits fall exclusively on the head of surveys. His assistants José Fuchs and Humberto Beghin shared with the firemen, laborers and cooks. In this story, all the credit goes to those in charge of drilling and in their renowned telegram where they announce to Geminas (Mining and Geology Division) that they have found "*kerosen of the best quality*" at a depth of 539 m.

On the other hand, the enormous efforts to purchase drilling rigs of greater power and depth, to systematically survey the geology, and to study the oil sources, kept the results of the drilling in a state of alert. Already in 1895, Eng. Hermitte had advised the "*Compañía Mendocina de Petróleo S. A.*" in its exploration work in Cacheuta. He had advised that they should continue with the surveys (Riccardi 2018). This is how Eng. Hermitte, a few hours after receiving Fuchs' telegram, had already issued the decree that allowed the State to reserve the area of oil discovery in 1907. It was at Hermitte's suggestion that the national government made the reservation in the

Fig. 11.2 Work associated with the beginning of the drilling led to the oil discovery well in Comodoro Rivadavia in December 1907. Although it was originally done to search for water, they were part of a program for the knowledge of subsoil resources supervised by Hermitte (Turic and Ferrari 2000)

case by decree of December 15, 1907: "*The denunciation of mining belongings and granting of search permits in the port of Comodoro Rivadavia, Territory of Chubut, is prohibited within a radius of 5 km leagues, all direction, counting from that of the population*". Today there is a certain consensus that Eng. Hermitte was prepared with his drilling program for an eventual discovery given that since 1902 he was in charge of studying, as head of the Commission for the Study of Water and Coal Deposits, the potential of probable oil deposits.

The General Direction was in charge not only of directing the drilling, but also of the first geological studies that were carried out by Stappenbeck who was sent to Comodoro Rivadavia to supervise the drilling in 1908. He made an excellent study of the groundwater potential of the region, established the geology of the different sequences, and analyzed the oil horizon, assigning it to the "*variegated sandstones*" of Carlos Ameghino (Hechem 2016), currently considered the Chubut Group. He mentions the presence of oil shows in the Chico River, being one of the first works related to the hydrocarbon potential of this basin (Stappenbeck 1908, 1909).

Given the decline in production in 1910, the General Direction of Petroleum Exploration of Comodoro Rivadavia was created, placing it in charge of Engineer Luis A. Huergo. This is the first precedent of an official institution that is responsible for the direct exploitation of the resource. In the following three years and under the strong direction of Eng. Huergo, new drilling rigs were brought in, an aqueduct was built to supply water to the drilling wells, tanks were made to store the oil, a dock was built to ship the production, and a whole series of measures rapidly increased production. Eng. Huergo was confident in the importance of the field and in 1913 he asked Congress for a strong increase in funds to achieve self-sufficiency in oil. Unfortunately, he died that same year without seeing his development program become a reality.

In those years, geological studies of the site were carried out in parallel by other institutions or companies. Walter Schiller, who worked on several occasions starting in 1910 (Fig. 11.3), considered that the traps were anticlines characterized by gentle dipping domes (Hechem 2016), visible in the outcropping Tertiary and that were transmitted with few variations towards the subsoil (Schiller 1919, 1925), not paying attention to an unconformity that separated the Tertiary layers from the Cretaceous ones.

In 1916 and 1917 Juan Keidel worked in the Sierras de San Bernardo ranges, creating one of the first maps of the folded layers in the Senguer elbow, belonging to the orogen that he called Patagónides (Keidel 1917).

Ricardo Wichmann carried out studies on some 60 wells in the reservoir, agreeing that the oil was in the layers with *Lahillia luisa* from the upper Senonian. In his opinion, the layers of the reservoir were in an almost horizontal position and the trap was given by the lenticularity of the reservoirs (Wichmann 1921).

Windhausen was subsequently sent to Patagonia to support the studies on the Comodoro Rivadavia oil fields (Fig. 11.4), which he carried out between 1918 and 1923. His studies extended to the Sierras de San Bernardo and allowed him to propose again the structural control of the deposits in contrast to Wichmann's previous hypothesis (Windhausen 1921, 1923).

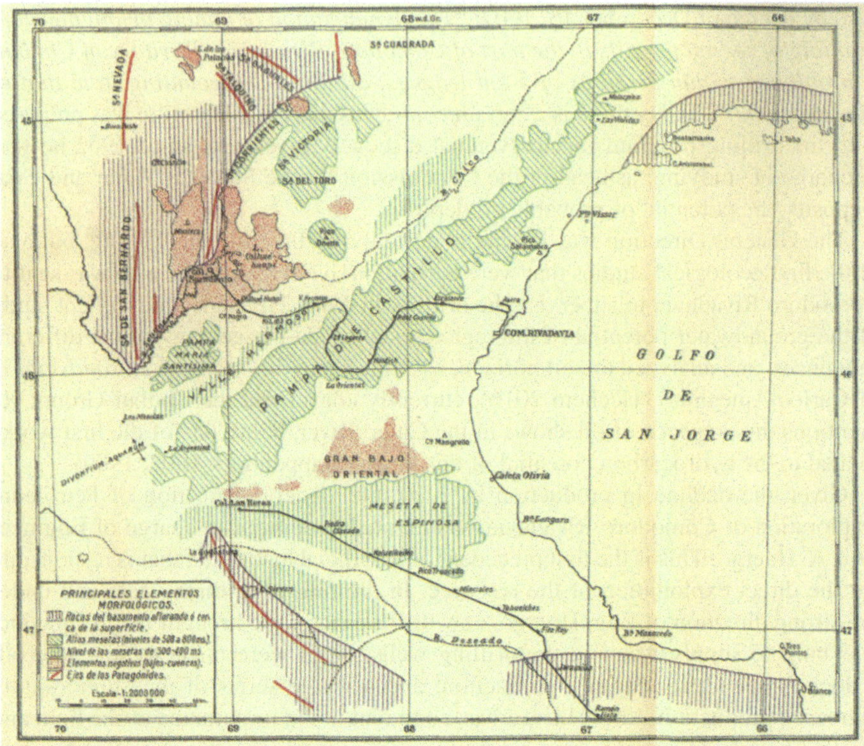

Fig. 11.3 Main elements of the San Jorge basin with indications of the main anticlines according to Windhausen (1923). In this work he shows the presence of a gentle anticline parallel to the coast north of Comodoro Rivadavia that affects the oil formation, confirming Schiller's proposal of 1919

This led to one of the first geological discussions about the petroleum system, on the one hand, those who believed that trapping was "*a tectonic issue*" with those who believed that it was "*a facies issue*" (Hechem 2015, 2016).

One of the fundamental works on the San Jorge Gulf basin in that decade is the complete study published by Windhausen (1924). In this work, he establishes the stratigraphy, the characteristics of the oil horizon, the structure, and the tectonic, and paleogeographic evolution of the different units, complemented with an evaluation of previous geological studies. It establishes the detailed stratigraphy of the basin, highlighting the location and facies of the petroleum formation, the development of anticlines, and the unconformity that separates the Cretaceous layers from the Tertiary deposits that Keidel (1917) had already attributed to the formation of the Patagónides. He completed this work while he was on commission at the recently created YPF, but he ended up publishing it in the *Annals of the Academy of Sciences* in Córdoba.

This comparison only highlights Windhausen's solid academic training and his clear ideas on the geology of the Gulf of San Jorge basin. His ideas and convictions

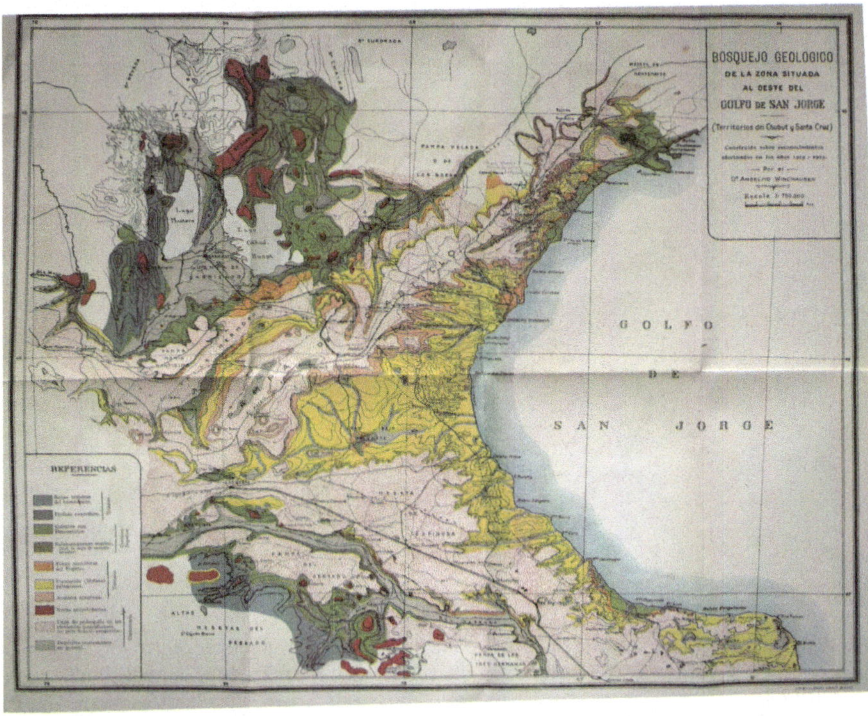

Fig. 11.4 Geological map of the San Jorge basin and the Sierra de San Bernardo done by Windhausen between 1919 and 1923 (Windhausen 1924)

collided with another strong personality such as that of General Mosconi, and in 1923 he had to leave the nascent state oil company where he had been sent on a commission by the Direction of Mines and Geology.

11.3 The Discovery of Oil in Plaza Huincul

In previous chapters, we had analyzed the works of Keidel (1913) and Windhausen (1914) who evaluated the geology of the oil manifestations in the region of what we know today as the Huincul ridge, where they recommended drilling based on its oil potential. Based on the studies carried out, Eng. Hermitte asked Keidel to choose a favorable structure for drilling (Figs. 11.5 and 11.6). This drilling, which began in February 1916, finally found oil in October 1918 between 603 and 608 m deep (Turic and Ferrari 2000). The drilled location was a compromise between the distance to the railroad tracks and the presence of an anticlinal structure (Fig. 11.7). The Direction of Mines, Geology and Hydrology drilled six more wells in the region, of which five

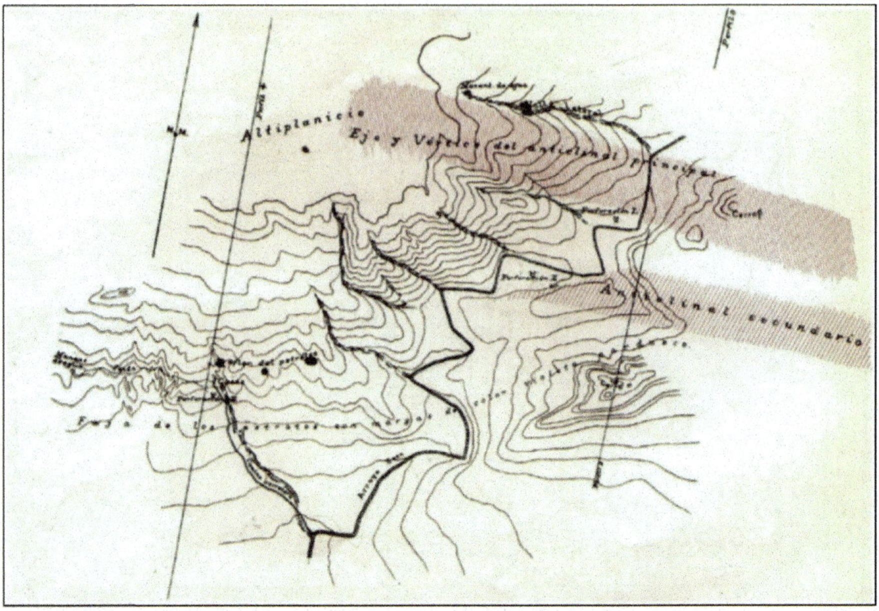

Fig. 11.5 East–west structures of the Challacó region where two anticlines of that direction stand out, associated with manifestations of oil on the surface that are part of the current Huincul ridge (Keidel 1913)

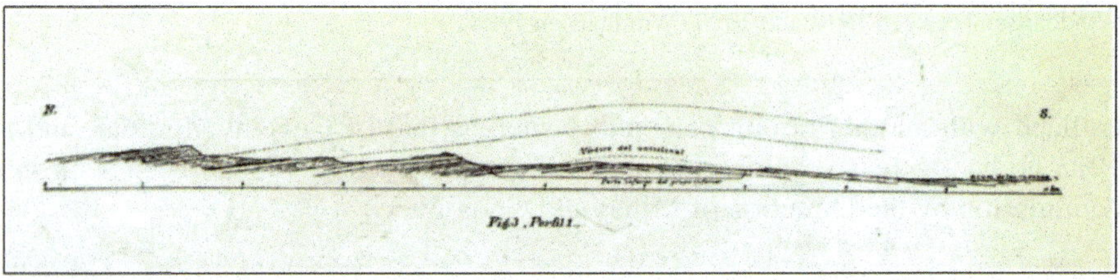

Fig. 11.6 North–south structural section surveyed by Keidel (1913) in the Challacó region, today recognized as the Huincul ridge. Location on the map in the previous figure

were productive. In December 1922, more than 4,500 cubic meters of oil had been produced before it was founded by YPF.

11.4 The Creation of YPF

In September 1910, Law 7059 was finally enacted, authorizing the executive branch to reserve 5,000 hectares for oil exploitation and exploration in Comodoro Rivadavia. A few months later, the Administrative Commission was created to carry out work

Fig. 11.7 Transfer of the first well from the railroad tracks to its drilling site in Plaza Huincul in 1917; drilling of well No. 1 supervised in 1918 by Juan Keidel, Enrique Cánepa, and Juan Soufal. *Source* Guevara (1927), and Giordano (2002)

in a new department, the General Direction of Petroleum Exploitation in Comodoro Rivadavia.

In December 1910, Eng. Hermitte placed Eng. Luis A. Huergo in charge of exploitation in Comodoro Rivadavia. This, despite the lack of financial support, gives a great boost to the work. 100,000 pesos were needed to acquire materials for the aqueduct, and Huergo, under his responsibility, decided to buy them, ensuring that, if the Government did not recognize the expense, he would pay for them out of his pocket (Yacimientos Petrolíferos Fiscales 1932, p. 33). His management made it possible to significantly increase production, which went from 2,082 to 20,732 cubic meters between 1911 and 1913. Given the results obtained, he requested Congress for 2,000,000 pesos, promising oil self-sufficiency if they allocated the necessary funds. Unfortunately, he died in 1913, leaving a great work behind.

The Administrative Commission finally built the aqueduct, hired tankers (Fig. 11.10) for its transportation and drilled numerous wells, leaving in 1922, 89 wells in production with a total of 343,000 cubic meters of oil.

The first oil specialists who arrived at Comodoro Rivadavia were geologists and engineers who emigrated from Russia after 1920 (Fig. 11.8), such as Vladimiro Vinda, Theopent Nikulin, Alejandro Piatnitzky, Alejandro Stessin, Traian Serghiescu, Demetrio Chahnazaroff and Nicolás Koukharsky, among many other operators and specialists of this nationality that settled in that city (see Tchoumatchenco et al. 2018). Among them, Vinda (1877–1930) stood out, who participated in numerous field commissions and analyzed more than 600 wells in the Gulf of San Jorge basin (Vinda 1928).

President Hipólito Yrigoyen (1852–1933) became aware of the importance of oil supply for the country. After he visited the Comodoro Rivadavia fields in 1919, he presented a bill declaring that all oil fields were state property and established that the executive branch could exploit or explore them. Although this project was not approved, it continued to struggle to achieve a law that would authorize exploitation

Fig. 11.8 YPF geologists in the field in 1927. From left to right N.N., Alejandro Stessin, Alejandro Piatnitzky, Vladimiro Vinda, N.N. and Enrico Fossa Mancini (Hechem 2016)

and administration by the state, until on June 3, 1922, the creation, by decree of the executive branch, of the *Direccción General de Yacimientos Petrolíferos Fiscales* (YPF), within the Ministry of Agriculture, with a hierarchy equal to that of the Direction of Mines, Geology and Hydrology. We had seen that that same year Hermitte had resigned as director of Geology and Mines and was replaced by José María Sobral.

With the assumption of Marcelo T. de Alvear to the presidency in 1922, Colonel Enrique Mosconi was appointed director of YPF. He had studied engineering at the Faculty of Exact, Physical, and Natural Sciences, graduating in 1903 as a civil engineer.

Mosconi began its activities at YPF by taking control of the Plaza Huincul and Comodoro Rivadavia deposits, with a series of measures to achieve greater efficiency in its production and transportation (Fig. 11.9). His provisions made it possible to achieve an integration of production, distillation, and sale that had great development during his direction (see details in Yacimientos Petrolíferos Fiscales 1972). He turned YPF into a self-sufficient company that was financed with profits from oil extraction during his 8 years in charge of managing this company (Mosconi 1936). In 1924, he began the publication of the *Boletín de Informaciones Petroleras* (Petroleum Information Bulletin), which for decades has been an important means of publicizing new finds, stratigraphic reviews, and exploration and exploitation news in the geological community.

When Mosconi took over YPF he had only one exploration geologist, Anselmo Windhausen, who worked at Comodoro Rivadavia. To conduct the geological activities, he later hired Guido Bonarelli on the recommendation of José M. Sobral who was serving as Director of Mining. Bonarelli had left the Direction of Mining and Geology in 1922 to return to Italy, after his important investigations in northern

Fig. 11.9 Enrique Mosconi (1877–1940). *Source* Wikimedia

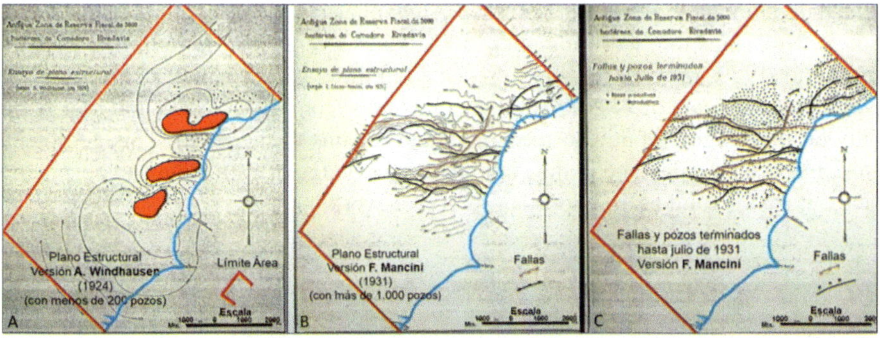

Fig. 11.10 Different interpretations of the structural control of Comodoro Rivadavia over time (after Calegari 2022a). **a** Anticlines with 200 wells from Windhausen (1924); **b** Structural map, and **c** Gentle structures controlled by faults based on 1000 wells of Fossa Mancini (1931)

Argentina, which highlighted the oil potential of the sub-Andean region. He arrived in the country again in 1923 and joined as head of the Geology Division, in charge of YPF exploration. Upon his arrival, he planned the exploration not only of Patagonia and Neuquén, but also of the Cuyo region and northern Argentina. He invited Egidio Feruglio (1897–1954), who arrived in 1925 and took charge as Head of the Geology Section of Comodoro Rivadavia. Due to an altercation with the administrator, he was sent to Buenos Aires and only returned to Comodoro Rivadavia in 1927 (Hechem 2016). After his return, in the following decades, he carried out titanic work in the geological study of Patagonia.

In 1927, invited by Bonarelli, Enrico Fossa Mancini (1884–1950) arrived, who had obtained a doctorate from the University of Pisa in 1913 and already had a notable geological career in Italy. He had participated in the First World War where he stood out as an observer in captive balloons (AAPG Memorial 1953) and had carried out oil studies in Venezuela (Frenguelli 1950; Urbani 1991; Riccardi 2016). Upon his arrival at Comodoro Rivadavia he took charge of the Geological Commission, being one of the first to recognize the structural control by east–west extensional faults of the gentle folding structures (Fig. 11.10), improving the interpretation of Windhausen.

Upon Bonarelli's return to Italy, Fossa Mancini became head of the Geology Division (Fig. 11.11), a position he held from 1928 to 1939 (see Sassaroli 2008; Riccardi 2016). During this period, he developed various deposits in the Neuquén and San Jorge basins, as well as in Cuyo and northwest Argentina. In those years joined YPF María Casanova (1899–1947) (see Calegari 2022b), Vicente Franceschi, Danilo Ramaccioni, and Ivo Conci (Anónimo 1951), who together with Feruglio and Fossa Mancini would constitute the "*Italian school*" of YPF that so much contributed to the geological knowledge of the San Jorge Gulf basin and other regions of the country.

The work of the different commissions in Patagonia was supervised for many years by Egidio Feruglio (Fig. 11.12), who carried out important field tasks, together with geologists such as Juan Brandmayr, Danilo Ramaccioni, Ivo Conci (1901–1951), Alejandro Piatnitzky (1879–1959) (see Archangelsky and Riccardi 1989), among others. He brings this work together in his "*Geological Description of Patagonia*" (Feruglio, 1949–1950). This work presented in three volumes and accompanied by two maps at a scale of 1: 2,000,000 encompasses from 32° to 55°S and covers not only all of Patagonia but also the immediate regions to the north. This geological work has been a milestone in the geological knowledge of Patagonia and despite the years, it is still valid as an important reference work.

Fig. 11.11 Enrico Fossa Mancini (1884–1950) and in the field observing fractures of the Chubut Group in 1928 (Herrero Ducloux 1953; Fossa Mancini 1934)

Fig. 11.12 Egidio Feruglio in the field in 1932 with N.N, his colleagues Juan Brandmayr and Alejandro Piatnitzky; Portrait of Feruglio preserved in the photo library of the *Musei Civici* of Udine. *Source* Courtesy of Edgardo Rolleri and del Bianco (1955)

11.5 The First Argentine Petroleum Geologists

At the end of the 1920s, the rector Ricardo Rojas of the University of Buenos Aires proposed to General Mosconi to create an institution at this university to prepare technicians and professionals for the oil industry. Thus, it was born the Petroleum Institute of the University of Buenos Aires, which began operations at the beginning of 1930, in the Faculty of Exact, Physical and Natural Sciences. His students were civil engineers, chemists and geologists who spent a year studying specialization, graduating as petroleum engineers. These courses were taken by a group of young geologists, trained in the country at the universities of Buenos Aires, La Plata and Córdoba, through an aggressive YPF scholarship plan proposed by Windhausen, already a professor at the National University of Córdoba, generated since 1932, an explosion of vocations to work in the oil industry (Martino and Ramos 2018). The first scholarship recipients, such as I. Osvaldo Bracaccini, Tomás Suero, Enrique Herrero Ducloux and Joaquín Daniel, among others, began a tradition of field internships for 2 months in Comodoro Rivadavia, where they were prepared for professional work. They arrived by boat to that town and were received by instructors such as Enrico Fossa Mancini, who introduced them to the geology of the basin and the different operations that were carried out. This custom was followed for decades, as reported by Rodolfo Moreno, Pedro Lesta, and Edgardo Rolleri, among others (Ramos 2022).

In those years Fossa Mancini had an outstanding task, first as head of the Geology Division in charge of YPF exploration, later as a professor at the Petroleum Institute of the University of Buenos Aires, and in his last years as a professor at the University of La Plata. In this institution he directed 23 doctoral theses, highlighting among his disciples Teruggi, Bassi, Methol, Mauri, among many others (Riccardi 2016). He died in 1950, the victim of an unexpected traffic accident.

It is interesting to note that the development of new wells to the east of the Comodoro Rivadavia field along the coast at the end of the 20 s inspired Juan J. Nágera on a visit to this city to propose his "*Mar Libre*" doctrine. Observing that the oil fields were extending offshore, he became aware of the potential that our continental shelf had and proposed in 1927 to extend our sovereignty to the limit with the continental slope (Fig. 11.13). His doctrine served as the basis for years later to establish sovereignty over the "*Argentine Epicontinental Sea*" through a law, a concept coined by Nágera (1927).

In the following years, the young generation of Argentine geologists took control of geological studies and hydrocarbon exploration in the different basins of the country.

Among the first stands out I. Osvaldo Bracaccini (1913–1979), graduated from the National University of Córdoba and trained under the guidance of teachers such as Bodenbender and Windhausen, who joined YPF in 1935 (Fig. 11.14). At the first "*YPF Geologists Meeting*", the young Bracaccini presented in 1938 his novel structural interpretations and fracture analyzes carried out in the north of Mendoza. This inclination towards structural geology was a constant in his studies of the Cuyo region and northern Argentina and culminated in his 1945 synthesis of the structural factor in the country's oil accumulations (Bracaccini 1945). In 1944, he became head of Geology and in 1948 he became the Exploration Manager. In those years, he completed his tectonic investigations in the Precordillera of San Juan with a complete review of its geology (Bracaccini 1950). His work in managing and training the new generations at YPF was fundamental (Ramos 1979; Calegari 2023).

Tomás Suero (1915–1963) had joined YPF in 1937, finishing his doctoral studies in 1939, at the National University of La Plata under the direction of Joaquín Frenguelli. His classic studies of Cerro Lotena allowed him to identify the tectonics

Fig. 11.13 The first wells made in the *restinga* in the late 20s and early 30s that allowed the oil deposits to be extended on the coast (courtesy of Luis Stinco 2022)

Fig. 11.14 Osvaldo Bracaccini, Tomás Suero, Abel Herrero Ducloux, and Carmelo De Ferrariis of the first generations of Argentine geologists awarded scholarships by YPF as students who produced important advances in the knowledge of our regional geology. *Source* A.G.A. archives

of the Upper Jurassic and Supracretaceous in the Neuquén basin. After his advanced studies in the United States, he returned to the country working in the Cuyana basin and in the northern region, which he combined with other studies in the Neuquén basin (Rolleri 1964). His main contributions were made in the western sector of the province of Chubut where he identified for the first time the late Paleozoic marine deposits of the Tepuel basin (Suero 1948), which he later integrated into his late Paleozoic geosyncline that extended to Santa Cruz and the Malvinas Islands (Suero 1962). He had an extraordinary teaching job at YPF and at the universities of La Plata and Buenos Aires.

In the Mendoza sector, the research of Eduardo Trümpy (1903–1966), a Swiss geologist who made numerous reports on the geology of Mendoza, stood out. He was one of the first to ascend to the headwaters of the Tunuyán River where he built a detailed geological map (Pascual 1949). Most of his contributions have remained as unpublished reports, with the exception of a few that were published in the Petroleum Information Bulletin of those years (Trümpy 1943; Trümpy and Lehz 1937).

The stratigraphy of the Neuquén basin known by the studies of Windhausen and Keidel had a notable boost through the studies carried out by Charles E. Weaver (1880–1958) between 1922 and 1925, published in his fundamental work on "*Paleontology of the Jurassic and Cretaceous of West-Central Argentina*" (Weaver 1931). This work published by the University of Washington in Seattle in 1931 introduces a modern stratigraphy with a nomenclature through the definition of formations for the first time in the basin (Lazo 2009). These studies were complemented by the various works of Pablo Groeber in the mountain region mentioned in the previous chapters, among which his "*Fundamental Lines of the Geology of Neuquén*" stands out (Groeber 1929). These were the bases for the survey and exploration work carried out in this basin by YPF.

Among the numerous YPF geologists who worked in that basin, Joaquín Daniel, Carmelo de Ferrariis, and Abel Herrero Ducloux were the first YPF scholarship recipients and the first three doctors from the National University of La Plata under the direction of Keidel in 1938. Abel Herrero Ducloux (1915–1965) graduated with a thesis in the Quebrada de Humahuaca, joined YPF as a geologist student in 1933, and as soon as he finished his doctorate he began working in Neuquén in 1938 (Yrigoyen 1965). After a period of specialization at the Colorado School of Mines in 1940, he produced a series of reports on the Neuquén basin that culminated with a fundamental work, his "*Contribution to the geological knowledge of Neuquén extrandino*" (Herrero Ducloux 1946), a work of synthesis that, together with those of Weaver and Groeber, became classic contributions that for decades were required consultation for several generations.

In the Neuquén basin, Carmelo I.C. de Ferrariis (1912–1998) had important contributions as Head of the Neuquén District from 1941. His most important work was to identify and characterize the Huincul Ridge through his classic work on "*Age of the Arc or ancient Ridge of Eastern Neuquén, according to the stratigraphy of the immediate area*" (De Ferrariis 1947). His interpretation of the ridge based on numerous structural cross sections was for decades a paradigm that allowed us to understand the paleogeography of the basin. He also carried out surveys in Santa Cruz and Tierra del Fuego, working for more than 40 years at YPF. He spent his last years teaching at the National University of La Plata (Stipanicic 2000).

These pioneering YPF explorers were followed by a second legion of geologists who continued systematically surveying the geology of the different basins. Many of them benefited from the scholarships awarded by YPF to complete their studies and came from different national universities.

In the San Jorge Gulf basin, Roberto Ferello (1922–1987) stood out (Fig. 11.15), replacing Tomás Suero in Geological Commission 3, where he worked as an assistant. Ferello had graduated in 1946 from the University of Buenos Aires with a thesis on "*Geological study in the Piedra del Águila region (Neuquén)*" under the direction of Pablo Groeber, where he documents and describes an interesting Liassic flora. He developed intense fieldwork in the Sierras de San Bernardo in the central part of Chubut, where he surveyed in detail the stratigraphy and structure of this region. His precise geological maps allowed him to identify the Bernárdides as a large structural unit in 1950, thus characterized by their topographic expression and tectonic characteristics (Ferello 1950, in Ramos 2022, p. 1388). His work in the San Jorge Gulf basin served as the basis for current interpretations of its structure.

Another geologist who started at YPF in those years was Marcelo R. Yrigoyen (1924–1997). He entered as a scholarship recipient in 1942, selected by the demanding Dr. Bracaccini, and received his doctorate from the National University of La Plata in 1948, under the direction of Pablo Groeber, with a thesis on the geology of the Buta Ranquil region in Neuquén. He began as an assistant at Geological Commission 6 of southern Mendoza under the direction of Pedro Criado Roque, whom he succeeded shortly after. During those years and until 1957, he actively worked on surveys of the Tertiary deposits of Mendoza, as well as on the Jurassic Cretaceous sequences. His planchette surveys and numerous reports have remained unpublished. Years later, now outside YPF, he published a complete synthesis of these deposits in his "*Stratigraphic Problems of the Tertiary of Argentina*" (Yrigoyen 1970), which was followed by other important works on Tertiary synorogenic deposits, largely initiated during their work at YPF.

Fig. 11.15 Roberto Ferello, Marcelo Yrigoyen, Pedro Stipanicic, and Alberto Mingramm, a second generation of geologists with YPF scholarships and specialized in the Petroleum Institute of the University of Buenos Aires. *Source* A.G.A. archives

Pedro N. Stipanicic (1921–2008) also worked in Geological Commission 6 under the supervision of Yrigoyen, who graduated from the University of Buenos with a thesis directed by Pablo Groeber on the Triassic deposits of the San Juan Precordillera. He joined YPF in 1947 in the paleontology section, to later carry out stratigraphic field studies on the Triassic and Jurassic, actively collaborating in those classic works carried out in collaboration with Groeber (Groeber and Stipanicic 1953; Groeber 1953). He left YPF in 1953, but in those years, he published numerous results of research and work carried out on the stratigraphy and paleontological content of the Mesozoic sequences (Riccardi 2009).

Alberto R. Mingramm (1921–1994), who graduated from the National University of Córdoba in 1948, also stood out in that generation of geologists. Although his formal thesis sponsor was Otto Schlagintweit, his study on the Ñorquín region in the Neuquén basin was supervised by Pablo Groeber. He joined YPF in 1944 as a fellow and began working on stratigraphic and tectonic studies of the Jurassic and Cretaceous basins of San Juan, Mendoza, and Neuquén. His studies carried out in Geological Commission 6, under the supervision of Yrigoyen, contributed substantially to Groeber's Mesozoic work already mentioned (Stipanicic and Hunicken 1994). Mingramm most transcendental contribution is the tectonic characterization of the Subandean Sierras of Salta and Jujuy, where he presented in 1969 the first balanced structural sections of our country (Ramos 1996). He left YPF in 1970 continuing in the oil industry until his retirement, when he became a professor of Structural Geology at the universities of Buenos Aires and later on in Córdoba.

Although a certain injustice is committed by not highlighting other important contributions made to Argentine geology by YPF geologists, we have restricted attention in this stage of knowledge consolidation to those that we have considered fundamental for its initial development.

Among the new generations of geologists trained at YPF, the figure of Miguel A. Uliana (1944–1997) stands out (Fig. 11.16) in the field of sedimentology and sequential stratigraphy (Legarreta 1998; Biddle and Mitchum 1998). This geologist trained at the National University of La Plata and the Petroleum Institute of the University of

Buenos Aires, in his 30 years of professional work, first at YPF and then at various international companies, made important contributions to the knowledge of our sedimentary basins. He trained alongside leaders of sequence stratigraphy and basin analysis such as Peter Vail and Robert Mitchum, specialists at the frontier of knowledge. He passed on his teachings to his colleagues with maximum generosity and through classic reviews and syntheses that have become fundamental milestones of our geology. Among them, his pioneering works on "*Seismic stratigraphy of carbonate depositional sequences, Upper Jurassic-Lower Cretaceous, Neuquén Basin*" and "*Mesozoic-Cenozoic paleogeographic and geodynamic evolution of southern South America*" stand out, where he lays the foundations for a stratigraphic and paleogeographic evolution modern of those basins. His clear concepts on Mesozoic tectonics in "*Mesozoic extension and the formation of Argentine sedimentary basins*" and in turn the dominant Andean structural style of our basins in "*Inversion structures and hydrocarbons occurrences in Argentina*" produced notable changes and new interpretations in geology Argentina. His early and unexpected death found him in full scientific production (see Legarreta 1998).

However, Uliana's proposals, followed by her colleagues and co-workers, have taken deep roots in an entire generation, as can be seen in the works of Leandro Legarreta, Carlos Gulisano, Estanislao Kozlowski, Daniel Depllapé, among many others who have continued analyzing sedimentary basins through sequential stratigraphy. A good example of these are the modern interpretations of the progradation of the depositional sequences of the Vaca Muerta Formation identified by Uliana almost 40 years ago and which are more valid today than ever.

Fig. 11.16 Miguel A. Uliana introduced sequence stratigraphy in our sedimentary basins. *Source* A.G.A. archives

11.6 Final Considerations

In this chapter, we have attempted to characterize how oil exploration has contributed substantially to the knowledge of our geology, especially when YPF consolidates itself as the national company in charge of the exploration and exploitation of oil and gas under the direction of Mosconi.

The process through which *Yacimientos Petrolíferos Fiscales* integrated a solid team of Argentine geologists, capable of replacing foreigners, was very effective. It was based on selecting the best geology students at the universities of Buenos Aires, La Plata and Córdoba as scholarship holders, exposing them to campaign tests, and incorporating those with the best average into the company. His first tasks consisted of geological surveys using a planchette, which often included the preparation of the topography, without the aid of photography or images, through continuous fieldwork in the geological commissions. These studies covered large areas, with months of fieldwork and a few monthly office days, which allowed several generations of geologists to be trained in the beginning under the so-called Italian school that had geologists and teachers such as Bonarelli, Fossa Mancini, and Feruglio. Another key piece in the training of these geologists starting in 1930 was the Petroleum Institute of the University of Buenos Aires, where graduates from different universities specialized not only in geology but also in geophysics and other disciplines associated with oil exploration.

This system produced geological surveys of large areas of our sedimentary basins, many times in competition with the regular surveys carried out by the National Direction of Mining and Geology, as seen in the geological sheets raised at a scale of 1: 100,000 by Eduardo L Padula, Oscar L. Baulíes, Pedro Criado Roque, Carlos A. Di Persia, among several others, have remained unpublished YPF reports.

In the past 50 years, knowledge of our basins and their oil potential, much of it originated by geologists trained at YPF, has emerged through the work of hundreds of professionals active in the oil industry. Numerous private companies, both Argentine and international, have made important contributions in recent years and have created a rich history showing the evolution of concepts in hydrocarbon exploration. The philosophy of exploration in those years, the history of productive basins, offshore prospecting and the discoveries of unconventional deposits, have a plethora of geologists who have contributed substantially to our history of Argentine geology.

The Argentine Institute of Oil and Gas, under the coordination of Gustavo Vergani and the sponsorship of YPF, has presented the chronicle and history of those years in two important volumes, whose analysis due to its dimensions is beyond the objectives of the present history.

There is no doubt that not only in these initial decades did YPF train numerous generations of geologists who have contributed substantially to the geological knowledge of our country. This YPF school continues to be an important institution over the years that continues to contribute directly or indirectly to the advancement of our geology.

References

AAPG Memorial (1953) Enrico Fossa Mancini (1884–1950). American Association of Petroleum Geologists, Bulletin 37(9):2245

Alonso RN (2015) El petróleo y el polaco Zuber. Diario El Tribuno, 14 de diciembre 2015, Salta

Alonso RN (2016) Historia del petróleo del norte argentino: desde Pablo Sardicat de Soria a Francisco Tobar. Mundo Gráfico, 160 p, Salta

Alonso RN (2020) Los manaderos de petróleo. Diario El Tribuno, 11 de mayo 2020, Salta

American Association of Petroleum Geologists (1953) Enrico Fossa-Mancini (1884–1950) Memorial. American Association of Petroleum Geologists, Bulletin 37(9):2245–2246

Anónimo (1951) Doctor Ivo Conci. Revista de la Asociación Geológica Argentina 6(2):113

Archangelsky S, Riccardi A (1989) Alejandro Matveievich Piátnitzky. Revista de la Asociación Geológica Argentina 44(1–4):442–447

Biddle KT, Mitchum RM (1998) Miguel A. Uliana (1944–1997). American Association of Petroleum Geologists, Bulletin 82(11):2149–2150

Bracaccini IO (1938) Estudios estructurales en el Terciario del Norte de Mendoza. Resumen, Boletín de Informaciones Petroleras 171:41–41, Buenos Aires

Bracaccini IO (1945) Acerca de los movimientos intertriásicos de Mendoza Norte. 1ra. Reu nión de Comunicaciones, Instituto Panamericano de Ingeniería de Minas y Geología (IPIMIGEO) p. 1–26, Buenos Aires

Bracaccini IO (1950) Investigaciones tectónicas en la Precordillera Sanjuanina. Boletín de Informaciones Petroleras 301:1–34, Buenos Aires

Brackebusch L (1883) Estudios sobre la formación petrolífera de Jujuy: Informe oficial. Boletín de la Academia Nacional de Ciencias 5:137–184, Córdoba

Calegari RJ (2022a) El Doctor Fossa Mancini, un actor importante en la exploración de hidrocarburos. Actas del 21° Congreso Geológico Argentino 1435–1436, Puerto Madryn

Calegari RJ (2022b) María Casanova, la primera petrógrafa de YPF. Actas del 21° Congreso Geológico Argentino 1433–1434, Puerto Madryn

Calegari RJ (2023) El Dr. Bracaccini, de alumno-geólogo a Sub Administrador General de YPF (1932–1955). 6° Congreso Argentino y 1° Congreso Latinoamericano de Historia de la Geología, Resúmenes, pp 61–69, Tucumán

De Ferrariis CIC (1947) Edad del Arco o Dorsal antigua del Neuquén Oriental, de acuerdo a la estratigrafía de la zona inmediata. Revista de la Sociedad Geológica Argentina 2(3):256–283

Del Bianco D (1955) La vita operosa di Egidio Feruglio/Michele Gortani. Atti del 1° Convegno Friulano di Scienze Naturali, 16 p, Udine

Ferello R (1950) Levantamiento geológico de la región al oeste del lago Musters Anticlinal VI. YPF (unpublished report), Buenos Aires

Fossa Mancini E (1934) Las primeras exploraciones de YPF en Cañadón Perdido, Escalante, Manantiales Behr, Cañadón Lagarto y Pampa María Santísima. Boletín de Informaciones Petroleras 2(114):9–36

Frenguelli J (1950) Profesor Doctor Enrique Fossa Mancini (1884–1950). Revista de la Asociación Geológica Argentina 5(2):77–81

Giordano HM (2002) La trascendencia del pozo N°1 en Neuquén. Petrotecnia 43(4):74–76, Buenos Aires

Groeber P (1929) Líneas fundamentales de la geología del Neuquén, sur de Mendoza y regiones adyacentes. Dirección Nacional de Geología y Minería, Publicación 58:1–109, Buenos Aires

Groeber P, Stipanicic PN (1953) Triásico. In Groeber, P., Mesozoico, Geografía de la República Argentina. Sociedad Argentina de Estudios Geográficos GAEA 2(1):13–141

Groeber, P (1953) Jurásico. Con la colaboración de Stipanicic, P.N. and Mingramm, A.R.G. In Groeber, P., Mesozoico, Geografía de la República Argentina. Sociedad Argentina de Estudios Geográficos GAEA 2(2):1–347

Guevara, C (1927) Historia del yacimiento fiscal de Plaza Huincul. Boletín de Informaciones Petroleras. Año IV(40):1175–1793, Buenos Aires

Halperín F (2019) Exploración: El puente desde las ideas a la energía. YPF, 194 p, Buenos Aires
Hauthal R (1895) Nota sobre algunas observaciones geológicas en la provincia de Mendoza. Revista del Museo de La Plata 7:9–96
Hechem JJ (2015) Cien años de modelos geológicos en la Cuenca del Golfo San Jorge. Revista de la Asociación Geológica Argentina 72(1):1–11
Hechem JJ (2016) Los primeros geólogos de YPF en Comodoro Rivadavia y sus aportes en la región del Golfo San Jorge. Revista del Museo de la Plata, La Historia de la Geología en el Bicentenario de la Argentina, Número Especial 1:119–129
Herrero Ducloux AP (1946) Contribución al conocimiento geológico del Neuquén extrandino. Boletín de Informaciones Petroleras 23(266):245-281, Buenos Aires
Herrero Ducloux AP (1953) Enrico Fossa Mancini (1884–1950). Bull Am Assoc Pet Geol 37(9):2245–2246
Keidel J (1913) Informe geológico sobre el yacimiento petrolífero de Challacó (Territorio del Neuquén), Talleres Peuser, 47 p, Buenos Aires
Keidel J (1917) Über das patagonische Tafelland und ihre Beziehungen zu den geologischen Erscheinnungen in den argentinischen Andengebiet und Littoral. Zeitschrift der Deutsche Akademie Wissenschaft 3(4):219–245
Lazo DG (2009) Vida y obra de Charles Edwin Weaver (1880–1958) y su paso por la cuenca neuquina en los años '20. Revista de la Asociación Geológica Argentina 65(2):241–247
Legarreta L (1998) Miguel Ángel Uliana (1944–1997). Revista de la Asociación Geológica Argentina 53(1):137–140
Martino R, Ramos VA (eds.) (2018) El Nacimiento de la Geología en la enseñanza universitaria. Revista de la Facultad de Ciencias Exactas, Físicas y Naturales de la Universidad Nacional de Córdoba, vol. 5, Suplemento 1, 163 p, Córdoba
Miecznik JB (2021) Profesor Rudolf Zuber—ojciec polskiej geologii naftowej *(El padre de la geología petrolera de Polonia).* Przeglad Geologiczny 69(8):528–543, Warszawie
Mosconi E (1936) El petróleo argentino, 1922–1939. Librería El Ateneo, 245 p, Buenos Aires
Nágera JJ (1927) Mar Libre: Doctrina. Facultad de Humanidades y Ciencias de la Educación, Revista Humanidades 14:95–102, Buenos Aires
Pascual R (1949) Observaciones geológicas en la Alta Cordillera de Mendoza, Departamento de Tunuyán, entre el Mesón de San Juan y el río Palomares, al oeste del Alto Río Tunuyán. Universidad Nacional de La Plata, unpublished PhD Thesis, La Plata
Ramos VA (1979) Dr. Inocencio Osvaldo Bracaccini (1913–1979). Revista de la Asociación Geológica Argentina 34(3):249–252
Ramos VA (1996) Las primeras secciones balanceadas de Argentina: obra de un pionero. Revista de la Asociación Geológica Argentina 51(1):92–95
Ramos VA (2022) La evolución tectónica de la provincia del Chubut. In: Giacosa R (ed) Geología y Recursos Naturales de la Provincia del Chubut, XXI° Congreso Geológico Argentino, Relatorio: 1372–1397, Puerto Madry
Riccardi AC (2016) Enrico Fossa Mancini: significación y trascendencia de su obra geológica. Revista del Museo de La Plata, La Historia de la Geología en el Bicentenario de la Argentina, Número Especial 1:257–284
Riccardi AC (2018) El desarrollo histórico de las exploraciones petroleras en la Argentina. Academia Nacional de Ciencias de Buenos Aires, Comunicación, El Guion Ediciones, 5–41, Buenos Aires
Rolleri EO (1964) Tomás Suero (1915–1963). Revista de la Asociación Geológica Argentina 19(1):5–8
Sassaroli S (2008) Vita e opere di Enrico Fossa Mancini. In: Sassaroli S (ed) Enrico Fossa Mancini e la storia natural dell´Appennino, Liceo Classico Statale Vittorio Emanuelle II, Jesi. Ed. Abbatelli, pp 15–31
Stappenbeck R (1908) Geologische Beschreibung der Umgebung des Sees Musters in Patagonie. Akademie der Wissenschften 117:1243–1249, Wien

Stappenbeck R (1909) Informe preliminar relativo a la parte Sudeste del Territorio del Chubut. Anales del Ministerio de Agricultura de la Nación, Sección Geología, Mineralogía y Minería 4(1):1–20, Buenos Aires
Stipanicic PN (2000) Carmelo C.I. de Ferrariis. Notas necrológicas. Revista de la Asociación Geológica Argentina 55(1–2):139–140
Suero T (1948) Descubrimiento del Paleozoico Superior en la zona extra-andina. del Chubut. Boletín de Informaciones petroleras 287:31–48. Buenos Aires
Suero T (1962) Paleogeografía del Paleozoico Superior en la Patagonia (República Argentina). Revista de la Asociación Geológica Argentina 16(1–2):35–42
Schiller W (1919) Geologie und Erdöl con Comodoro Rivadavia (Patagonien). Geologische Rundschau 10(1):14–31, Leipzig
Schiller W (1925) Estratigrafía, tectónica y petróleo de Comodoro Rivadavia. Anales del Museo de La Plata 2 (Entrega 1ª): 9–57
Tchoumatchenco P, Riccardi AC, Durand Delga M, Alonso R, Wiasemsky M, Boltovskoy D, Charrier R, Minina E (2018) Geologists of Russian origin in Latin America. Revista del Museo de La Plata 3(2):223–295
Trümpy E (1943) Discordancias en Mendoza. Boletín de Informaciones Petroleras 20(223):39–42, Buenos Aires
Trümpy E, Lehz R (1937) División estratigráfica de los terrenos aflorantes en la región comprendida entre Luján de Cuyo, Potrerillos y Tupungato. Boletín de Informaciones Petroleras 14(152):39–56, Buenos Aires
Turic MA, Ferrari JC (2000) La exploración de petróleo y gas en la Argentina: el aporte de YPF. YPF S.A., 178 p, Buenos Aires
Urbani F (1991) Enrico Fossa Mancini (1884–1950): su vida y obra en Italia, Venezuela y Argentina. Revista de la Asociación Geológica Argentina 46(3–4):353–357
Vergani G (coord) (2022) Crónicas de exploración y desarrollo de hidrocarburos. 50 años de conceptos e historias en primera persona. Instituto Argentino del Petróleo y el Gas, vol 1, 806 p; vol 2, 786 p
Vinda V (1928) Comodoro Rivadavia y las exploraciones petrolíferas en la región vecina al Golfo San Jorge. Informe Yacimientos Petrolíferos Fiscales (inédito), 77 p, Comodoro Rivadavia
Weaver CE (1931) Paleontology of the Jurassic and Cretaceous of west-central Argentina. University of Washington, Memoirs 1:1–595, Seattle
Wichmann R (1921) Estudio geológico de la zona de reserva de la explotación nacional de petróleo en Comodoro Rivadavia (Territorio Nacional del Chubut). Boletín de la Dirección General de Minas, Geología e Hidrología (Serie B Geología) 25:1–18, Buenos Aires
Windhausen A (1914) Contribución al conocimiento geológico de los territorios de Río Negro y Neuquén, con un estudio de la región petrolífera de la parte central del Neuquén (Cerros Lotena y Covuncó). Anales del Ministerio de Agricultura de la Nación. Sección Geología, Mineralogía y Minería 10(1):1–69, Buenos Aires
Windhausen A (1921) Ein Blick auf Schichtenfolge und Gebirgsbau im südlichen Patagonien. Geologische Rundschau 12(3–5):109–137
Windhausen A (1923) Cambios en el concepto de las condiciones geológicas del yacimiento petrolífero de Comodoro Rivadavia. Boletín de la Academia Nacional de Ciencias 27:1–8, Córdoba
Windhausen A (1924) Líneas generales de la constitución geológica de la región situada al oeste del golfo de San Jorge. Informe sobre reconocimientos efectuados en los años 1919–1923. Boletín de la Academia Nacional de Ciencias 27:167–320, Córdoba
Yacimientos Petrolíferos Fiscales (1932) Desarrollo de la industria petrolera fiscal 1907–1932. Editorial Jacobo Peuser, 429 p, Buenos Aires
Yacimientos Petrolíferos Fiscales (1972) Cincuentenario de YPF, una empresa estatal al servicio del país, 1922–1972. Talleres Gráficos Sacerdoti, 191 p, Buenos Aires
Yrigoyen M (1965) Dr. Abel Pedro Alberto Herrero Ducloux (1915–1965). Revista de la Asociación Geológica Argentina 20(4):399–402

Yrigoyen MR (1970) Problemas estratigráficos del Terciario de Argentina. Ameghiniana 6(4):315–329

Yrigoyen, MR (1983) Reseña sobre los conocimientos y la explotación de los hidrocarburos en Argentina antes de 1907. Petrotecnia, ediciones de marzo y abril de 1983 (reprinted in February 2007, pp 16–36)

Zuber R (1887) Z podrózy do poludniowej Ameryki (Tresc wykladu wygloszonego na posiedzeniu Towarzystwa im. Kopernika 26 pazdziernika 1887). De un viaje a América del Sur (El contenido de la conferencia pronunciada en la reunión de la Sociedad Copérnico el 26 de octubre). Kosmos 12:409–414

Zuber R (1890a) Estudio geológico del cerro de Cacheuta y sus contornos, República Argentina. Boletín de la Academia Nacional de Ciencias 10:448–472, Córdoba

Zuber R (1890b) Informe sobre el petróleo de Laguna de la Brea (provincia de Jujuy, República Argentina). Boletín de la Academia Nacional de Ciencias 10:442–447, Córdoba

Zuber R (1890c) Szesc tygodni w Kordylierach (Seis semanas en la Cordillera). Kosmos 15:189–217

Zuber, R (1892) Informe sobre los terrenos petrolíferos del Departamento de San Rafael (Provincia de Mendoza). Boletín de la Academia Nacional de Ciencias 12:370–375, Córdoba

Zuber R (1895) O wêglu kamiennym w Kordylierach argentyñskich (Sobre el carbón en la Cordillera Argentina). Kosmos 20:517–519

Chapter 12
Teaching and Research of Geology in Universities

Abstract We had analyzed the beginning of the teaching of Geology in 1865 at the University of Buenos Aires. Years later, in 1906, with the creation of the National University of La Plata, the teaching of geological sciences began at the Museum of La Plata, with Walther Schiller and Santiago Roth, with courses in Geology and Paleontology. In the 1930s, Joaquín Frenguelli and Enrico Fossa Mancini joined, which coincided with an explosion in the enrollment of Geology students given the boost that oil exploration produced in the country. The creation of the Faculty of Natural Sciences and Museum in 1949 consolidated the teaching together with new professors such as Mario Teruggi, Rosendo Pascual, and Alfredo Cuerda. In Buenos Aires, the school of geology was consolidated in 1922 with the arrival first of Hans Keidel as professor and later with Pablo Groeber. Both contributed to the training of human resources, among which Horacio Harrington stood out with his studies of the Flora of *Glossopteris* and his connection to the Gondwana continent, who was a strong promoter of Continental Drift. Harrington's return to the country after his second doctorate at Oxford University allowed the curricula to be updated. Together with Armando Leanza, they created the Institute of Geology of the University of Buenos Aires, which concentrated on excellent research and teaching. At the University of Córdoba, where the Doctorate in Natural Sciences was founded in 1876, it gradually disappeared until its reinstatement in 1918. Bodenbender started teaching again in 1919 and together with Oskar Schmieder, Roberto Beder, and Anselm Windhausen consolidated the courses in Natural Sciences. In the 1940s, Juan Olsacher and Otto Schlagintweit were in charge of teaching.

Keywords Teaching in La Plata · Córdoba · First professors · Teaching consolidation · Research

In previous chapters, it had been analyzed how the teaching of geology at the University of Buenos Aires began in 1865 through the classes of Pellegrino Strobel. At the same time, we had seen how President Sarmiento, under the supervision of Germán Burmeister, in 1869 had German scholars brought in to begin the teaching of Exact and Natural Sciences at the University of Córdoba. This attempt ended with the

V. A. Ramos, *History of the Geological Sciences in Argentina*, The Latin American Studies Book Series, https://doi.org/10.1007/978-3-031-83139-3_12

creation of the National Academy of Sciences due to the impossibility of these scholars being able to have the doors of the university open to teaching. The transcendent management of the expert Moreno in the Museo de la Plata was cut short when the museum became dependent on the brand-new National University of La Plata in 1905 (Fig. 12.1). These examples show the difficulties that existed at the beginning of the teaching of geology both in Córdoba as in La Plata. In this chapter, we will try to reconstruct how the teaching of geology began and developed in these three universities, which for several decades concentrated its teaching and the training of its first geologists.

12.1 The Beginning of Teaching in la Plata

The National University of La Plata was founded in 1905, to which institutions such as the Faculty of Agronomy and Veterinary Medicine and the Museum were annexed, which led to the resignation of the expert Moreno as its director. The annexation of the Museum was linked to the establishment of several schools in its facilities starting in 1906, including Geology. The schools of Geology, Biology, and Anthropology merged into the School of Natural Sciences in 1912, which had few students, unlike the others such as Chemistry and Pharmacy which had a greater number of students and remained in the Museum until 1920, when they moved to their own building. In 1919, the Faculty of Natural Sciences and Museum was created (Riccardi 2018a; Teruggi 1977).

The first two professors at the Geology school were Walther Schiller (1879–1944) and Santiago (Kaspar Jacob) Roth (1850–1924), who began their classes in 1906 for Natural Sciences students (Fig. 12.2).

Schiller was a professor of Mineralogy until 1915 when he returned to Germany to participate in the First World War as a *kriegsgeologe* (military engineer) in the German Army Staff. He returned to the country in 1919 and worked from 1921 as a professor of Physical Geography until 1944 when he died trying to climb Aconcagua. In the first years, he was commissioned by the General Direction of Mines to work on the survey of the High Cordillera of northern Mendoza (see Chap. 9). Shortly after the discovery of oil in 1907 in Comodoro Rivadavia, he carried out detailed structural studies that made it possible to identify the structural control of the reservoirs. He summarizes the oil exploration work of those years in a complete synthesis on "*Stratigraphy, tectonics and petroleum of Comodoro Rivadavia (Chubut)*", published years later (Schiller 1925).

The government of the province of Buenos Aires entrusts the Museum Institute with the geological survey of its territory, with Schiller beginning important studies in the region of Tandilia and Ventania in those years (Fig. 12.3). He presented the studies on the thrust structure of the Sierras de la Ventana in another extraordinary synthesis published in 1930, with important maps and structural sections (Schiller 1930, 1932, 1938) More than 50 years must have passed for the thrusts postulated by Schiller to be accepted by the community (Varela et al. 1986).

Fig. 12.1 Museo de la Plata in 1905 when the schools of Geology, Biology and Anthropology were founded, which along with others operated in its facilities. *Source* Museo de La Plata archives

Fig. 12.2 Walther Schiller and Santiago Roth, the first professors of geology and paleontology at the School of Geology of the National University of La Plata. *Source* Actis (1972) and Scheinsohn (2020)

Fig. 12.3 Schiller with the Ducós family during his work in the Sierra de la Ventana. *Source* Courtesy Centro Cultural Argentino de Montaña

His studies on the structure of the Sierras de Tandil, where he postulates the existence of late Paleozoic thrusts that delimit the different mountain blocks, associated with the contraction of Ventania, still await recognition, although more and more evidence of Permian blind thrusts is emerging. that controlled the present structure (Ramos and De Barrio 2022).

Santiago Roth, as head of the Paleontology section and professor of that discipline until 1924, was in those years dedicated mainly to the study of the Cenozoic deposits of the Pampas region (Fig. 12.4). His work on "*Origin and age of the Pampas formation of the Argentine Republic*" and his "*Geological investigations in the Pampas plain*" are considered classic syntheses of knowledge about this unit and its paleontological content (Torres 1927; Gertrud 1951; Bond 1999).

Various professors from other institutions collaborated with the teaching of different subjects such as Juan J. Nágera, Juan Keidel, Pablo Groeber, Horacio Harrington, among others, whose research and contributions were published mainly in their workplaces (Riccardi 2016a, 2018a).

In the 1930s joined Joaquín Frenguelli (1883–1958) and later Enrico Fossa Mancini who had retired from YPF, who at the time were full-time professors at the institution (Fig. 12.5).

Fossa Mancini's arrival coincided with the beginning of the YPF scholarships, which marked a numerical and conceptual leap in the training of geologists at the National University of La Plata. The first geologists to graduate were Joaquín Daniel, Carmelo de Ferrariis, and Abel Herrero Ducloux in 1938 (Teruggi 1994). Both Frenguelli and Fossa Mancini supervised dozens of doctoral theses, the first preferably in Neuquén and the second on petrographic topics.

Fig. 12.4 Roth in the Pergamino area, province of Buenos Aires, exhuming a glyptodont. *Source* Gertrud (1951)

Fig. 12.5 Joaquín Frenguelli and Enrico Fossa Mancini, were two professors who marked an era and trained the first geologists of La Plata. *Source* Riccardi (2018a, b)

Frenguelli's research focused on the Cenozoic stratigraphy of the Pampean region where he made important contributions, especially by recognizing the position of loess in glacial-interglacial sequences (Riccardi 2016b). His synthesis of the Neozoic for the Geography of the Argentine Republic (Frenguelli 1957a, b), together with his numerous contributions to the knowledge of Argentine diatoms and the stratigraphy and paleontology of Gondwana, remain fundamental works of reference.

In the late 1930s and early 1940's, there was an outbreak in enrollment, considered by Teruggi (1977) a period of hatching between 1935 and 1958. In these years more than 100 to 200 students registered in natural sciences, with a clear majority in geology (Riccardi 2018a). The creation of the Faculty of Natural Sciences and Museum in 1949 marked an increasing dedication to teaching to the detriment of research. In that decade, Mario E. Teruggi (1919–2002) and Rosendo Pascual (1925–2012) stood out for the significance of their research and their training in human resources, who created an excellent school of Petrology and Sedimentology and Vertebrate Paleontology with numerous and outstanding disciples (Fig. 12.6).

The school of Petrology, including both igneous and sedimentary rocks, had its greatest exponent in Mario Teruggi. He graduated as a geologist in 1945 and received his doctorate in 1946 at the Faculty of Natural Sciences and Museum under the direction of Fossa Mancini. He did postgraduate studies at the Royal School of Mines at Imperial College London and spent his entire teaching career in La Plata. He supervised 25 doctoral theses and had numerous management tasks (Ametrano et al. 2003). His excellent book on "*Eruptive rocks under the microscope, Their systematics and nomenclature*" (Teruggi 1951). It was a classic for decades for petrography students. His "*Sedimentological Lexicon*", co-authored with Félix González Bonorino, was another transcendental work for teaching with more than ten reprints. His geological research work was dedicated to Tandilia where he made important contributions to its

Fig. 12.6 Mario E. Teruggi, Rosendo Pascual, and Alfredo J. Cuerda, three prominent professors from the Faculty of Natural Sciences and Museum. *Source* Rapela and Depetris (2016), Tonni et al. (2013a, b), and Cingolani (2009a)

petrology and tectonic interpretation (Teruggi et al. 1989). However, his most significant contributions were in the field of sedimentology, where he created a school that over time remains one of the main ones in the country with disciples such as Renato Andreis, Mario Iñiguez, and Luis Spalletti, teachers of numerous generations. He made fundamental contributions to the Cenozoic sedimentology mainly of the province of Buenos Aires (see complete bibliography in Spalletti 2016). His literary contributions cannot be ignored, both his novels and his essays on lunfardo as urban speech.

Rosendo Pascual was born in Mendoza and since his adolescence, he was interested in the mountains (Fig. 12.7). He participated as a member of the Club Andino Mendoza in the rescue of Walter Schiller in 1944 (Personal Comm. 2009). He carried out all his scientific and academic activity at the National University of La Plata where he graduated as a doctor in Natural Sciences in 1949 with a geological thesis in the Alta Cordillera de Mendoza carried out under the direction of Fossa Mancini and the supervision of Pablo Groeber. He was a professor of Vertebrate Paleontology for decades, where he was in charge of that division of the Museum (Tonni et al. 2013a, b). He created a school with numerous disciples who continue to transmit his main achievements, including establishing an evolutionary succession of South American mammals associated with geological changes in ecosystems, such as the rise of the Andes and their important climatic restrictions. Hundreds of works mark his profound mastery of the biological and geological processes that dominated the last fifty years of the discipline (see Tonni 2005, Brandoni de Gasparini and Vucetich 2012).

The beginning of the teaching of geology at this university has been focused on a series of professors who have stood out in the first decades, training the first graduates in Geology and creating schools of excellence that transcend their career. However, other teachers collaborated in the training of those first classes of students, one of the largest on a national scale, who were characterized by their commitment to teaching and research (see a complete review of teaching at this university in Riccardi (2018b). We would not like to forget Ángel V. Borrello (1918–1971), Alfredo Cuerda (1920–2009), and Guillermo Furque (1921–1999) who graduated from this institution in

Fig. 12.7 Rosendo Pascual carried by his thesis partner Vicente Armando and Dr. Horacio Camacho when receiving the 2009 Pellegrino Strobel Prize at the University of Buenos Aires

1942 and 1947. The first, through years of teaching and numerous publications, introduced the geosynclinal theory, which achieved great popularity with his book "*Los geosynclinales de la Argentina*" (Borrello 1969), in addition to various studies on the Precordillera of San Juan and Mendoza. Guillermo Furque as a professor of General Geology and Alfredo Cuerda in Historical Geology made outstanding contributions in the early Paleozoic of our country (Cingolani 2001, 2009a, b).

Sedimentology had one of its main promoters in Renato (Tito) Andreis (1935–2009), a disciple of Teruggi who had graduated in 1961 (Fig. 12.8). In 1975, as a professor, he began to teach special courses after returning from his postgraduate studies in Europe, which modernized this discipline. The military coup of 1976 took him temporarily to Brazil, to return to teaching with democracy in 1983, as a teacher in Buenos Aires. His teachings and his research, shared with a host of young researchers, updated sedimentology at both universities and through his active participation in the Argentine Association of Sedimentology of which he was president (Pazos 2009).

12.2 The First Geologists Graduated in Buenos Aires

The teaching of Geology at the University of Buenos Aires began in 1865 with Pellegrino Strobel (Fig. 12.9). It continued basically as a support subject for the studies of Engineering, Chemistry, and Natural Sciences until 1914 (see beginnings

Fig. 12.8 Renato (Tito) Andreis, maximum referent in sedimentology at the end of the last century. *Source* Courtesy of Ricardo Palma

in Chap. 5). From that year onwards, the first geologists graduated: Franco Pastore in 1914, Juan J. Nágera in 1915, Luciano Catalano in 1917, and Edelmira Mórtola in 1921, all of whom had Prof. Enrique Hermitte as their thesis director. In this first batch of graduates, Franco Pastore and Edelmira Mórtola, the first male and female geologists to graduate in our country, stood out in their teaching work at the Faculty of Exact, Physical and Natural Sciences. Although Catalano obtained a doctorate in Chemistry, he always worked in mineral deposits in a discipline he called Chemical Geology or Geological Chemistry.

In those years, oil had been discovered in Comodoro Rivadavia and Plaza Huincul under the direction of Eng. Hermitte through the General Direction of Mines and Geology, and that slowly attracted Natural Sciences students interested in Geology. The first graduates worked at that institution and taught on an honorary basis. From 1920 Franco Pastore (1885–1958) began to teach his courses in Petrography and later in Mineralogy, being appointed Full Professor in 1924, secondary by Edelmira Mórtola in Mineralogy. Franco Pastore carries out numerous investigations in the Sierras Pampeanas of San Luis, his native land, and in those of Córdoba. He presents a synthesis of the magmatic and metamorphic evolution of the crystalline basement in 1926, on the occasion of his incorporation into the National Academy of Exact, Physical and Natural Sciences, where based on his field observations he concludes that the granites of the Sierras Pampeanas are of early Paleozoic age (Pastore 1929). He made various geological sheets and made a conspicuous *Geological Description of the Province of Córdoba*, with a map at a scale of 1:1,000,000 and structural sections (Pastore 1938), which updated the first provincial map made by Bodenbender at the beginning of the century. On the 50th anniversary of the Argentine Scientific Society, he presents a complete synthesis of the history of Argentine geology (Pastore 1925).

Starting in 1922, changes were noted in the teaching of geology. Juan Keidel retired from the General Direction of Mines and Geology and began to teach Physical Geography first and then Geology and Paleontology at the university (Fig. 12.10).

Fig. 12.9 Faculty of Exact, Physical and Natural Sciences of the University of Buenos Aires where the teaching of geology began in 1865, and Argentine geologists graduated uninterruptedly for more than one hundred years in the *Manzana de las Luces* (Square of the Enlightenment). *Source* A.G.N. archives

During his 20 years of teaching until his retirement in 1942, he did important research and published various works of great importance, among which stand out "*On the paleogeographic development of the large geological units of Argentina*" and "*The relations between South America and South Africa revealed by geological research in the southern mountains of Buenos Aires*" (Keidel 1925, 1929), among many others. He supervised several doctoral theses, including that of Horacio J. Harrington (1910–1973), defended in 1933, which marks an important milestone in geological research at the university. It can be said that his thesis on "*The flora of Glossopteris in the Sierras Australes of Buenos Aires*" produces a substantial change in the research that was done in the doctoral theses. This thesis can be considered the first to be developed at the frontier of knowledge at that time, and its results had strong international significance. Harrington received the 1934 Strobel Prize. In 1933 he participated in a study mission to the States Island with Dr. Alberto Castellanos, whose geological results he later made known (Harrington 1943). He received a scholarship from the University of Oxford where he obtained his Doctorate in Philosophy with the thesis on "*Some Ordovician trilobites from Northern Argentina*" in 1936. Upon his return to the university, he was appointed Professor of Geology, later replacing Keidel when he retired in 1942 (Ramos 2016).

In 1931 Erwin Kittl began teaching, teaching Geology for the Engineering degree. This Austrian professor works at the same time in the *Dirección Nacional de Vialidad*, pouring his experience into his famous work on "*Derrumbamientos, deslizamientos*

Fig. 12.10 Geology professors in the first decades of the twentieth century: Enrique Hermitte (1871–1955), Juan Keidel (1877–1954), Erwin Kittl (1890–1983), and Pablo Groeber (1885–1964). *Source* Biblioteca Digital F.C.E.N

y torrentes en caminos de montaña en la República Argentina" (Slumps, landslides, and floods in mountain roads of the Argentine Republic), which becomes a classic work of Applied Geology to Engineering published by the faculty (Kittl 1939). He was also a professor of Geology and Mineralogy in the Natural Sciences doctorate until his resignation in 1948.

12.3 The Consolidation of Teaching and Research in Buenos Aires

Pablo Groeber joined the university as a professor of Physical Geography in 1933 and served until his retirement in 1952. Groeber directed 23 doctoral theses and published numerous publications sharing his experience of more than 40 years of studying Argentine Geology. He had a predominant role in scientific research (Groeber 1953, 1959; Groeber et al. 1953) and the training of disciples, being considered one of the great masters of Argentine geology. He and Juan Keidel were the first two honorary members of the newly created Argentine Geological Society in 1945.

In 1938, at the request of Harrington, who had rejoined after his second thesis submitted at Oxford University in 1936 (Fig. 12.11), the study plan was modified, creating a geology orientation within the doctorate in Natural Sciences, separating Biology, despite maintaining a series of common subjects to both orientations. This involves substantial changes in the program, increasing the number of geological subjects and the duration of the degree by one year, ending the old concept of a broad-based naturalist.

At the end of 1944, the Institute of Geology was founded at the faculty, with Harrington being its first director between 1944 and 1948. In this last year, Dr. Armando F. Leanza (1919–1975) joined the faculty, who was appointed professor of Paleontology and director of the Institute from 1948 to 1952. Lorenzo Aristarain

Fig. 12.11 The directors of the Institute of Geology of the Faculty of Exact, Physical, and Natural Sciences: Dr. Horacio J. Harrington (1944–1948); Dr. Armando F. Leanza (1948–1952), and Dr. Lorenzo Aristarain (1953–1957). *Source* Biblioteca Digital F.C.E.N

(1926–2013) later took charge of the direction of the institute until its dissolution in 1957 (Ramos 2018).

The studies carried out by Harrington and Leanza at that time marked an important milestone in the geological research of the faculty, which was truncated in 1953 with the resignation of Harrington for not adhering to and affiliating with the government party (Riccardi 2008). His research was published in a series of publications, both local and abroad, culminating in a reference work that marked a milestone in the knowledge of the early Paleozoic faunas of our country: "*Ordovician trilobites of Argentina*" published by the University of Kansas in 1957 (Fig. 12.12). In this book, in addition to a systematic description of a complete collection of trilobites, Harrington in his introduction makes an excellent analysis of the geology of the early Paleozoic and a robust paleogeography of the Ordovician of northwest Argentina. The stratigraphic profiles and the collections studied had an important contribution from the fieldwork of YPF geologists.

In those years, the syntheses carried out by Pablo Groeber with the collaboration of Pedro Stipanicic and Alberto Mingramm, his doctoral thesis students, stand out, presented in a complete work on Geography of the Argentine Republic, where he wrote the four volumes of the Mesozoic, which are still in our days are an obligatory reference for the Mesozoic studies (Fig. 12.13).

This work was an update of the geological knowledge of our country, where in addition to the Mesozoic, a series of contributions from prestigious geologists of the time stood out. Thus, Juan Keidel wrote The Precambrian and The Paleozoic (Keidel 1947a, b) where he establishes the stratigraphy and a supplement to The Precambrian and Paleozoic folding, where he describes the structure and tectonic phases (Keidel 1957). This work is completed with the last part of the Neozoic volume by Joaquín Frenguelli from 1957 and a supplement by this author on the large physical units of the Argentine territory (Frenguelli 1946). The latter is complemented by Egidio Feruglio with the orographic systems of Argentina (Feruglio 1946), where he describes in

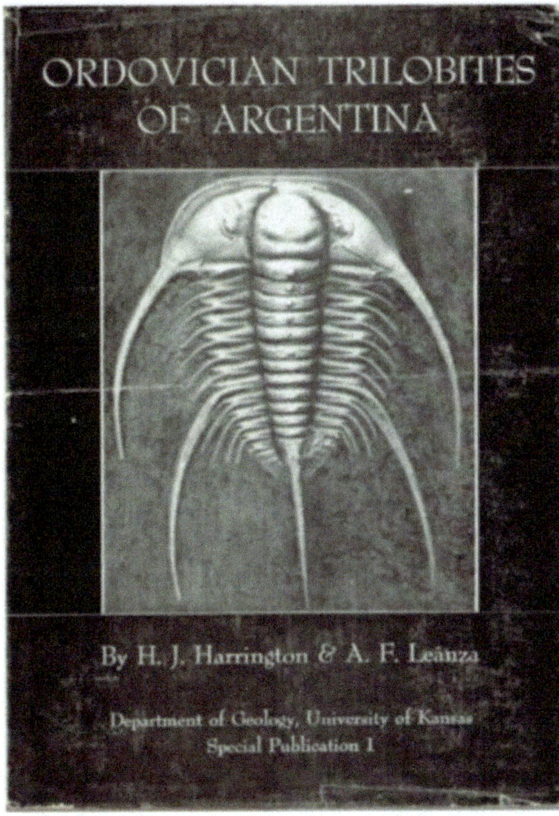

Fig. 12.12 Reference book on Ordovician faunas by Harrington and Leanza (1957), which in addition to the paleontological description has an excellent synthesis of the Argentine lower Paleozoic carried out by Horacio Harrington. *Source* FCEN Library, University of Buenos Aires

Fig. 12.13 Geography of the Argentine Republic edited by the Argentine Society of Geographic Studies where Pablo Groeber oversaw four volumes dedicated to the Mesozoic that bring together the knowledge of the time in more than 700 pages. *Source* Groeber (1953, 1959)

detail the characteristics of the mountain and central region, which culminates with the glaciers of the Cordillera Argentina: *Distribution of glaciers in the mountain range* (Feruglio 1957). Of these important contributions, the Mesozoic stands out for its development and the contributions it makes, for which Groeber received the *Francisco Pascasio Moreno Prize* from the Argentine Society of Geographic Studies in 1953.

In those years, Félix González Bonorino (1918–1998) also stands out, a disciple of Pablo Groeber, who received his doctorate in 1942. He produced an important change in the teaching and research of the petrology of igneous and metamorphic rocks. As Teruggi (1977) established, in the first decades of the twentieth century the German school was dominant in the geology of the different institutions of the country. González Bonorino marked a significant break in Argentine scientific evolution, introducing modern concepts from the North American school both in his courses on the *Evolution of Igneous Rocks* taught in Buenos Aires and La Plata, and in his research in the Sierras Pampeanas (Teruggi and Llambías 2000). He had trained in his postdoctoral studies with petrologists of the stature of Norman Bowen and Arthur Buddington from the universities of Chicago and Princeton, introducing fractional crystallization and the application of the Bowen Series to understand petrological processes in the country (Ramos 2023).

González Bonorino stood out both in the field and in the office, with extensive geological campaigns in the Sierras Pampeanas, where through the survey of numerous geological sheets as part of his work in the National Direction of Geology and Mines, he covered more than 20,000 square kilometers of that region. His contributions to the knowledge of igneous and metamorphic processes were published in both national and foreign magazines. His concern for teaching bequeathed us didactic works such as his *Principles of Mineral Optics, Optical Mineralogy,* and the *Sedimentological Lexicon*, the latter carried out with the collaboration of Mario Teruggi (González Bonorino 1954, 1976; González Bonorino and Teruggi 1952).

12.4 The Beginning of Teaching in Córdoba

The beginnings of the teaching of Geology date back to the Jesuit period, despite not being established as a science. The Philosophy courses that began shortly after the creation of the university in 1621 covered concepts associated with the earth and approached in a scholastic manner following Aristotelian principles (García Castellanos 2000). These concepts slowly evolved with the teaching of two prominent professors, the Jesuits José Sánchez Labrador (1717–1798) and Tomás Falkner (1702–1784). The first is in his book Paraguay Natural where he makes valuable observations on *Land, waters and air* and the second is in his book *Description of Patagonia and the adjoining parts of South America*, in which he makes important paleontological observations (see Chap. 2).

Charles III's royal order of 1767 expelling the Jesuits led to the Franciscans taking over the university. Although the erudition and knowledge of these friars were not as advanced as that of the previous ones, the teachings of Friar Cayetano Rodríguez (1761–1823) stand out at the university, who years later had an outstanding activity with the patriots. In his course on General Physics and Particular Physics in 1782, he specified that Physics "*the third part of Philosophy…has as its object the study of things in nature*" (Furlong 1952).

The concepts taught were elementary and must wait until after the nationalization of the university in 1854 when the Faculty of Exact and Philosophical Sciences was finally created in 1864. This faculty took a while to function, but courses, where Terrestrial Physics was taught, have been recognized which, starting in 1869, followed the text of Eujenio Sismonda (1815–1870). He was a professor of Natural History at the University of Turin and his book from the late 1850s had been translated and augmented by Juan Ramorino (1840–1876), a professor at the University of Buenos Aires since 1867 (Fig. 12.14).

It is interesting to note that the text mentions that the "*Elements of Terrestrial Physics by Eujenio Sismonda* (have been) *translated, considerably enlarged and arranged for the preparatory courses of the Argentine Republic by J. Ramorino,*

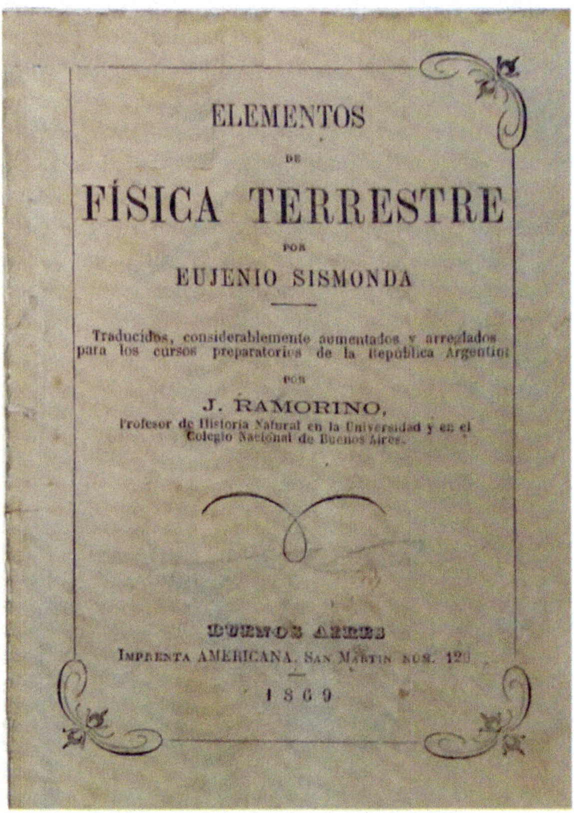

Fig. 12.14 Cover of the book on Terrestrial Physics by Eujenio Sismonda translated by Juan Ramorino in 1869. *Source* Biblioteca Nacional, Buenos Aires

professor of Natural History at the University and at the National College of Buenos Aires".

The content of the book deals with the Earth's crust, igneous, metamorphic and sedimentary rocks, mountain systems, and fossils of different ages. This treatise is complementary to the book *Rudiments of Mineralogy* written by Ramorino (1869), both printed in the same year at the Imprenta Americana in Buenos Aires.

However, we must wait for the arrival in 1871 of Alfred Stelzner (1840–1895) to the National Academy of Sciences in Córdoba to begin geological investigations (see Chap. 6). The problems between the Academy and the University were only overcome in 1874 when German scholars were allowed to teach. Stelzner's stay was three and a half years, being succeeded by Ludwig Brackebusch (1849–1906) who taught Mineralogy and Geology from 1874 to 1888. Upon his return to Germany Guillermo Bodenbender (1857–1941), taught from 1885 those disciplines for many years until his retirement in 1918. These courses were basically taken by engineering students. For more than forty years, Bodenbender carried out numerous investigations, highlighting his studies of the Paganzo deposits, late Palaeozoic to Triassic continental sequences of the Sierras Pampeanas.

Due to a lack of interest from potential students, the Doctorate in Natural Sciences of the Faculty of Mathematical and Physical Sciences, which had been founded in 1876, gradually disappeared until its reinstatement in 1918. That year it had José María Sobral as professor of Mineralogy and Geology, who was only in charge of dictation for a few months. Bodenbender returned in 1919 to take charge of the classes, until a competition was called and Oskar Schmieder (1891–1980), a geographer who graduated from the University of Heidelberg in 1915, was appointed professor, who taught Mineralogy until 1926 (Fig. 12.15).

We must wait for a new proposal from 1926 to reinstate the Doctorate in Natural Sciences in the faculty, which from 1935 was called by its current name of Faculty of Exact, Physical and Natural Sciences. On this occasion, Anselmo

Fig. 12.15 Professors Oskar Schmieder, Anselmo Windhausen, and Roberto Beder restarted the Doctorate of Natural Sciences at the Faculty of Mathematical and Physical Sciences in the 1920s. *Source* A.N.C. archives

Windhausen (1882–1932) and Roberto Beder (1885–1930) are appointed, who must take charge of teaching all subjects during an institutional crisis that was trying to close this doctorate (Martino 2018). Both made important studies and contributions to Argentine geology (see Chap. 6 and subsequent).

During this period as a professor in Córdoba Windhausen wrote a work until now unmatched is a complete text of Argentine Geology by a single author where he describes and analyzes the stratigraphic and structural aspects and presents his tectonic interpretations in support of continental drift and the exotic nature of Patagonia (Fig. 12.16). A jury that included Dr. Bernardo Houssay awarded him postmortem the second National Science Prize for his "*Argentine Geology*".

In addition to his academic tasks, Windhausen had the initiative in 1931 for YPF to generate a scholarship plan for geology students to increase the vocations for these studies in the national universities of the time. This scholarship program was successful and benefited several generations of Argentine geologists.

Both Beder and Windhausen died early, one from an accident in 1930 and the other from a cerebral embolism in 1932. This was a hard blow for the Geology career, given that upon Beder's death the young Juan A. Olsacher (1903–1964), who was his teaching assistant, had to take charge of teaching Mineralogy and Petrography. Olsacher, who had studied at the *Technische Hochschule* in Dresden in 1926, received his doctorate in 1930 in Córdoba. With the death of Windhausen, Olsacher had to take sole charge of teaching almost all the subjects in the Geology curriculum between 1932 and 1939 (Fig. 12.17). For more than three decades he was in charge of teaching Mineralogy, directing 35 doctoral theses and producing numerous crystallographic and mineralogical contributions (Leanza 1964; Martino 2018).

At the beginning of 1943 Otto Max Gabriel Schlagintweit (1881–1956) joined as a professor to teach Petrography (Fig. 12.18). This German geologist, a favorite

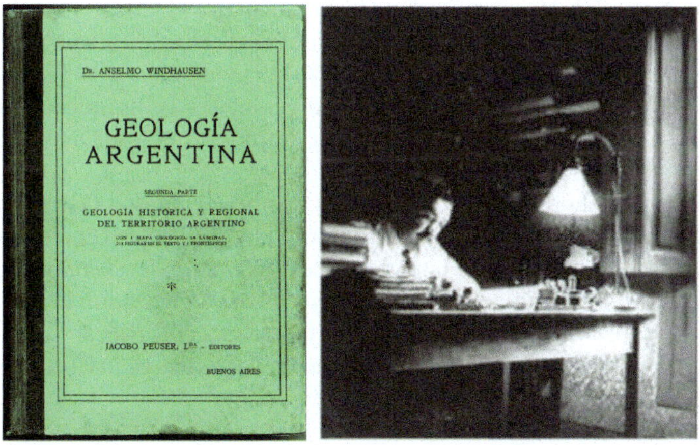

Fig. 12.16 Historical and Regional Geology of the Argentine Territory written by Anselmo Windhausen and in his years in Córdoba dedicated to producing this monumental work of 1931. *Source* FCEN Library and A.N.C. archives

Fig. 12.17 Juan Olsacher in 1962 and one of his most transcendent works. *Source* A.N.C. archives

disciple of Gustav Steinmann, was selected by his teacher to accompany him to Peru, where he actively collaborated in his work on the *Geology of Peru*. He interrupted his work to participate in the First World War and then dedicated himself to petroleum geology. He worked for several years in oil exploration in Colombia and Venezuela, and then worked in Northern Argentina between 1930 and 1938 for YPF (Schlagintweit 1937). He was able to establish the stratigraphic position of the Zapla iron deposit and reconstruct the paleogeography of the glacial horizon in northern Argentina. He then went to Bolivia hired by YPFB, where he made the first correlations between the Miraflores Limestone in Bolivia with the Dolomitic Calcareous Horizon of Argentina. He was contacted by Olsacher in 1943 to offer him the position of professor in Córdoba, where he worked until his death (see details in Enrique A. Sanmartino 1956) and biographical data in García Castellanos 1956). He directed more than 25 doctoral theses with renowned geologists as disciples such as Pedro Lesta, Florián Wetten, Marcelo Pensa, Héctor Martínez Cal, José Digregorio, among others.

In those years, shortly after receiving his doctorate, Telasco García Castellanos (1912–2008) joined teaching, and had an important academic and geological activity, highlighting his contributions to history such as *History of the teaching of geological science in the National University of Córdoba* and the organization in this university of the first symposia on the history of science (Depetris 2008; Riccardi 2020). He was for many years president of the National Academy of Sciences in Córdoba (Depetris 2019).

One cannot fail to mention Carlos E. Gordillo (1925–1984), one of Olsacher's beloved disciples (Fig. 12.19), who created a true school of petrology in Córdoba through more than 30 years of teaching and research. He graduated from this university in 1953 and through his research elucidated the evolution of the basement of

Fig. 12.18 Professors Otto Schlagintweit and Telasco García Castellanos were prominent teachers of the 1940s. *Source* A.N.C. archives

the Sierras Pampeanas, especially of metamorphic rocks and their minerals, such as cordierite. He did postgraduate studies with famous German petrologists through two von Humboldt scholarships (see Caminos 1985).

An outstanding disciple of Olsacher was Hebe Dina Gay (1927–2018), who graduated from Córdoba in 1950, did postgraduate studies at the *Università degli Studi* in Milan and who assisted him at the Museum. In addition, for more than forty

Fig. 12.19 Carlos Gordillo in his old Petrology and Chemistry laboratory located in the Academia building and Hebe Gay, two professors who created a school with numerous disciples at the National University of Córdoba. *Source* Courtesy of Roberto Martino

years, she taught Mineralogy, devoting all her efforts to the study of mineral species in Argentina. She dedicated her last 25 years fully to the University's Mineralogy Museum, created by Alfred Stelzner (Lira 2019).

12.5 Final Considerations

The origins of the teaching of Geology in La Plata, Córdoba, and Buenos Aires were completely different in their development and in the times when it was carried out. However, a period is observed where the three universities had an explosion in enrollment associated with YPF's interest in promoting the vocation of potential students for Geology through a vigorous program of scholarships for their study. Starting in 1930, at the initiative of a Windhausen proposal formally carried out through the authorities of the National University of Córdoba, an increase in the admitted and graduated geologists was observed in all three. This motivated the need to update the study plan to meet the requirements that the new graduates needed to have. These modifications with a small difference in time were made in the three universities.

Another common factor that they have had in the first decades of the twentieth century, already highlighted by Teruggi (1977), is the strong German influence that prevailed in the teaching of the subjects and their professors. Perhaps Buenos Aires and to a lesser extent La Plata, were the first to open to the new ideas and hypotheses that arose in the field of petrology with a strong influence from the studies and teaching carried out in both institutions by González Bonorino. These changes are also recognized in stratigraphy, which through the teachings of Horacio Harrington opens to the concepts of facies, formational units and the acceptance of a common stratigraphic lexicon. The North American influence is also identified in these concepts.

In all of them, the presence of two or three great leaders of Geology is recognized who carried the responsibility of advancing teaching in difficult times, especially during the 1920s and the crisis of the 1930s that inflicted strong financial strain on the available budgets. The vocation and passion they put into it made it possible to sustain the few vocations of those years.

A notable difference is the break that occurred in Buenos Aires after the "*night of the long sticks*" *(Noche de los bastones largos)* of July 1966 in its teaching body and in the level of teaching. Although the temporary loss of autonomy affected all study houses, the changes have not been as drastic in the other institutions.

The following chapters will analyze the subsequent evolution of these institutions and the impact that these years have had on the development of research and subsequent teaching.

References

Actis OC (1972) Una Empresa al servicio del País. Cincuentenario de YPF (1922–1972). Editado por YPF, Imprenta Sacerdoti, 191 p, Buenos Aires

Ametrano S, Bermúdez A, Delpino D (2003) Dr. Mario Egidio Teruggi. Profesor Emérito UNLP. (1919–2002). Nota necrológica, Revista de la Asociación Geológica Argentina 58(1):130–131

Bond M (1999) Santiago Roth. Fundación Museo de la Plata, Revista Museo 13:33–37, La Plata

Borrello AV (1969) Los geosinclinales de la Argentina. Anales de la Dirección Nacional de Geología y Minería 14:1–188, Buenos Aires

Brandoni de Gasparini ZN, Vucetich MG (2012) Rosendo Pascual: 15/7/1925–23/12/2012. Anales de la Academia Nacional de Ciencias Exactas, Físicas y Naturales 64:6–7

Caminos R (1985) Carlos E. Gordillo (1925–1984), Nota Necrológica. Revista de la Asociación Geológica Argentina 40(1–2):131–135

Cingolani CA (2001) Alfredo José Cuerda, homenaje al cumplir 80 años. El Museo y su gente, Fundación Museo de La Plata, Revista del Museo 15:31–37, La Plata

Cingolani CA (2009a) Prof. Dr. Alfredo José Cuerda (1920–2009). Revista de la Asociación Geológica Argentina 64(4):713–721

Cingolani CA (2009b) Prof. Dr. Alfredo José Cuerda (1920–2009), Nota Necrológica. Ameghiniana 45(2):443–444

Depetris PJ (2008) Dr. Telasco R. García Castellanos (1912–2008), Necrológica. Revista de la Asociación Geológica Argentina 63(3):470–471

Depetris PJ (ed.) (2019) La Academia Nacional de Ciencias, 150 Años acompañando la Ciencia Argentina. Academia Nacional de Ciencias, 400 p, Córdoba

Feruglio E (1946) Los sistemas orográficos de la Argentina. Geografía de la República Argentina, Sociedad Argentina de Estudios Geográficos GAEA 4:5–542, Buenos Aires

Feruglio E (1957) Los glaciares de la Cordillera Argentina. Geografía de la República Argentina, Sociedad Argentina de Estudios Geográficos GAEA 7(primera parte): 5–86, Buenos Aires

Frenguelli J (1946) Las grandes unidades físicas del territorio argentino. Geografía de la República Argentina, Sociedad Argentina de Estudios Geográficos GAEA 5:1–114, Buenos Aires

Frenguelli J (1957a) Neozoico. Geografía de la República Argentina, Sociedad Argentina de Estudios Geográficos GAEA 2(tercera parte): 1–115, Buenos Aires

Frenguelli J (1957b) El glaciarismo cuaternario. Geografía de la República Argentina, Sociedad Argentina de Estudios Geográficos GAEA 2 (tercera parte): 117–218, Buenos Aires

Furlong G (1952) Nacimiento y desarrollo de la Filosofía en el Río de La Plata. Editorial G. Kraft, 1536–1810, Buenos Aires

García Castellanos T (2000) Historia de la enseñanza de la ciencia geológica en la Universidad Nacional de Córdoba, Argentina. Boletín de la Academia Nacional de Ciencias 64:5–16, Córdoba

Gertrud W (1951) Santiago Roth 1850–1924: ein Berner als wissenschaftlicher Pionier in Südamerika. Berner Zeitschrift für Geschichte und Heimatkunde 13:19–39, Berna

González Bonorino F (1954) Principios de óptica mineral. Instituto Nacional de Investigaciones en Ciencias Naturales y Museo Bernardino Rivadavia, 147 p, Buenos Aires

González Bonorino F (1976) Mineralogía óptica. Temas de Eudeba, 342 p, Buenos Aires

González Bonorino F, Teruggi ME (1952) Léxico Sedimentológico. Publicación Extensión Cultural y Didáctica, Museo Argentino de Ciencias Naturales Bernardino Rivadavia 6:1–160, Buenos Aires

Groeber P (1953) Ándico. Geografía de la República Argentina, Sociedad Argentina de Estudios Geográficos GAEA 2(primera parte): 349–541, Buenos Aires

Groeber P (1959) Supracretácico. Geografía de la República Argentina, Sociedad Argentina de Estudios Geográficos GAEA 2 (segunda parte): 1–165, Buenos Aires

Groeber P, Stipanicic P (1953) Triásico. Geografía de la República Argentina, Sociedad Argentina de Estudios Geográficos GAEA 2 (primera parte): 9–131, Buenos Aires

Groeber P, Stipanicic P, Mingramm A (1953) Jurásico. Geografía de la República Argentina, Sociedad Argentina de Estudios Geográficos GAEA 2(primera parte): 143–347, Buenos Aires
Harrington HJ (1943) Observaciones geológicas en la Isla de los Estados. Museo Argentino de Ciencias Naturales, Anales 41:29–52
Harrington HJ, Leanza AF (1957) Ordovician Trilobites of Argentina. University of Kansas, Department of Geology, Special Publication 1:1–276, Lawrence
Keidel J (1925) Sobre el desarrollo paleogeográfico de las grandes unidades geológicas de la Argentina. Sociedad Argentina de Estudios Geográficos, GAEA, Anales 4:251–365, Buenos Aires
Keidel J (1929) Las relaciones entre Sud América y Sud África reveladas por la investigación geológica en las sierras australes de Buenos Aires; Extensión Universita ria (Conferences), Universidad Nocional de La Plata, año 1929, n° 3, La Plata
Keidel J (1947a) El Precámbrico. Geografía de la República Argentina, Sociedad Argentina de Estudios Geográficos GAEA 1(primera parte): 47–126, Buenos Aires
Keidel J (1947b) El Paleozoico. Geografía de la República Argentina, Sociedad Argentina de Estudios Geográficos GAEA 1(primera parte): 127–304, Buenos Aires
Keidel J (1957) Los plegamientos precámbricos y paleozoicos. Geografía de la República Argentina, Sociedad Argentina de Estudios Geográficos GAEA 1(suplemento): 5–218, Buenos Aires
Kittl E (1939), Derrumbamientos, deslizamientos y torrentes en caminos de montaña de la República Argentina; Publicación de la Facultad de Ciencias Exactas, Físicas y Naturales, Serie B 16, 91 p., Buenos Aires
Leanza AF (1964) Prof. Dr. Juan Augusto Olsacher (1903–1964). Revista de la Asociación Geológica Argentina 19(4):195–200
Lira R (2019) Hebe Dina Gay (1927–2018), Necrología. Revista de la Asociación Geológica Argentina 76(2):166–172
Martino RD (2018) Breve reseña del nacimiento de la enseñanza en Ciencias Geológicas en la Universidad Nacional de Córdoba y de sus primeros 70 años (1870–1940). In Ramos VA; Martino RD (eds.) El nacimiento de la Geología en la enseñanza universitaria, Revista Facultad de Ciencias Exactas, Físicas y Naturales 5(Suplemento 1):17–22, Córdoba
Olsacher J (1962) Ciento cincuenta años de geología en la Argentina. Anales de las Primeras Jornadas Geológicas Argentinas 1:21–36
Pastore F (1925) Nuestra Mineralogía y Geología durante los últimos cincuenta años. Cincuentenario de la Sociedad Científica Argentina (1872–1922), Evolución de las ciencias en la República Argentina 6:1–47, Buenos Aires
Pastore F (1929) Conocimientos sobre la composición y orogenia del macizo cristalino central de Argentina. Anales de la Sociedad Científica Argentina 108:275–286, Buenos Aires
Pastore F (1938) Descripción Geológica de la Provincia de Córdoba. Aguas Minerales de la República Argentina 4:1–47, Buenos Aires
Pazos P (2009) Dr. Renato Rodolfo Andreis (1935–2009), Necrológica. Revista de la Asociación Geológica Argentina 65(1):230–235
Ramorino J (1869) Rudimentos de Mineralogía. Imprenta Americana, Buenos Aires. 107 p, Buenos Aires
Ramos VA (2016) La primera clase de Exactas y el inicio de la enseñanza de la Geología. In Ramos VA (ed.) 150 Años de Exactas, Eudeba, 2–49 p., Buenos Aires
Ramos VA (2018) El desarrollo de la Geología en la Universidad de Buenos Aires. Revista Facultad de Ciencias Exactas, Físicas y Naturales, Nueva Serie 5(Suplem. 1):23–32, Córdoba
Ramos VA (2023) Félix González Bonorino: el geólogo que cambió la historia. Fundación Azara, Universidad Maimónides, Revista de Historia Natural 13(1):13–43
Ramos VA, de Barrio RE (2022) La estructura de Tandilia: Schiller y su libreta de campaña. 21° Congreso Geológico Argentino, Actas: 1297–1298, Puerto Madryn
Rapela CW, Depetris PJ (2016) Geochemistry in Argentina: from pioneers to the present. Environ Earth Sci 75(524):15

Riccardi AC (2008) Horacio J. Harrington: significación y trascendencia de su obra geológica. In: Aceñolaza G (ed) Historia de la Geología Argentina, InsuGeo Serie Correlación Geológica 24:259–270, Tucumán

Riccardi AC (2016a) Geographical and geological explorations of the La Plata Museum 1884–1905. In: Mayer W, Clary RM, Azuela LF, Mota TS, Wołkowicz S (eds) History of geoscience: celebrating 50 years of INHIGEO, vol 442. Geological Society, Special Publications, London, pp 1–12

Riccardi AC (2016b) Life and geological studies of Joaquín Frenguelli. In: Mayer W, Clary RM, Azuela LF, Mota TS, Wołkowicz S (eds) History of geoscience: celebrating 50 years of INHIGEO, vol 442. Geological Society, Special Publications, London, pp 1–13

Riccardi AC (2018a) El desarrollo histórico de las exploraciones petroleras en la Argentina. Academia Nacional de Ciencias de Buenos Aires, Comunicación, El Guion Ediciones, 5–41, Buenos Aires

Riccardi AC (2018b) Origen y desarrollo de la enseñanza de la Geología en la Universidad Nacional de La Plata. In: Martino R, Ramos VA (eds) El nacimiento de la geología en la enseñanza universitaria, Revista Facultad de Ciencias Exactas, Físicas y Naturales 5(Suplemento 1):33–47, Córdoba

Riccardi AC (2020) Telasco García Castellanos: Promotor de la Geología y de las ciencias. Revista Facultad de Ciencias Exactas, Físicas y Naturales 7 (Suplemento 1):95–114, Córdoba

Sanmartino EA (1956) Dr. Otto Schlagintweit, Noticia necrológica. Revista de la Facultad de Ciencias Exactas, Físicas y Naturales Año 18(3–4):321–328

Scheinsohn V (2020) Cómo construir una carrera de naturalista en la Argentina de fines del siglo XIX: Santiago Roth y la red suiza del Museo de La Plata. In: Dávila L, Arenas P (eds) El americanismo germano en la antropología argentina de fines del siglo XIX al siglo XX, Ediciones CICCUS, 103–131, Buenos Aires

Schiller W (1925) Estratigrafía, Tectónica y Petróleo de Comodoro Rivadavia (Chubut), Sección Mineralogía y Geología, Anales Museo de La Plata (Segunda Serie) 2:2–56

Schiller W (1930) Investigaciones geológicas en las montañas del Sudoeste de la Provincia de Buenos Aires. Anales del Museo de La Plata, 2, Parte 4:1–101

Schiller W (1932) Complicaciones tectónicas (cobijaduras) en las sierras de Tandil (Provincia de Buenos Aires). Revista del Museo de La Plata 32:299–306

Schiller W (1938) Cobijaduras tectónicas en el Paleozoico de las Sierras de la Tinta (Provincia de Buenos Aires). Notas del Museo de La Plata, n.s. 3, Geología 5:35–46

Schlagintweit O (1937) Observaciones estratigráficas del Norte Argentino. Boletín de Informaciones Petroleras 14(156):1–49, Buenos Aires

Spalletti LA (2016) Mario E. Teruggi: un puente entre la ciencia y el humanismo. In: Riccardi A (ed) La Historia de la Geología en el Bicentenario de la Argentina, Revista del Museo de la Plata, vol 1, Número Especial: 302–315

Teruggi ME (1951) Las Rocas Eruptivas al Microscopio. Su sistemática y su Nomenclatura. Publicación del Instituto Cultural y Didáctica, Museo Argentino de Ciencias Naturales Bernardino Rivadavia 3:410 p. Buenos Aires

Teruggi ME (1977) Cien años de geología en el Museo de La Plata. Obra del Centenario del Museo de La Plata. I, Reseña Histórica: 59–73, La Plata

Teruggi M, Leguizamón MA, Ramos VA (1989) Metamorfitas de bajo grado con afinidades oceánicas en el basamento de Tandil: sus implicaciones geotectónicas, provincia de Buenos Aires. Revista de la Asociación Geológica Argentina 43(3):66–374

Teruggi ME (1994) Museo de la Plata 1888–1988, una centuria de honra. 3ra. Edición, Fundación Museo de La Plata, 155 p

Teruggi ME, Llambías EJ (2000) Félix González Bonorino (1918–1998). Revista de la Asociación Geológica Argentina 55(4):424–427

Tonni EP (2005) El último medio siglo en el estudio de los vertebrados fósiles. Asociación Paleontológica Argentina. Publicación Especial 10, 50º Aniversario, pp 1–13, Buenos Aires

Tonni EP, Brandoni de Gasparini ZN, Cione AL, Scillato Yané G, Vucetich MG (2013a) Prof. Emérito Dr. Rosendo Pascual. 15/7/1925–23/12/2012. Revista de la Asociación Geológica Argentina 70(3):436–442

Tonni EP, Brandoni de Gasparini ZN, Vucetich MG, Cione AL, Scillato Yané G (2013b) Prof. Emérito Dr. Rosendo Pascual (1925–2012). Ameghiniana 50(3):377–378

Torres LM (1927) Doctor Santiago Roth (1850–1924). Revista del Museo de La Plata 30:165–169, La Plata

Varela R, Leone EM, Manceda R (1986) Estructura tectónica en la zona del Cerro del Corral, Sierras Australes de Buenos Aires. Revista de la Asociación Geológica Argentina 41:256–261

Chapter 13
The National Direction of Mining and Geology

Abstract The 1940s were very difficult years, with a low budget and limited technical staff, and at times only three geologists working. Numerous geological maps at a scale of 1:200,000 with their corresponding text were produced at the end of the decade through surveys carried out by Horacio Harrington, Félix González Bonorino, Christian Petersen, Raúl Dessanti, César R. Vilela, Enrique De Alba, Oscar Ruiz Huidobro, Juan C. Turner, Guillermo Furque, Eduardo Methol and Rafael González, among others. These geological sheets began to be published along with the arrival of a host of young geologists. This second generation of geologists was mostly graduates of the universities of La Plata and Buenos Aires, such as Oscar Ruiz Huidobro, Juan Carlos M. Turner and Guillermo Furque, Eduardo Methol and Eduardo Holmberg. During their time, they compiled numerous geological maps on a scale of 1:200,000, which marked the beginning of the systematic survey of the country. The creation of the National Direction of Geology and Mining in 1956 gave a new impetus to geological studies under the direction of Roberto Tezón.

Keywords Systematic surveys · Geological maps · González Bonorino · Turner · Tezón director

In the previous chapters had been analyzed the beginnings of the institution and the geological contributions made from 1904 to the crisis of 1930, which led to the departure of José M. Sobral as director of the institution. After a temporary interregnum of Luis M. Vicat, the industrial engineer Tomás M. Ezcurra took over the Direction of Mines and Geology in 1932 at that time belonged to the Ministry of Agriculture of the Nation. This engineer graduated from the University of Buenos Aires in 1924 had a good management record at YPF. Where he held positions in Comodoro Rivadavia, Plaza Huincul and Salta. In 1927, he had made trips to the United States to learn about the oil equipment in use for its possible acquisition by YPF. His appointment as director of Mines lasted until 1943, when he resigned after 11 years of positive management.

V. A. Ramos, *History of the Geological Sciences in Argentina*, The Latin American Studies Book Series, https://doi.org/10.1007/978-3-031-83139-3_13

13.1 Direction of Mines and Geology of the Ministry of Agriculture

During his direction, some geological maps were published, especially the sheets drawn up in previous years by Franco Pastore in the ranges of San Luis and Córdoba, which, with the exception of Hoja 20i of Córdoba (Pastore 1932), had to wait until that the texts were completed by Oscar Ruiz Huidobro, Eduardo Methol and Rafael González years later (Pastore and Ruiz Huidobro 1952; Pastore and Methol 1953; Pastore and González 1954). In those years, the geological map of Hoja 30c Puntilla de Huincán was published by Pablo Groeber (1937), who also made a series of provincial maps at a scale of 1,000,000 such as those of Neuquén, Mendoza, and La Rioja with their corresponding texts for the series Mineral Waters of the Argentine Republic (Groeber 1938, 1939, 1940).

The director Tomás Ezcurra complained in 1936 about the meager production of the institution, which was attributed to the "*insufficient resources, insufficient technical personnel and, why not say it, insufficient concern of the Public Powers about these matters*" (Zappettini 2004). He complained that the staff was made up of three geologists, two assistant geologists, and two mining engineers, that when compared to the beginnings of the institution in the first years of the century, the lack of funds for personnel and field surveys was evident.

It must be recognized that despite the meager resources in those years, geological services were provided for the construction of various dams, the trans-Andean tunnel, and the location of the Uruguayana-Paso de los Libres bridge, along with support for various exploration and prospecting mining activities.

In 1934 Victorio Angelelli (1908–1991) joined the institution, who in those years provided important services until 1944 (Fig. 13.1). He was born in Ancona, Italy, and came with his parents in 1910 to settle in San Juan. He graduated as a chemical engineer at the University of San Juan and did further studies at the famous Mining Academy of Freiberg where he graduated as a mining engineer in 1934 (Brodtkorb and Schalamuk 1992). He made numerous mining and mineralogical studies, but the most important contribution of those years was "*The mineral deposits and application rocks of the Argentine Republic*", a complete work that in its 400 pages provides a complete inventory of the known deposits (Angelelli 1941). This work received that same year the Eduardo Holmberg Award from the National Academy of Exact, Physical and Natural Sciences.

The Engineer Tomás Ezcurra participated in the First Pan-American Congress of Mining Engineering and Geology held in Santiago de Chile in 1942 (Fig. 13.2), along with other prominent geologists such as Horacio Harrington, Pablo Groeber, and Esther W. de Carral Tolosa who presented interesting works at the meeting. One of the contributions that stands out in this event is that of Sobral (1942), who, already working for YPF, made a complete survey of the west of the Pampa and the east of Mendoza, describing the geology and evaluating their oil potential.

Concerning the mining studies undertaken by the repartition, the first discovery of uranium mineralizations in Argentina in Calamuchita in 1938 had an important

Fig. 13.1 The Eng. Victorio Angelelli when he joined the National Academy of Sciences in Córdoba, and working in the field in northwest Argentina. *Source* A.G.A. archives

Fig. 13.2 The Argentine delegation to the First Pan American Congress of Mining Engineering and Geology in Santiago, Chile, 1942. In the center, first row, the head of the delegation, Ing. Tomás Ezcurra; to his right is Eng. Hermitte and at the end to the right is Egidio Feruglio; between the first two in the second row Horacio Harrington and to his left Pablo Groeber. In the second photo in the center, the Vice President of Congress Tomás Ezcurra. *Source* Allbala (1942)

significance, a discovery that would be followed by those of uranium mineralizations in La Rioja, San Luis and, the most notable of all, the one corresponding to the Don Otto mine, in Salta.

Engineer Tomás Ezcurra was in charge of management until June 1943, when he resigned following the overthrow of President Ramón S. Castillo.

13.2 From the Direction of Mines, Geology and Hydrogeology to the Direction of Mining

The revolution of 1943 coincided with the decline of the conservative governments and marked the beginning of a development policy linked to the numerous industries that were established because of the Second World War and the processes of import substitution with local products. It was one of the best moments of the Argentine economy, with full employment and excellent exports (Zappettini 2004).

These political changes led to the Direction of Mines, Geology, and Hydrogeology ceasing to depend on the Ministry of Agriculture after many years, passing under the orbit of the Ministry of Industry and Commerce. Upon Ezcurra's resignation, Augusto Tapia took over as temporary director. In the first years, this industrial impulse meant that the objectives of the old Direction of Mines and Geology were focused mainly on mining, as is evident in the successive names it had as the General Direction of Mining Industry (1948–1951) and the National Direction of Mining (1951–1955) where geology was a little more relegated. The directors who followed Augusto Tapia's interim term in 1944 were not geologists and varied from military engineers to surveyors, generally with little background.

In those years, some geological surveys of the chart at a scale of 1:200,000 were published. New geologists were slowly joining the Direction of Mines such as Horacio Harrington, Félix González Bonorino, Christian Petersen, Raúl Dessanti, César R. Vilela, Enrique De Alba, Oscar Ruiz Huidobro, Juan C. Turner, Guillermo Furque, Eduardo Methol and Rafael González, among others (Fig. 13.3). These regional geologists between the 1940s and early 1950s surveyed numerous geological sheets that remained unpublished for many years.

Horacio J. Harrington (1910–1973) completed his second doctorate at the University of Oxford in 1936 (Turner 1974) and joined the Department of Mines and Geology that same year (Fig. 13.4). Although he has an outstanding teaching activity,

Fig. 13.3 Horacio Harrington, Félix González Bonorino, and Christian Petersen in their time as geologists at the National Direction of Mines and Geology. *Source* Biblioteca Digital F.C.E.N

Fig. 13.4 The opening event of the Pan American Institute of Mining Engineering and Geology meeting in Buenos Aires under the presidency of Perón. Military, civil, and ecclesiastical authorities are present, the latter represented by Bishop Santiago Luis Copello. In the third row behind Perón, Dr. Telasco García Castellanos represents the National Academy of Sciences. In the lower right corner Dr. Harrington president of IPIMIGEO of Argentina. *Source* Courtesy Harrington's family

the lack of adequate remuneration as there was no exclusive dedication and there were no subsidies to do field research, forced teachers to have stable jobs in other institutions. In those years Harrington did commendable work. Finally, he drew up the maps of the Hojas 33 m Sierras de Curamalal and 34 m Sierra de la Ventana and wrote a complete text that was published a little later where, among other novelties, he announced the presence of the *Eurydesma* fauna in the region (Harrington 1947). He carried out detailed geological investigations in the Sierras of Villavicencio and Malpaís (Harrington 1941), along with more applied studies in different sectors (Turner 1974).

From those years is his precise description of the San Juan earthquake of 1944, published in great detail (Harrington 1944), one of the few studies carried out where the areal distribution of the damage and the intensity of the earthquake could be evaluated, through of an isosist map. He organized the first meeting in Buenos Aires of the Pan-American Institute of Mining Engineering and Geology, where he presented the development of the Gondwana System in Uruguay (Harrington 1945), along with other papers that were published that year.

His merits are recognized by the National Academy of Exact, Physical and Natural Sciences, which incorporated him as a full member in 1948. The last of his works in the Direction of Mines and Geology is the survey of Hoja 22c, Ramblón, which appeared published years later (Harrington 1971). In this sheet, the distribution of the Paleozoic units of the Precordillera was shown for the first time from the Cerrito Salinas in the eastern foothills to the Uspallata valley, where it recognizes

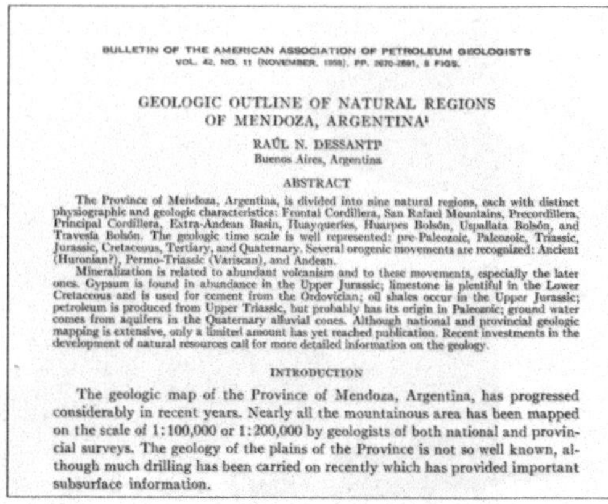

BULLETIN OF THE AMERICAN ASSOCIATION OF PETROLEUM GEOLOGISTS
VOL. 42, NO. 11 (NOVEMBER, 1958), PP. 2670-2691, 8 FIGS.

GEOLOGIC OUTLINE OF NATURAL REGIONS OF MENDOZA, ARGENTINA[1]

RAÚL N. DESSANTI[2]
Buenos Aires, Argentina

ABSTRACT

The Province of Mendoza, Argentina, is divided into nine natural regions, each with distinct physiographic and geologic characteristics: Frontal Cordillera, San Rafael Mountains, Precordillera, Principal Cordillera, Extra-Andean Basin, Huayquerías, Huarpes Bolsón, Uspallata Bolsón, and Travesía Bolsón. The geologic time scale is well represented: pre-Paleozoic, Paleozoic, Triassic, Jurassic, Cretaceous, Tertiary, and Quaternary. Several orogenic movements are recognized: Ancient (Huronian?), Permo-Triassic (Variscan), and Andean.

Mineralization is related to abundant volcanism and to these movements, especially the later ones. Gypsum is found in abundance in the Upper Jurassic; limestone is plentiful in the Lower Cretaceous and is used for cement from the Ordovician; oil shales occur in the Upper Jurassic; petroleum is produced from Upper Triassic, but probably has its origin in Paleozoic; ground water comes from aquifers in the Quaternary alluvial cones. Although national and provincial geologic mapping is extensive, only a limited amount has yet reached publication. Recent investments in the development of natural resources call for more detailed information on the geology.

INTRODUCTION

The geologic map of the Province of Mendoza, Argentina, has progressed considerably in recent years. Nearly all the mountainous area has been mapped on the scale of 1:100,000 or 1:200,000 by geologists of both national and provincial surveys. The geology of the plains of the Province is not so well known, although much drilling has been carried on recently which has provided important subsurface information.

Fig. 13.5 Raúl N. Dessanti when he joined the Direction of Mines, and one of his fundamental works. *Source* Biblioteca Digital F.C.E.N

the Cortaderas and Alojamiento facies, today interpreted as Ordovician subduction complexes (Haller and Ramos 1984).

Raúl N. Dessanti (1913–1987) together with Harrington was one of the first Argentine regional geologists of the Direction of Mines and Geology (Fig. 13.5). He studied at the University of Buenos Aires through a scholarship from *Yacimientos Petrolíferos Fiscales* and presented his thesis in 1938 under the direction of Juan Keidel entitled *On the stratigraphy and tectonics of Cerro La Cal, province of Mendoza.* In his early years he worked at YPF until in 1944 he joined the Direction of Mines, where he became Head of the Geological Service (Caminos 1987). In those years he raised the geological sheets 27 c Cerro Diamante, 29 b Bardas Blancas, and 28 b Malargüe, at the same time he coordinated the first Geological Map of the Argentine Republic on a scale of 1: 2,500,000 published by the Directorate of Mines in 1950 (Fig. 13.6).

Among his most notable contributions is the discovery of marine Carboniferous in the Uspallata and San Rafael mountains. He was an enthusiastic member of the Argentine Geological Association since its founding and its president between 1955 and 1957. He was a professor at the universities of Buenos Aires and La Plata in the 1950s and from 1964 until his retirement in 1979 he was a full-time professor of Argentine and Historical Geology at the Universidad Nacional del Sur. One of his most important contributions can be considered his paper published in the Bulletin of the American Association of Petroleum Geologists on *Geologic outline of natural regions of Mendoza* where he makes an interesting division of the main geological provinces, their stratigraphy and structure, based on his surveys carried out both in YPF and in the Direction of Mines (Dessanti 1958). He was a Guggenheim Fellow in 1961 carrying out studies at the State University of Pennsylvania and upon his return, he drew attention to the need to implement Photogeology, at the same time he published one of the first geochronological studies on the granites and metamorphic rocks of the Frontal Ciordillera (Dessanti and Caminos 1967).

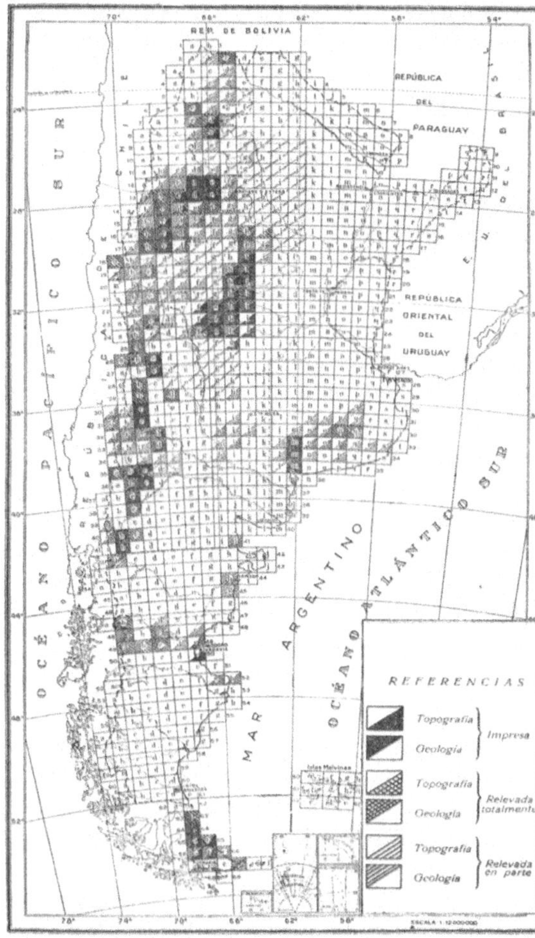

Fig. 13.6 Status of the survey of Geological Sheets at a scale of 1:200.00 at the beginning of 1950. *Source* Dessanti (1954)

González Bonorino (1918–1998) had joined the Direction of Mines in 1941, publishing his observations in the Patagonian Mountains in the Hoja 41b Río Foyel, which was his doctoral thesis directed by Pablo Groeber (González Bonorino 1944a). From those years is his complete description of almost a hundred rock samples collected by Egidio Feruglio from the Bariloche region (González Bonorino 1946) and the work with Christian Petersen on the igneous rocks of Cholila (Petersen and González Bonorino 1947). He makes known for the first time the ignimbrites, a rock that despite its abundance had gone unnoticed, which had a great impact due to its novelty (González Bonorino 1944b). He began working in the Sierras Pampeanas in 1945 with the survey of Sheets 12 d (Capillitas), 12 e (Aconquija), 13 c (Fiambalá), 13 d (Andalgalá), 13 e (Villa Alberdi) and 14 f (Catamarca) in the provinces of Catamarca and Tucumán. These extensive surveys with several months in the field that covered more than 20,000 km^2 of the Sierras Pampeanas were published years later by the Direction of Mines (González Bonorino 1950a, b, 1951, 1972, 1978). He published his observations from those years in a series of contributions to the journal of the Argentine Geological Association, of which he was founder and president (see Ramos 2023).

Christian S. Petersen (1912–1962) studied at the University of Buenos Aires, defending his doctorate at the end of 1942 under the direction of Pablo Groeber. The previous year he had joined the Direction of Mines and Geology with González Bonorino, and his thesis on *Geological Studies in the Middle Chubut River Region* was published in 1946 by this department (Petersen 1946). His great contribution has been his first geological surveys on the Island of Tierra del Fuego where he organized several expeditions until 1949 (Petersen 1948; Petersen and Methol 1949). Seven geologists participated in these expeditions, as well as surveyors who did the cartography. Guillermo Furque, Horacio Camacho, Raúl Sister, Eduardo Methol, Juan C.M. Turner, Adolfo Stegman and Luis Lambert participated in those memorable campaigns. He resigned from the Direction of Mines, dedicating himself to teaching at the University of Buenos Aires and developing a private commercial printing press dedicated to the natural sciences where classic manuals such as *Geology applied to Engineering*, among others, were published (Methol 1963).

Enrique de Alba (1920–2000) had joined the Direction of Mines as a geologist assistant in 1944 while he was pursuing his doctorate at the University of Buenos Aires. He received his doctorate in 1947 with the "*Geological Study of the Sierra de Maz, province of La Rioja*" directed by Horacio Harrington. From the beginning, his regional works were dedicated to the Famatina System, where he surveyed sheets 16c Villa Unión and 16d Chilecito, both in La Rioja between 1946 and 1949, although they were published years later. Once these sheets were completed, he was commissioned to survey Sheet 41j Sierra Grande in Río Negro. He carried out this task between the years 1950 and 1951. His most significant geological contribution was his syntheses of the Famatina System on the first and second Regional Geology Symposia of 1969 and 1976 (De Alba 1972, 1979). Subsequently, he carried out several applied geology studies for the Salto Grande, Paraná Medio, and Yaciretá dams (González Díaz 2000). He carried out important management tasks in the division until his retirement as National Director of Geology and Mining.

In previous chapters, we had seen that there was some competition in the geological surveys carried out by the Direction of Mines and Geology with those of YPF, which used the same distribution of geological sheets in those years. An exception to this competition was that of Tomás Suero (1915–1963) who had carried out the survey of Sheet 36 c Cerro Lotena as a YPF geologist and used it as a doctoral thesis at the University of La Plata under the direction of Joaquín Frenguelli in 1939 (Fig. 13.7). This work was published by the National Direction of Mining as part of the program of geological sheets of the institution (Suero 1951).

César R. Vilela (1921–2016) studied and received a doctorate in Natural Sciences at the University of La Plata in 1945 with a "*Geological study of the Ramos anticline area in the Sierra de San Antonio, Salta*" under the direction of Joaquín Frenguelli. As soon as he graduated, he worked at the Standard Oil Company of New Jersey, where he spent several years studying the Sierras Subandinas, which allowed him

Fig. 13.7 Angular unconformity in the Cerro Lotena between the Tithonian (Vaca Muerta Formation) and the *Strata with dinosaurs* (Neuquén Group) described by Suero in his doctoral thesis (Suero 1951). *Source* Courtesy Osvaldo Carbone

to complete his doctoral thesis. He joined the Direction of Mines in 1949, mapping several geological sheets in northwest Argentina. Among them, 6c San Antonio de los Cobres, 7d Rosario de Lerma, 9e Amblayo stand out, as well as others further south such as Hoja 48c, Lago Musters, Chubut (Fig. 13.8). The geological sheets were published later (see Ramos 2018), but their findings and contributions such as the occurrence of lacustrine sediments in the Calchaquí Valleys, the eruptive periods of the Puna de Atacama, and the discovery of the Dolomitic Calcareous Horizon in the Puna Salto-jujeña, were published in the Journal of the Argentine Geological Association (Vilela 1951, 1953a, b). He worked at the institution until 1959.

Oscar Ruiz Huidobro (1917–2004) was born in Tucumán, but he completed his secondary studies in Buenos Aires (Fig. 13.9). Before starting university, he spent five years traveling through the province of Tucumán and northwest Argentina. He entered the University of Buenos Aires in 1941, graduating with a doctorate in Natural Sciences in 1948 (Tófalo, 2004). His doctoral thesis dealt with a *Geological Study of the region of the Quitilipi and Pirgua hills, province of Salta.* He had joined the Direction of Mines the previous year where he works as a regional geologist. From that time are his geological surveys of sheets 12c Laguna Helada, 8e Chicoana, 7e Salta, and 11e Santa María, all of them in northwest Argentina surveyed between 1949 and 1955, but published years later (Ruiz Huidobro 1960, 1968, 1972, 1976). In 1957 he dedicated himself fully to hydrogeology, carrying out studies in the Water Resources Division of the U.S. Geological Survey in 1961. Simultaneously, he developed teaching activities at the University of Buenos Aires, where he began

Fig. 13.8 Cesar R. Vilela at the time when worked in the Direction of Mines, and on one of his expeditions to the Iruya region in 1949. *Source* Ramos (2018)

teaching Hydrogeology for the first time, a discipline that he was in charge of for several decades until his retirement in 1992.

Juan Carlos M. Turner (1918–1979) was the regional geologist of the Directorate of Mines who surveyed more than 50,000 km^2 of our mountain ranges at different latitudes, through more than 15 geological sheets. By far he has been the one who has carried out the most field survey work in this institution. Born in Buenos Aires into an English family, he was absolutely bilingual, but despite this he took care to speak and write Spanish devoid of Anglicisms, both in his daily dealings and in writing, leaving us a complete English–Spanish geological technical dictionary (Turner 1971), which has been a classic of use not only in the country but abroad.

Turner defended his doctoral thesis at the University of Buenos Aires in 1946 with a study on the *Geological Survey of the area north of Cerro Tocino in the Sierra de Famatina*, work carried out under the direction of Horacio Harrington. As soon

Fig. 13.9 Oscar Ruiz Huidobro young and at mature age, and Juan Carlos M. Turner, two geologists who surveyed important sectors of northwest Argentina through their geological sheets. *Source* Biblioteca Digital F.C.E.N

as he graduated, he went to do his second doctorate at the University of Cambridge in the United Kingdom between 1948 and 1950, obtaining his PhD with a study of *The graptolithic faunas of South America*, which still remains a continental reference study of these early Paleozoic faunas (Camacho 1979). He had joined the Direction of Mines as a geologist assistant in 1942 and as soon as he received it, he began surveying geological sheets (Fig. 13.10). His first maps were 15c Vinchina and 15d Famatina, both in the province of La Rioja, surveyed between 1947 and 1948, as a continuation of his thesis work. Starting in 1949, it began its surveys in northwest Argentina with sheets 2c Santa Victoria and 2b La Quiaca, which were followed in 1952 by 13b Chaschuil, 7c Nevado de Cachi, and 11d Laguna Blanca, to culminate with sheets 1a-b Santa Catalina and 2a San Juan de Oro in the Puna at the end of that decade. In 1956 and 1958 he mapped sheet 37a-b Junín de los Andes and later 36a Aluminé, in Neuquén. He returned to Puna in 1960 to survey sheet 3a-b Pirquitas and closed his surveys in the 1970s with sheets 44c Tecka and 44 d Colán Conhué, in Chubut. Many of these changes from one region to another were conditioned by the prior availability of the necessary topographic survey.

In parallel with these works, he made important contributions such as *The Cambrian of Northern Argentina* and *The Andes of Northwestern Argentina*, along with the publication of the Second Symposium on Regional Geology of the National Academy of Sciences in 1976, where he served as editor while writing the chapters corresponding to the Puna and the Eastern Cordillera, updating those he had written in the first Regional Geology symposium in 1969 (Alonso 2013; Turner 1970, 1972a, b, 1979). His teaching activity was carried out at both the University of Buenos Aires and La Plata, where he supervised bachelor's and doctoral theses. He was widely recognized by being incorporated into both the National Academy of Exact, Physical and Natural Sciences and the National Academy of Sciences as an academic.

Guillermo Furque (1921–1999), originally from San Juan, studied at the National University of La Plata, graduating in 1946 as a doctor in Natural Sciences with a

Fig. 13.10 Juan Carlos M. Turner taking a break in the field and saddling his horse in 1965 in Pocoya, Jujuy. He was always proud of his saddle with double girth, waistband, and *baticola* covered with two to three furs woven in Famatina and awarded with a thick double lamb fur and an authentic *quirquincho* overlay

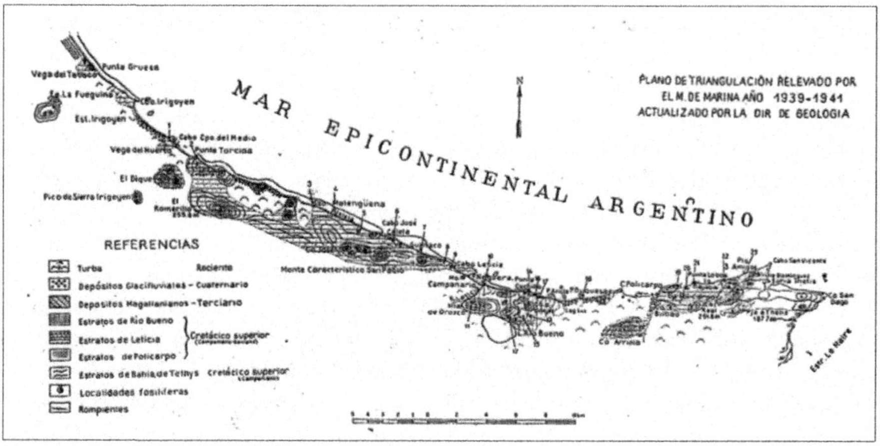

Fig. 13.11 Survey of the Atlantic coast of Tierra del Fuego carried out by Guillermo Furque in 1948. *Source* A.G.A. archives

thesis on the *Study of the geology and sites around Cerro Blanco, Sierra de la Huerta, San Juan* under the direction of Pablo Groeber. He joined the Direction of Mines and his first job consisted of participating along with other geologists led by Christian Petersen in the first expedition to Tierra del Fuego of the Direction of Mines in 1948. Furque was responsible for studying the Atlantic coast of the Mitre Peninsula from Cape San Pablo to Cape San Vicente (Fig. 13.11). An epic legend that is transmitted orally in the institution tells that Furque, armed with a shotgun and a packet of salt, traveled alone for several weeks on this rugged coast, fording rivers such as the Policarpo and others, exploring for the first time its Tertiary outcrops, collecting fossils that, studied by Camacho, were published the following year (Furque and Camacho 1949). Of the geologists who participated in these expeditions, the most valuable observations were made by Furque, who later published them in the Journal Journal of the Argentine Geological Association.

As a regional geologist, he was responsible for surveying sheet 17b Guandacol starting in 1949, continuing with sheet 16b Cerro La Bolsa, and years later sheet 18c Jáchal, culminating with sheet 19c Ciénaga de Gualilán, all of them in the San Juan Precordillera. Furthermore, he ventured into the Sierras Australes of Buenos Aires in the early 1960s where he raised sheets 33n Coronel Pringles and 34n Sierra de Pillahuincó, and years later in the Patagonian Andes, he mapped sheet 58b Lago Argentino (see Furque 1963, 1972, 1973, 1979a, b,1983). In his last years, he compiled the Jachal sheet at a 1:250,000 scale, based on previous studies and new fieldwork. He made important contributions to regional geology and the early Paleozoic orogenies, defining the Guandacol phase in the middle Ordovician (Caballé and Cuerda 2000).

Eduardo J. Methol (1922–1977) died early in an airplane accident, while he was working as a geologist for the Direction of Mines (Fig. 13.12). He graduated from the University of La Plata in 1946 with a thesis on the *Economic geological study of*

Fig. 13.12 Eduardo Methol and Eduardo Holmberg, two regional geologists from the Direction of Mines and Geology. *Source* Segemar archives

Cerro Champaquí, Córdoba, under the direction of Enrique Fossa Mancini. He had joined as an assistant geologist in 1944 and alternated his first years after his thesis finishing some geological sheets of the Sierra de Córdoba with the expedition to Tierra del Fuego under the direction of Christian Petersen starting in 1948. In Tierra del Fuego he studied together with Raúl Sister the gold alluvium of San Sebastián and collaborated with Petersen on the final report of the studies (Petersen and Methol 1949). In the Sierras of Córdoba he continued the work of Franco Pastore by raising the sheet 18i Deán Funes; writing the text of the 19 h Capilla del Monte whose map Pastore had drawn up in 1924–1928; the 21i Alta Gracia previously surveyed as part of her doctoral thesis by Verena Kull between 1944 and 1945, which Methol completed in 1954; the 22 h surveyed by Remigio Rigal between 1936 and 1938, adjusting it to the new topography and updating the text. In all these updates he carried out fieldwork and new petrographic studies.

Beginning in 1961 after fellowship training in photogrammetry and photogeology at the U.S. Geological Survey he began a new stage as a photogeologist in the Direction of Mines (Camacho 1977). In 1965 he worked for three years on the Valcheta Plan for geological-mining prospecting in the province of Río Negro, being in charge of the geological photointerpretation and whose field verification was carried out by Eduardo Holmberg and Fernando Sesana. In 1966, the departure of Henry S. Römer (1926–2019), professor of Photogeology at the University of Buenos Aires, allowed him to continue teaching that subject, which he later also taught at the University of La Plata. His early death cut short his academic and professional career.

Eduardo Holmberg (1915–1979) was another illustrious regional geologist who joined the Direction of Mines in 1944 as a student. He was educated in a cultured family, under the influence of his grandfather Eduardo L. Holmberg, who instilled in him his love for nature. He received his doctorate in 1946 from the University of Buenos Aires with a thesis on a structural geological study of the Cerro Bola region, Mendoza directed by Pablo Groeber. Immediately afterward he began his regional surveys in the Neuquén basin where he raised several geological sheets. The first was the 33d Auca Mahuida sheet surveyed in 1945, followed by the 32d Chachahuén, 29d Cerro Nevado and 33c Buta Ranquil, which ended in 1954. In turn, there was an unpublished sheet 32b Loncopué, with which they culminate his regional geological studies in the south of Mendoza and north of Neuquén. These surveys consumed on average more than six months of field per sheet, which explains the detailed description of the different units (De Alba 1980).

13.3 National Direction of Geology and Mining

The overthrow of Perón in 1955 brought back Eng. Tomás Ezcurra as interim director, followed by Roberto V. Tezón in 1956. This was a positive change for geological studies, given that the new institution, the National Direction of Mining and Geology became dependent on the brand-new Ministry of Mining.

Roberto V. Tezón (1921–1973) graduated from the University of La Plata at the end of 1946 and received his doctorate in 1948 under the direction of Pascual Sgrosso (Fig. 13.13). His doctoral thesis in Natural Sciences with a specialization in Geology dealt with the *Los Arboles—Los Ratones District* in Catamarca receiving the highest marks (Methol 1974). His entry into the Direction of Mines as a geologist assistant allowed him to carry out the work of his doctoral thesis, carrying out a detailed economic geological study of the deposits of lead, silver, zinc, tungsten, tin and copper in that mining district.

In the beginning he did various mining works, among which his studies of the Sierra Grande iron in Río Negro, and the survey of the gold, silver and manganese

Fig. 13.13 Roberto Vitelmo Tezón as director of the National Direction of Geology and Mining and the brothers William and Walter Stoll mining engineers in 1936 at the Independence Mine in Alaska. Walter Stoll trained an entire generation of mining geologists from Argentina and Chile. *Source* Segemar archives

deposit of "*Farallón Negro*" and the *Agua de Dionisio District* of Catamarca stood out. It was trained under the direction of Walter C. Stoll (1915–1978), an American geologist whom the Direction of Mines hired in 1948. The mining engineer Stoll was the son of Walter W. Stoll, one of the most famous Alaskan miners in the first decades of the twentieth century. Walter and his younger brother William had studied Mining Engineering at the University of Washington, where he graduated in 1939. He worked at the USGS and in 1949 he came to Argentina. He did important services to the country by studying the *Farallón Negro* districts, the Andacollo mines in Neuquén, *Mogote Blanco* in La Rioja, among many others. His greatest contribution was teaching an entire generation of young mining geologists in his ten years in Argentina where he formed a school of mining exploration and evaluation studies. Jorge Valvano, Hugo Bassi, Raúl Sister, Oscar Reverberi, Héctor de la Iglesia, Mario Klein, Susana Bockmann, among several others, stood out. In addition, he also taught Economic Geology at the University of Buenos Aires and later at the University of Chile.

Tezón held important management positions as National Director between 1957 and 1964, putting Walter Stoll in charge of Mining Exploration. Among his achievements is the prioritization of the old Direction of Mines and the development of the project that created the Subsecretariat of Mining. Additionally, he took management courses at the U.S.G.S. and the University of Madison in 1960. He was concerned with perfecting the technical level of his geologists and mining engineers, sending 17 technicians to the United States, 20 to France and Germany, and to other countries. He obtained a budget to increase the number of professionals, fieldwork, and to improve the salaries, promoting geological research (Methol 1974).

During Tezón's administration, the *Cordillerano Norte Plan* was put into operation, later followed by the NOA-I, which included the survey of vast sectors of northwest Argentina, for which he managed the survey of 200,000 square kilometers of aerial photographs of the provinces of Catamarca, Santiago del Estero, Salta, Jujuy and Tucumán. He is also responsible for the programming of the Patagonia-Comahue Plan, one of the most important regional surveys in the country. He did not neglect the academic part, supporting the organization of the First Argentine Geological Conference in San Juan in 1960 and being its first president. These days were the beginning of the Argentine Geological Congresses, an old desire of the Argentine Geological Association. Its objective was always to promote Geological Sciences in general towards the great context of the country's economic development.

Among the new geologists who entered the Direction of Mines in those years, Jorge Polanski (1892–1975) stands out, who arrived in the country in 1947 (Fig. 13.14). This Ukrainian geologist, who considered himself Polish, studied at the University of Lviv after having participated in the First World War. He worked in the Polish geological survey and emigrated after the Second World War. He worked for a few years in the Provincial Direction of Mining in Mendoza and then joined the National Direction of Mining in 1951. As a regional geologist, he had to survey the Main and Frontal Cordilleras of Mendoza, through sheets 25a San José Volcano and 24a-b Cerro Tupungato, whose fieldwork was carried out during the years 1952 and 1957 with nine months of working in the field. During his work in the Provincial

Fig. 13.14 Jorge Polanski upon his arrival in the country and in a field camp in the San Rafael block with his disciples Enrique Nuñez, Néstor Porro, Francisco Fidalgo, and Emilio González Díaz. *Source* González Díaz (1975)

Direction of Mining in 1950 he ventured into the San Rafael Block with his 26c sheet La Tosca in Mendoza, but his masterpiece can be considered to be the *Stratigraphy, neotectonics, and geomorphology of the Pedemontane Pleistocene between the Diamante and Mendoza rivers* published in the Journal of the Geological Association (Polanski 1962).

Other important contributions have been *Rapid flows of rocky debris in arid and volcanic zones* (Polanski 1966) and *Carbonic and Permian of Argentina* (Polanski 1970), two Eudeba manuals with wide impact. Among his contributions to Argentine Geology, we must highlight the identification of mass removal phenomena that for years were confused with glacial moraines, which caused strong controversies with Pablo Groeber. At the same time, he introduced the teaching of Geomorphology, replacing the traditional Physical Geography that had been taught for decades, and the first dictation of Quaternary Geology (González Díaz 1975, 2011), where he made important contributions and created a school both in Buenos Aires and in La Plata with disciples of the stature of Emilio González Díaz and Francisco Fidalgo.

13.4 Final Considerations

The 40 s and 50 s, especially after 1943, marked a change between conservative governments with agricultural-livestock priorities, to governments that attempted industrial development. This development, partly driven by the restrictions of the Second World War, showed the need to know and mobilize our mining resources. To this end, they were conscious of recognizing the value of geological knowledge in the evaluation of resources, promoting surveys, regional prospecting plans, acquisition of aerial photographs of extensive areas of the country and intense geological activity.

When you compare the success that the Direction of Mines had in the first years under the leadership of Enrique Hermitte, with a large staff of German professionals, with the difficult years of the 20 s and 30 s, where the directors complained since they only had three geologists in the entire institution, the great differences in the production of geological knowledge can be understood.

In the 1940s, the Direction of Mines notably increased the quantity and quality of its geologists, partly helped by the increase in geological vocations produced by YPF scholarships. Some of its geologists benefited from these scholarships as students, even though they did not subsequently carry out petroleum studies. When one sees that in the team of geologists, we had personalities such as Harrington, Dessanti, González Bonorino, Petersen, Vilela, Turner, Furque, Polanski, among many others, it is understood how those years were covered with intense geological studies in the northwest of Argentina, the central region of the Precordillera, Frontal Cordillera and Principal Cordillera, and part of Patagonia. These studies involved months of fieldwork, and due to the lack of access roads, much of the work was carried out on horseback or mule, which explains the knowledge acquired from remote places in the mountain range. We are aware that we have shown a group of the most outstanding, leaving dozens of geologists who contributed in the most diverse ways to the institution.

That generation of Argentine regional geologists, formed for the most part by Groeber and Keidel, marked a fundamental milestone in the knowledge of vast regions of Argentine Geology and developed a school of field geology where several subsequent generations of young students were trained who would take the post of those surveys. It was a period almost as successful as Hermitte's management.

References

Allbala A (1942) Front Matter. Anales del Primer Congreso Panamericano de Ingeniería de Minas y Geología, Tomo Primero, 1–83, Santiago de Chile

Alonso RN (2013) Vida y obra de Juan Carlos M. Turner. In: Alonso RN (ed) 3er. Congreso Argentino de Historia de la Geología, Actas 29–42, Salta

Angelelli V (1941) Los yacimientos minerales y rocas de aplicación de la República Argentina. Boletín de la Dirección de Minas y Geología 50:1–400, Buenos Aires

de Brodtkorb MK, Schalamuk IB (1992) Ingeniero Victorio Angelelli (1908–1991). Revista de la Asociación Geológica Argentina 46(3–4):358–360

Caballé M, Cuerda A (2000) Guillermo Furque (1921–1999), Notas necrológicas. Revista de la Asociación Geológica Argentina 55(1–2):141–142

Camacho HH (1977) Dr. Eduardo J. Methol (1922–1977). Revista de la Asociación Geológica Argentina 32:315–316

Camacho HH (1979) Dr. Juan Carlos Manuel Turner (1918–1979). Revista de la Asociación Geológica Argentina 35:290–292

Caminos R (1987) Raúl N. Dessanti 1913–1987. Revista de la Asociación Geológica Argentina 42(3–4):483–484

De Alba E (1972) El Sistema de Famatina. Primer Simposio de Geología Regional Argentina, Academia Nacional de Ciencias, pp 143–184, Córdoba

De Alba E (1979) El Sistema de Famatina. Segundo Simposio de Geología Regional Argentina, Academia Nacional de Ciencias 1:349–396, Córdoba

De Alba E (1980) Dr. Eduardo Holmberg (1915–1979). Revista de la Asociación Geológica Argentina 35(1):162–163

Dessanti RN (1958) Geologic outline of natural regions of Mendoza, Argentina. American Association of Petroleum Geologists 42(11):2670–2691

Dessanti R, Caminos R (1967) Edades potasio-argón y posición estratigráfica de algunas rocas ígneas y metamórficas de la Precordillera, Cordillera Frontal y Sierras de San Rafael, Mendoza. Revista de la Asociación Geológica Argentina 22:135–162

Furque G (1963) Descripción geológica de la Hoja 17b, Guandacol, Provincias de La Rioja y San Juan. Boletín del Servicio Nacional Geológico Minero 92:1–114, Buenos Aires

Furque G (1972) Descripción geológica de la Hoja 16 b, Cerro La Bolsa. Provincia de San Juan. Boletín del Servicio Nacional Minero Geológico 125:1–75, Buenos Aires

Furque G (1973) Descripción geológica de la Hoja 34 n, Sierra de Pillahuincó, Provincia de Buenos Aires. Boletín del Servicio Nacional Minero Geológico 141:1–70, Buenos Aires

Furque G (1979a) Descripción geológica de la Hoja 18 c, Jachal, Provincia de San Juan. Boletín del Servicio Geológico Nacional 164:1–87, Buenos Aires

Furque G (1979b) Descripción geológica de la Hoja 33 n, Coronel Pringles. Provincia de Buenos Aires. Boletín del Servicio Geológico Nacional 174:1–38, Buenos Aires

Furque G (1983) Descripción geológica de la Hoja 19 c, Ciénaga de Gualilán. Provincia de Sa Juan. Servicio Geológico Nacional, Boletín 193:1–126, Buenos Aires

Furque G, Camacho HH (1949) El Cretácico superior de la costa atlántica de Tierra del Fuego. Revista de la Asociación Geológica Argentina 4(4):264–297

González Bonorino F (1944a) Descripción geológica y petrográfica de la Hoja 41 b, Río Foyel (territorio de Río Negro). Boletín de la Dirección de Minas, Geología e Hidrología 56:1–124, Buenos Aires

González Bonorino F (1944b) Nota sobre la presencia de ignimbritas en la Argentina. Notas del Museo de La Plata, Geología 9(35):577–590

González Bonorino F (1946) Contribución a la petrología del noroeste de la Patagonia. Revista del Museo de La Plata, Nueva Serie, Sección Geología 4(20):1–87

González Bonorino F (1950a) Geografía y petrografía de las Hojas 12 d (Capillitas) y 13 d (Andalgalá). Boletín de la Dirección General de Industria Minera 70, 121 p, Buenos Aires

González Bonorino F (1950b) Descripción Geológica de la Hoja 13e, Villa Alberdi, Provincia de Tucumán. Carta geológico-económica de la República Argentina, escala 1:200. 000, Boletín de la Dirección Nacional de Minería 74:1–95, Buenos Aires

González Bonorino F (1951) Descripción Geológica de la Hoja 12e, Aconquija, Catamarca-Tucumán. Carta geológico-económica de la República Argentina, escala 1:200.000, Boletín de la Dirección Nacional de Minería 75:1–50, Buenos Aires

González Bonorino F (1972) Descripción Geológica de la Hoja 13c, Fiambalá, Catamarca. Carta geológico-económica de la República Argentina, escala 1:200.000, Boletín de la Dirección Nacional de Minería 127:1–86, Buenos Aires

González Bonorino F (1978) Descripción Geológica de la Hoja 14f San Fernando del Valle de Catamarca. Carta geológico-económica de la República Argentina, escala 1:200.000, Boletín del Servicio Geológico Nacional 160:1–86, Buenos Aires

González Díaz EF (1975) Dr. Jorge Polanski (1892–1975). Revista de la Asociación Geológica Argentina 30(4):388–391

González Díaz EF (2000) Enrique de Alba (1920–2000). Revista de la Asociación Geológica Argentina 55(3):284–285

González Díaz EF (2011) Dr. Jorge Polanski (1892–1975). Revista de la Asociación Geológica Argentina 68(3):357–364

Groeber P (1937a) Mapa geológico de la Hoja 30c Puntilla de Huincán, provincia de Mendoza a Escala 1: 200.000. Dirección Nacional de Geología y Minería (sin texto), Buenos Aires

Groeber P (1938b) Mapa geológico de la Gobernación del Neuquén, escala 1: 1.000.000. Territorio Nacional del Neuquén. En Aguas Minerales de la República Argentina. Ministerio del Interior, Comisión Nacional de Climatología y Agricultura 12:17–31, Buenos Aires
Groeber P (1939b) Mapa geológico de Mendoza. 2da. Reunión Ciencias Naturales (Mendoza), Physis 14(46):171–220, Buenos Aires
Groeber P (1940) Descripción geológica de la provincia de la Rioja (con Mapa geológico escala 1: 1.000.000 y texto). Ministerio de Agricultura de la Nación, 17 p, Buenos Aires.
Harrington HJ (1941) Investigaciones geológicas en la Sierra de Villavicencio y Mal País. Boletín de la Dirección de Minas y Geología 49:1–65
Harrington HJ (1944) El sismo de San Juan del 15 de enero de 1944. Corp. Prom. Intercambio, Publicación Especial, 30 p, Buenos Aires
Harrington HJ (1945) Algunas observaciones sobre el Sistema de Gondwana en el Uruguay. Primera Reunión de Comunicaciones IPIMIGEO, Secc. Arg. (Buenos Aires), pp 5–19
Harrington HJ (1947) Explicación de las Hojas geológicas 33m (Sierra de Curamalal) y 34m (Sierra de la Ventana), provincia de Buenos Aires. Boletín de la Dirección de Minas y Geología 61:1–43
Harrington HJ (1971) Descripción geológica de la Hoja 22c, Ramblón (Provincias de Mendoza y San Juan). Boletín de la Dirección Nacional de Geología y Minas 114:1–87
Methol EJ (1963) Cristian S. Petersen (1912–1962). Revista de la Asociación Geológica Argentina 18(1):5–7
Methol EJ (1974) Dr. Roberto Vitelmo Tezón (1921–1973). Revista de la Asociación Geológica Argentina 29(4):482–484
Pastore F (1932) Hoja 20i del Mapa Geológico de la Argentina. Región oriental media de la Sierra de Córdoba. Relevamiento geológico y explicación. Boletín de la Dirección de Minas y Geología 36:1–67, Buenos Aires
Pastore F, Ruíz Huidobro O (1952) Descripción geológica de la Hoja 24g. Saladillo (San Luis). Boletín de la Dirección Nacional de Minería 78:1–63, Buenos Aires
Pastore F, Methol EJ (1953) Descripción geológica de la Hoja 19i, Capilla del Monte (Córdoba). Boletín de la Dirección Nacional de Minería 79:1–70, Buenos Aires
Pastore F, González RRL (1954) Descripción geológica de la Hoja 23g, San Francisco (San Luis). Boletín de la Dirección Nacional de Minería 80:1–62, Buenos Aires
Petersen CS (1946) Estudios Geológicos en la Región del Río Chubut Medio. Boletín de la Dirección General de Minas y Geología 59:1–152, Buenos Aires
Petersen CS (1948) Nota preliminar sobre rasgos geológicos de la porción septentrional de Tierra del Fuego. Revista de la Sociedad Geológica Argentina 3(4):279–286
Petersen CS, Methol EJ (1949) Informe sobre los trabajos de relevamiento geológico efectuados en Tierra del Fuego entre 1945 y 1948. Dirección General de Industria Minera, unpublished report, 51 p, Buenos Aires
Petersen CS, González Bonorino F (1947) Observaciones geológicas en el Chubut occidental. Revista Sociedad Geológica Argentina 2(3):177–222
Polanski J (1962) Estratigrafía, neotectónica y geomorfología del pleistoceno pedemontano entre los ríos Diamante y Mendoza. Provincia de Mendoza. Revista de la Asociación Geológica Argentina 17(3–4):127–349
Polanski J (1966) Flujos rápidos de escombros rocosos en zonas áridas y volcánicas. Eudeba, 67 p, Buenos Aires
Polanski J (1970) Carbónico y Pérmico de la Argentina. Eudeba, 216 p, Buenos Aires
Ramos VA (2018) Cesar Reinaldo Vilela (1925–2017). Necrológica. Revista de la Asociación Geológica Argentina 75(2):309–311
Ramos VA (2023) Félix González Bonorino: el geólogo que cambió la historia. Fundación Azara, Universidad Maimónides, Revista de Historia Natural 13(1):13–43
Haller MA, Ramos VA (1984) Las ofiolitas famatinianas (Eopaleozoico) de las provincias de San Juan y Mendoza. 9° Congreso Geológico Argentino (S.C. Bariloche), Actas 2:66–83
Ruíz Huidobro O (1960) Descripción Geológica de la Hoja 8e, Chicoana, Provincia de Salta. Boletín de la Dirección Nacional de Geología y Minería 89:1–46, Buenos Aires

Ruíz Huidobro OJ (1968) Descripción geológica de la Hoja 7e, Salta (Prov. de Salta y Jujuy). Boletín de la Dirección Nacional de Geología y Minería 109:1–48, Buenos Aires
Ruíz Huidobro OJ (1972) Descripción geológica de la Hoja 11e, Santa María (Prov. de Catamarca y Tucumán). Boletín del Servicio Nacional Minero Geológico 134:1–72, Buenos Aires
Ruíz Huidobro OJ (1976) Descripción geológica de la Hoja 12e, Laguna Helada, Prov. de Catamarca. Boletín del Servicio Geológico Nacional 146:1–55, Buenos Aires
Sobral JM (1942) Geología de la región occidental del territorio de La Pampa (al oeste de Chadi-Leuvú) y zona vecina de la provincia de Mendoza. Anales del Primer Congreso Panamericano de Ingeniería de Minas y Geología, Geología, Primera parte 2:668–750, Santiago
Suero T (1951) Descripción Geológica de la Hoja 36 c, Cerro Lotena. Neuquén. Escala 1:200.000. Carta Ge, lógico-Económica de la República Argentina. Boletín de la Dirección Nacional de Minería 76:1–80, Buenos Aires
Tófalo RO (2004) Oscar Ruiz Huidobro (1917–2004). Revista de la Asociación Geológica Argentina 59(3):518–519
Turner JCM (1970) The Andes of Northwestern Argentina. Geologische Rundschau 59(3):1028–1063, Stuttgart
Turner JCM (1971) Diccionario Geológico Inglés-Español, Español-Inglés. Asociación Geológica Argentina. Serie B, Didáctica y Complementaria, 1:1–288, Buenos Aires
Turner JCM (1972a) Puna. In: Leanza AF (ed) Geología Regional Argentina, Academia Nacional de Ciencias, Publicación Especial, pp 91–116, Córdoba
Turner JCM (1972b) Cordillera Oriental. In: Leanza AF (ed) Geología Regional Argentina, Academia Nacional de Ciencias, Publicación Especial, pp 117–14, Córdoba
Turner JCM (1974) Profesor Horacio Jaime Harrington, 1910–1973. Revista Asociación Geológica Argentina 29(3):195–200
Turner JCM (coord) (1979) Geología Regional Argentina. Academia Nacional de Ciencias, Publicación Especial, 1691 p, Córdoba
Vilela CR (1951) Acerca del hallazgo del Horizonte Calcáreo Dolomítico de la Puna salto-jujeña y su significado geológico. Revista de la Asociación Geológica Argentina 6(2):101–107
Vilela CR (1953a) Acerca de la presencia de sedimentos lacustres en el Valle Calchaquí. Revista de la Asociación Geológica Argentina 7(4):219–227
Vilela CR (1953b) Los períodos eruptivos en la Puna de Atacama. Revista de la Asociación Geológica Argentina 8(1):5–36
Zappettini E (2004) Servicio Geológico Minero Argentino, 1904–2004, 100 Años al servicio del desarrollo nacional. Secretaría de Minería, Segemar 106 p, Buenos Aires

Chapter 14
Geology During the Five-Year Plan

Abstract The five-year plan of the Peronist government focused on geological aspects in the Bernardino Rivadavia Museum, where the Natural Sciences Research Institute was created. Agustín Riggi was its director from 1946 to 1955, a period when, through a generous budget, he set up 29 laboratories in the museum. These were years in which classic books on the inventory of mineral resources of Argentina were published, such as *Metalliferous Deposits* by Angelelli and *Solid Mineral Fuels* by Borrello. Textbooks on different aspects of geology were edited, such as the *Sedimentological Lexicon* and *Mineral Optics*, together with *Eruptive Rocks under the Microscope*, with the participation of Mario Teruggi and Félix González Bonorino, which were widely used by several generations. Other aspects were not neglected, such as *Argentine Antarctica, its Geography and its Geology* by Agustín Riggi, and Borrello's book on the *Geology of the Malvinas Islands*. During those years, Mario Teruggi stood out at the Institute, providing his experience in both sedimentary and igneous petrography in the training of his disciples. Esteban Boltovskoy one of our greatest specialists in current and past foraminifera, joined the Institute as a researcher in 1950, carrying out important scientific work for decades. Regional Geology was supported by the publication of Pablo Groeber's great work on *The High Andes Mountains*.

Keywords New institute · Riggi director · Laboratories · Fundamental publications · Teruggi ideas

The five-year plans of Peronism, during the decade 1946–1955, laid the foundations for economic planning by the State. This role meant a paradigm shift concerning the agro-export period, whose limitations were exposed after the crisis of 1929 (Marín 2019). The Natural Sciences Research Institute, through the actions of its new director, was part of these plans from both a political and budgetary point of view.

In previous chapters it had been seen as Germán Burmeister after his appointment as director of the Public Museum of Buenos Aires, which had begun in 1812 at the initiative of Bernardino Rivadavia and formally consolidated in 1823, turning it

V. A. Ramos, *History of the Geological Sciences in Argentina*, The Latin American Studies Book Series, https://doi.org/10.1007/978-3-031-83139-3_14

Fig. 14.1 Bernardino Rivadavia Argentine Museum of Natural Sciences in its current building inaugurated in 1937. *Source* M.A.C.N. archives

into a Museum of Natural History. This name was known until 1923, when it was assigned its current name the Bernardino Rivadavia Argentine Museum of Natural Sciences (Fig. 14.1). The museum with its different names was basically a place where collections were treasured, their results were studied and published.

In 1946, one month after General Juan D. Perón (1895–1974) assumed the presidency, important changes occurred in the museum. By decree of the national executive branch of July 1946, Doello Jurado's retirement was accepted and Agustín E. Riggi (1904–1979) was appointed as the new director to replace him. Riggi, highly committed to official government policies, began a new stage in the institution.

Agustín Riggi was one of the first five geologists to graduate from the University of Buenos Aires. He defended his thesis on "*Geology of the Sierra de las Tunas, province of Buenos Aires*" under the direction of Juan J. Nágera. His director had to challenge Juan Keidel to a duel, who had held the thesis as a member of the panel for several months in order to be able to defend it in December 1932. Edelmira Mórtola, Martín Doello Jurado, and Juan Keidel were at the examining table. In his early years, he was a geologist at the Direction of Mines and Geology, until 1940 when he joined the Bernardino Rivadavia Argentine Museum of Natural Sciences. At the same time from 1942, he was a professor at the University of La Plata, where he supervised some 15 doctoral theses until his resignation in 1955. He was serving as head of the geology section of the Museum when he was appointed director.

14.1 The National Research Institute of Natural Sciences and Annex Museum

Riggi's management produced important changes in the nature and organization of the Museum (Fig. 14.2). In his almost ten years of administration, he suppressed the sections of human sciences, which he transferred to the Ethnographic Museum of the University of Buenos Aires, and the Musicology Office became part of the National Direction of Culture, which in his opinion were not useful to the disciplines of Natural Sciences. These changes were aimed at converting the museum into a research institute in Natural Sciences (de Asúa 2012).

Riggi's main innovation was to convert the Museum into a Natural Sciences research institute with a strong emphasis on natural resource prospecting and applied research. The institute from a formal point of view was created by a decree of the executive power in 1948. The objectives of its new director were to separate research from the exhibition, where the institute would take charge of "*scientific and technical functions*", while the museum "*It will exclusively include the exposition part corresponding to popular culture*". Riggi, inspired by the wishes of the president of the nation and in mutual agreement with him and in line with the scientific policy of Peronism, privileged the study of natural resources, applied research, and technological application. In this way, he managed to incorporate the institute into the objectives of the Second Five-Year Plan.

To fulfill them he reorganized the structure of the institution into three departments, botany, zoology and geology, each of them with different sections. Until then, geology was almost absent from the museum. One of his first actions was to invite young geologists to join that department and collaborate in starting up the laboratories. In the first 3 years in office, the institute had a very generous budget, following the strong increase that the country had in its GDP. It had a special allocation to buy exhibition furniture, laboratory instruments, books, and publications, and carry out campaigns. This allowed him to create two new publications, the Magazine and the Communications of the Bernardino Rivadavia Argentine Museum of Natural

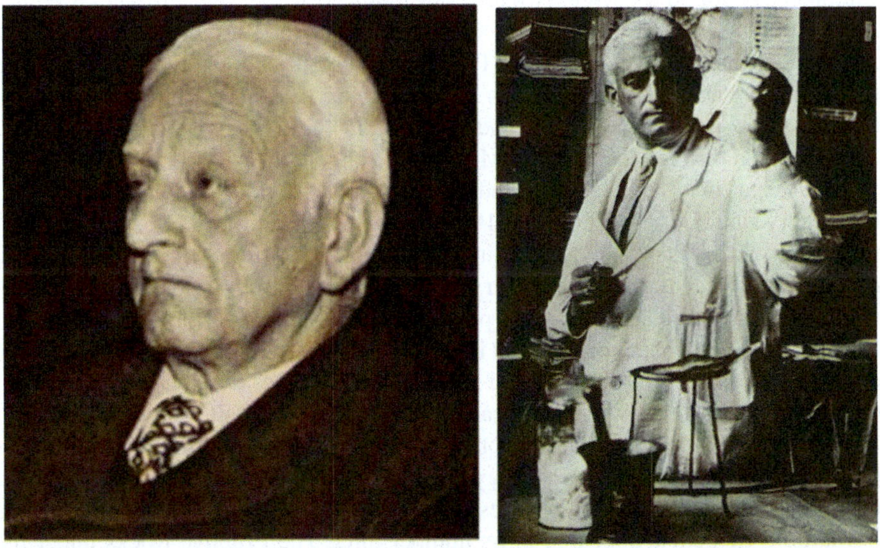

Fig. 14.2 Agustín E. Riggi (1904–1979) was the National Research Institute of Natural Sciences director between 1946 and 1955. *Source* M.A.C.N. archives and Abad de Santillán (1971)

Sciences. The budget covered the creation of 28 laboratories in which there was a clear predominance of mineralogy and geology instruments.

Riggi put Jorge Ricardo Cuomo (1919–2013) in charge of the Geology Department, who accompanied his entire management at the Institute, first as head of the department and later as vice director. Cuomo had completed a thesis on "*Geological-mining study of the sulfur deposits of the Cumbre Austral of the Alta Cordillera de Cori, Salta*", under the direction of Riggi in La Plata where he received his doctorate in 1954. He was always dedicated to mining, both in private companies and in public management, being Undersecretary of Mining of the Nation in 1982 and later president of the Argentine Mining Union.

14.2 The Geology Department of the New Institute

The modifications introduced by Riggi spontaneously coincided with the return to the country of Mario Teruggi, who after the Second World War obtained a scholarship from the British Council to develop advanced tasks in Petrology, Mineralogy and Applied Geology. He studied at the Royal School of Mines of the Imperial College of Science and Technology, University of London, returning to the country in the second half of 1946.

Upon his arrival, he was invited to join the institute, first as head of the Petrology Section and later as head of the Geology Department. Taking advantage of the new equipment acquired by Riggi, which even included new X-ray diffraction equipment, he carried out various studies, such as the discovery of mordenite, a rhombic zeolite from the basalts of Misiones. This along with other research was published in the Institute's new Communications series. During those years he began a series of studies on the basement of Tandil in which he again took advantage of its dissemination through the Institute Magazine (Spalletti 2016).

The Department of Geology had new graduates among whom María Clara Etchichury stood out. She entered in 1947 and in those years did her doctoral thesis at the University of Buenos under the direction of Franco Pastore, which she presented at the end of December 1951. Under the influence of Teruggi, she trained in sedimentology and years later became a professor of that discipline at the University of Buenos Aires. Esteban Boltovskoy (1912–1997), one of our greatest specialists in current and past foraminifera, joined the Institute as a researcher in 1950, carrying out important scientific work for decades (Boltovskoy and Boltovskoy 2022).

The ease that the budget allocations for publications gave him allowed Riggi to publish fundamental works on geology (Figs. 14.3 and 14.4). In the new Cultural and Didactic Extension Publication series, the Argentine Museum of Natural Sciences Bernardino Rivadavia appeared in its number three The Eruptive Rocks under the Microscope, Its Systematics and its Nomenclature (Teruggi 1951a), a textbook used for decades in the teaching of Petrology.

The first volume of the Institute's Journal featured articles such as "*The High Cordillera between latitudes 34° and 29°30',*" an excellent 350-page update by Pablo

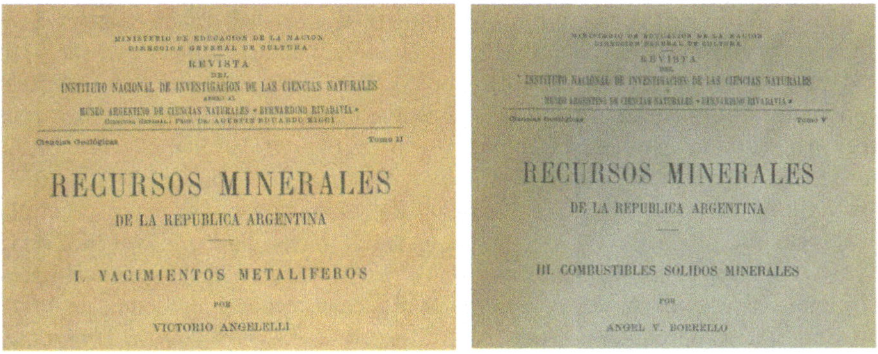
MINISTERIO DE EDUCACION DE LA NACION
DIRECCION GENERAL DE CULTURA
REVISTA
DEL
INSTITUTO NACIONAL DE INVESTIGACION DE LAS CIENCIAS NATURALES
MUSEO ARGENTINO DE CIENCIAS NATURALES «BERNARDINO RIVADAVIA»
Ciencias Geológicas — Tomo II
RECURSOS MINERALES
DE LA REPUBLICA ARGENTINA
I. YACIMIENTOS METALIFEROS
POR
VICTORIO ANGELELLI

MINISTERIO DE EDUCACION DE LA NACION
DIRECCION GENERAL DE CULTURA
REVISTA
DEL
INSTITUTO NACIONAL DE INVESTIGACION DE LAS CIENCIAS NATURALES
Ciencias Geológicas — Tomo V
RECURSOS MINERALES
DE LA REPUBLICA ARGENTINA
III. COMBUSTIBLES SOLIDOS MINERALES
POR
ANGEL V. BORRELLO

Fig. 14.3 Inventory of Mineral Resources published in the Institute Magazine, Metalliferous Deposits by Angelelli (1950), and Solid Mineral Fuels by Borrello (1956)

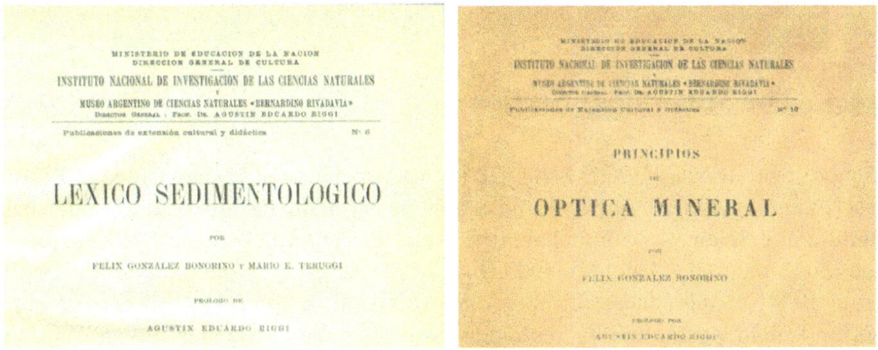
MINISTERIO DE EDUCACION DE LA NACION
DIRECCION GENERAL DE CULTURA
INSTITUTO NACIONAL DE INVESTIGACION DE LAS CIENCIAS NATURALES
Y
MUSEO ARGENTINO DE CIENCIAS NATURALES «BERNARDINO RIVADAVIA»
Director General: Prof. Dr. AGUSTIN EDUARDO RIGGI
Publicaciones de extensión cultural y didáctica — Nº 6
LEXICO SEDIMENTOLOGICO
POR
FELIX GONZALEZ BONORINO Y MARIO E. TERUGGI
PROLOGO DE
AGUSTIN EDUARDO RIGGI

INSTITUTO NACIONAL DE INVESTIGACION DE LAS CIENCIAS NATURALES
PRINCIPIOS
DE
OPTICA MINERAL
POR
FELIX GONZALEZ BONORINO

Fig. 14.4 Two texts published by the Institute in its series of cultural and didactic extension publications had a high impact on the community. *Source* González Bonorino and Teruggi (1952) and González Bonorino (1954)

Groeber on the geology and structure of the Principal Cordillera. It presents one of his latest contributions. It provides an integrative synthesis of his *Geological maps along the meridian 70°*, with new data and interpretations of the Andean uplift (Mescua et al. 2017). The institute also published fundamental works summarizing and inventorying mineral resources, following the precise political instructions issued by President Perón and taking into account that, as a result of World War II, there were severe restrictions on the supply of certain strategic minerals. The following volume features a complete inventory of the metalliferous deposits known at the time by Victorio Angelelli (1908–1991), a classic that remained a required reference for many years. This work updated the one he had prepared ten years earlier for the Direction of Mines (Angelelli 1941).It was published in the Journal of the Institute for Research in Natural Sciences as volume two of the Geological Sciences section (Angelelli 1950). In its series of Cultural and Educational Publications, the institute also focused on disseminating basic texts such as the *Sedimentological Lexicon*, written by Félix González Bonorino and Mario E. Teruggi, "*two of the country's most prominent petrologists*". This lexicon has had more than ten editions over the years and was considered the first of its kind published in Spanish in the world. Aware of the importance of this educational series, the institute aims to publish a new issue each year to meet the scientific and technical objectives of "*the aims and purposes of this Justicialist Government*" (Riggi's prologue in González Bonorino; Teruggi 1952). During its existence, it covered a variety of topics in the natural sciences, with the next issue in geological sciences being the popular treatise on *Principles of Mineral Optics* written by González Bonorino (1954). This book, too, had a significant impact, resulting in a series of subsequent reprints, and became a widely distributed textbook. It led to Eudeba introducing a revised and expanded version of *Optical Mineralogy* in its textbooks years later (González Bonorino 1976). This series of cultural and didactic extensions had in its number 4 "*Argentine Antarctica, its geography and its geology*", accompanied by complete geographical and geological maps, which gives foundations to the Argentine position in this region (Riggi 1950).

14.3 The Mineralogy Section

According to Riccardi (2015), this section was in charge of Romeo Croce (1901–?), a self-taught geologist who had studied Chemistry at the Faculty of Exact, Physical, and Natural Sciences, without having finished his doctoral thesis, in 1940. He joined the Museum as an assistant in 1923 (Fig. 14.5). He oversaw the mineralogy section from 1934, taking charge of the geology division in 1940. He had carried out a series of field works, mainly geological surveys in the Patagonian region. With the creation of the Institute, several of his studies were published in the Magazine and the Communications series. Little is known about Romeo Croce, who later dedicated himself to mineralogy, a discipline that he studied through his work at the Museum since at least 1930. Although he never graduated from the University, a series of contributions of a geological nature are known published by the Institute. Rodolfo Casamiquela relates that it was this fascinating mineralogist, whom he met in the late 1940s at Ingeniero Jacobacci, a small town in Patagonia, who encouraged him to study mining on the night shift at the Mariano Moreno School to become a mining expert (Casamiquela 1989).

In his work in Patagonia, he shows a direct interest in studying the porphyry complex currently considered to be of Jurassic age (Fig. 14.6). He describes its variations from the Atlantic coast to the Patagonian Mountains. He toured the Somuncura Massif from Valcheta to its northwestern sector where he described in detail the basement and its volcanic cover in the Gastre region and in the Sierra de Pire Mahuida that appear in the first volume of the Institute's Magazine. He presents interesting profiles of their relationships and structure. In addition, he studied the porphyry complex from Lake Ghío to the Sierra de Sangra, where he describes that it rests on powerful levels of conglomerates on the Altar de Dios hill, in the eastern foothills of that mountain range.

Fig. 14.5 Romeo Croce, second standing from the left, with Martín Doello Jurado sitting in the middle and other museum authorities in 1927. In those years he was already working as a geologist at the museum. *Source* M.A.C.N. archives

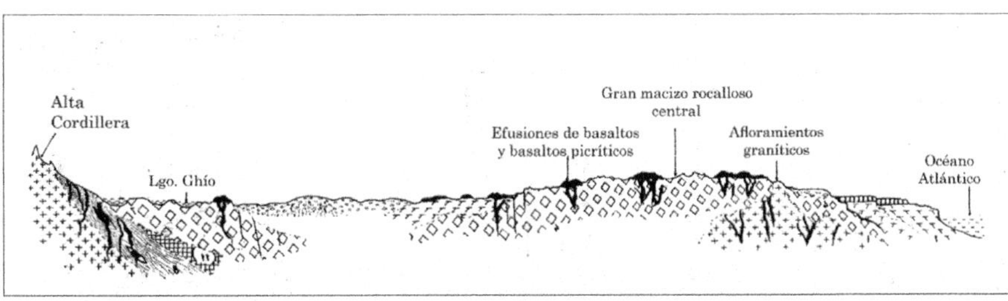

Fig. 14.6 Geological section of the Mesozoic rhyolitic porphyries from the High Patagonian Cordillera to the Atlantic Ocean showing the central rock massif, what is currently considered the Somuncurá Massif, and the relationships with the subjacent and overlying rocks. *Source* Croce (1954)

He also taught in the mining technical courses at the Mining School of the Argentine Chamber of Mining. In his publications he gives main importance to mineral deposits, describing various veins of fluorite both in Bajo de Valcheta and in the Gastre region (Croce 1950a, b, 1952). His last work presents a petrographic appendix written by Mario Teruggi (Croce 1956).

14.4 Final Considerations

The resignation of Agustín Riggi in 1955 associated with the coup d'état that deposed General Perón, marks the decline of the Institute's geological research and dissemination activities. Adolfo Dago Holmberg (1889–1980) was appointed director until 1958, a period during which it was unsuccessfully discussed whether the museum would become dependent on the University of Buenos Aires, as it was during a short period in the nineteenth century.

Finally, Mario Teruggi took over the interim management for a year until 1959 (Figs. 14.7 and 14.8). It is interesting to analyze his publication that year in Unesco, *The two natural science museums of Argentina* (Teruggi 1959), where he makes a comparison between the La Plata and Bernardino Rivadavia Museum. While the first fulfilled the mission of collecting, classifying, and researching, disseminating its results, with great weight in teaching, it stood out that Bernardino Rivadavia fulfilled only the first three, but was not involved in teaching within the museum. It is recognized that Agustín Riggi introduced the need for a museum to do research, an activity that continued after he left the institution until today. However, Holmberg's attempts to associate him with the University of Buenos Aires were not successful, despite González Bonorino's recommendations in that regard. Teruggi leaves the Bernardino Rivadavia Museum and concentrates his management activities at the La Plata Museum and teaching at the National University of La Plata.

When one analyzes those years of Agustín Riggi's management, one is struck by the capacity and budget that that institution had to publish fundamental works of Argentine geology, such as Groeber's extensive work on the Cordillera Principal and the complete inventories of Angelelli and Borrello on metalliferous and coals deposits respectively. One concern Riggi always had was to disseminate our geology

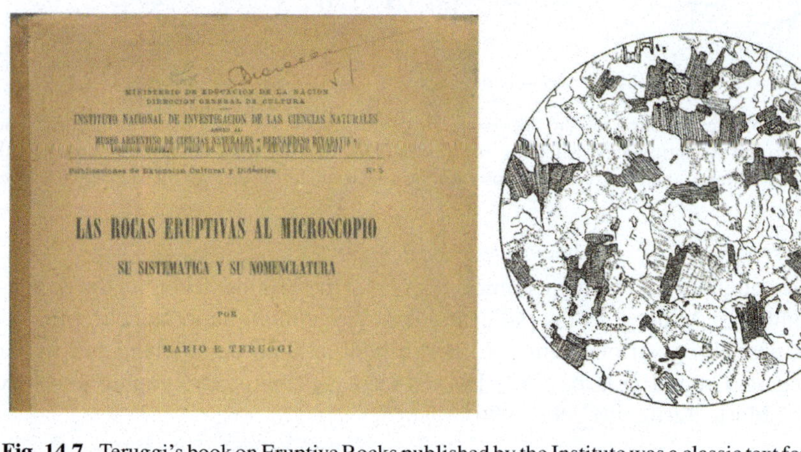

Fig. 14.7 Teruggi's book on Eruptive Rocks published by the Institute was a classic text for teaching several generations. It featured masterful drawings of petrographic sections of local rocks such as this tonalite from the Sierras de Córdoba. *Source* Teruggi (1951b)

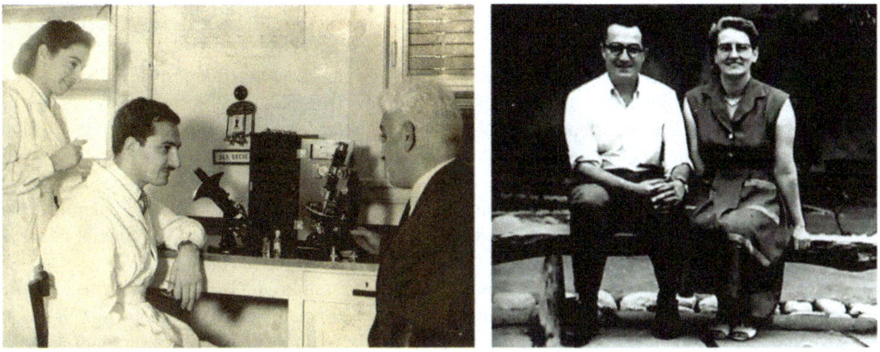

Fig. 14.8 The director of the Institute Agustín Riggi in the Petrography laboratory talking with María Clara Etchichury and Mario Teruggi; and years later Mario Teruggi with his wife Dr. Genoveva Dawson, a renowned researcher in Ethnobotany and Applied Botany. *Source* M.A.C.N. archives and Spalletti (2016)

and reaffirm sovereignty, such as his work on Argentine Antarctica (Riggi 1950) or Borrello's contribution to the Malvinas Islands geology promoted by the institute (Borrello 1963). This action was complemented by excellent textbooks on eruptive rocks, sedimentology and mineral optics, which were used for decades in universities, not only in Argentina but abroad. At the same time, he had an active policy of disseminating the geological research carried out by the institute which, although it has not had an impact on Argentine Geology, was due more to the limited personnel it had for these purposes than to the ability to publish them. Among them, we must claim Romeo Croce, a mineralogist and a little-known self-taught geologist, who has made interesting contributions to the knowledge of the volcanic rocks of northern Patagonia and their mineral occurrences.

References

Abad de Santillán D (1971) Historia Argentina. Tipográfica Editora Argentina, Tomo 5, 476 p, Buenos Aires

Angelelli V (1941) Los yacimientos minerales y rocas de aplicación de la República Argentina. Boletín de la Dirección de Minas y Geología 50:1–400, Buenos Aires

Angelelli V (1950) Recursos minerales de la República Argentina, I Yacimientos Metalíferos. Revista Museo Argentino de Ciencias Naturales Bernardino Rivadavia, Ciencias Geológicas 2:1–543, Buenos Aires

Boltovskoy A, Boltovskoy D (2022) Pioneers of plankton research: Esteban Boltovskoy (1912–1997). J Plankton Res 44(6):1–6

Borrello AV (1956) Recursos minerales de la República Argentina, V Combustibles sólidos minerales. Revista Museo Argentino de Ciencias Naturales Bernardino Rivadavia, Ciencias Geológicas 5:1–665, Buenos Aires

Borrello AV (1963) Sobre la Geología de las Islas Malvinas. Ministerio de Educación y Justicia. Ediciones Culturales Argentinas, 70 p Buenos Aires

Casamiquela RM (1989) El camino de la Fascinación. Ciencia Hoy, Entrevista 1(1):54–61, Buenos Aires
Croce R (1950a) Contribución al conocimiento de la geología y constitución de la Sierra Pire-Mahuida y sus estribaciones al NE. de Gastre, Chubut. (Estudio preliminar). Revista del del Instituto Nacional de Investigación de las Ciencias Naturales y Museo Argentino de Ciencias Naturales Bernardino Rivadavia, Ciencias Geológicas 1(2):66–86
Croce R (1950b) Algunas observaciones sobre los afloramientos graníticos, de rocas volcánicas antiguas y filonianas den Gastre, Chubut. Revista Instituto Nacional de Investigación de las Ciencias Naturales y Museo Argentino de Ciencias Naturales Bernardino Rivadavia. Ciencias Geológicas 1(3):90–115
Croce R (1952) Los afloramientos fluoríticos en las rocas cristalinas del Bajo de Valcheta (territorio de Río Negro). Comunicaciones del Instituto Nacional de Investigación de las Ciencias Naturales y Museo Argentino de Ciencias Naturales Bernardino Rivadavia. Ciencias Geológicas 10:1–22
Croce R (1954) Algunas observaciones realizadas en los bordes del gran macizo porfírico de Santa Cruz. Revista del Instituto Nacional de Investigación de las Ciencias Naturales y Museo Argentino de Ciencias Naturales Bernardino Rivadavia. Ciencias Geológicas 3(2):53–77
Croce R (1956) Formaciones características en las estructuras basales de la altiplanicie de Somuncurá en Río Negro (con un capítulo final abarcando el estudio de algunas rocas características por el doctor Mario Teruggi). Revista de la Asociación Geológica Argentina 11(3):158–201
De Asúa M (2012) Dos siglos y un museo. In: Penchaszadeh PE (ed) El Museo Argentino de Ciencias Naturales. 200 años, pp 13–70, Buenos Aires
González Bonorino F (1954) Principios de óptica mineral. Instituto Nacional de Investigaciones en Ciencias Naturales y Museo Bernardino Rivadavia, 147 p, Buenos Aires
González Bonorino F (1976) Mineralogía óptica. Temas de Eudeba, 342 p, Buenos Aires
González Bonorino F, Teruggi ME (1952) Léxico sedimentológico: Prólogo de Agustín Eduardo Riggi. Museo Argentino de Ciencias Naturales Bernardino Rivadavia 164 p, Buenos Aires
Marín M (2019) El efecto de los planes quinquenales del peronismo en la economía argentina. Facultad de Agronomía, Apuntes Agroeconómicos 13(9):1–7, Buenos Aires
Mescua JF, Giambiagi L, Ramos VA (2017) Los movimientos andinos de Groeber y la evolución tectónica de los Andes Centrales del sur. Revista de la Asociación Geológica Argentina 74(1):49–58
Riccardi AC (2015) Panorama histórico integral de las instituciones dedicadas a la geología en la Argentina, ca. 1880–1950. Manuscrito inédito preparado para el Diccionario histórico de las Ciencias de la Tierra en la Argentina (Archivo Histórico del Museo de La Plata) Prohistoria Ediciones
Riggi A (1950) La Antártida Argentina, su geografía y su geología. Publicación del Instituto Cultural y Didáctica, Museo Argentino de Ciencias Naturales Bernardino Rivadavia 4:32 p, Buenos Aires
Spalletti LA (2016) Mario E. Teruggi: un puente entre la ciencia y el humanismo. In: Riccardi A (ed) La Historia de la Geología en el Bicentenario de la Argentina, Revista del Museo de La Plata, Número Especial 1:302–315
Teruggi ME (1951a) Contribución a la petrografía del Partido de Tandil. El Cerro Noceti (Provincia de Buenos Aires). Revista del Instituto Nacional de Investigaciones Naturales, Geología 1(1):1–32
Teruggi ME (1951b) Las Rocas Eruptivas al Microscopio. Su sistemática y su nomenclatura. Publicación del Instituto Cultural y Didáctica, Museo Argentino de Ciencias Naturales Bernardino Rivadavia 3:1–410
Teruggi ME (1959) The two natural science museums of Argentina (Les deux principaux musees de sciences naturelles de l´argentine). Museum (UNESCO) 12(4):241–247, París

Chapter 15
Geology in *Fabricaciones Militares*

Abstract Restrictions on the import of raw materials as a consequence of the Second World War led to the creation of the General Directorate of *Fabricaciones Militares*. The studies began with the evaluation of the iron deposits of the Paleozoic basins of northern Argentina under the direction of Engineer Victorio Angelelli and the advice of Luciano Catalano. The iron ore of Zapla was exploited and the blast furnaces in Palpalá were built, which allowed the steel industry to begin in the region. The deposits of the Sierra Grande region, which were the most important at the time, were explored and evaluated. However, the exploration of strategic minerals was one of the most important activities carried out. They allowed the first deposits of porphyry copper to be discovered, such as Bajo de La Alumbrera, which was put into operation and allowed Argentina to enter into large-scale mining of disseminated metalliferous minerals. Among its geologists, Remo. R. Romani, Edgardo Menoyo, and Vicente Méndez stood out, with important contributions to the development of mining.

Keywords Second World War · New institutions · Strategic minerals · Iron potential · Disseminated copper · Large-scale mining

Restrictions on the import of raw materials and industrial products as a consequence of the start of the Second World War led President Roberto M. Ortíz to create the General Direction of *Fabricaciones Militares* as a state company dedicated to the generation of defense products that the industry private was not in a position to provide (Fig. 15.1). These objectives were expanded to the production of elements from the most diverse branches of activity from engines, and communications equipment, to railway cars, chemicals and steel products. Its creation had been promoted by Colonel Manuel Savio to contribute to industrial development in relation to national defense. It was formally institutionalized on October 9, 1941, through law 12.709 (Méndez 2004a).

In addition to this industrial activity, from its beginnings, it tried to produce those mineral raw materials considered essential to promote industrial development. With these objectives, he began explorations and exploitations aimed at obtaining copper, iron, manganese, tungsten, aluminum, and beryllium, among other substances.

V. A. Ramos, *History of the Geological Sciences in Argentina*, The Latin American Studies Book Series, https://doi.org/10.1007/978-3-031-83139-3_15

Fig. 15.1 Central Headquarters of the General Direction of Fabricaciones Militares in Buenos Aires where the Department of Geology and Mines was located, where the Army Higher Education Institute operated and is currently the headquarters of the National Defense University

15.1 The Northern Argentine Iron-Ore Basin

One of the first jobs that Colonel Savio undertook was the evaluation of the iron-ore basin in northwest Argentina (Fig. 15.2), where the occurrence of iron had been discovered in 1939. Two days before the creation of the DGFM, he received on October 7, 1941, a report that Luciano Catalano had prepared at his request, where he proposed a promising future for the sedimentary iron deposits in the Zapla region. A quick field visit allowed Catalano to estimate the presence of a potential deposit of 50 million tons with grades of 40 to 50% iron that would justify the installation of a blast furnace in Palpalá. The results of this report are complemented by Catalano the following year with a study of the limestone in the basement of the region where he estimates more than one hundred million tons of mineral that, due to its quality, could constitute an excellent flux that would allow the establishment of a steel industry in the region (see Catalano 1941, 1942, 1965).

Enthusiastic about these results, Savio led the new General Direction of Fabricaciones Militares to take charge of the exploration of the ferriferous occurrences covered by the provincial mining reserve of 1941 (Larra 1980) and requested the Direction of Mines to study the Zapla region, commissioning Eng. Angelelli for these works in 1942 (Fig. 15.3).

Fig. 15.2 Gen. Manuel N. Savio (1892–1948), a promoter of the exploitation of Zapla iron and founder of the Argentine steel industry. View of the first cast of pig iron in October 1945 at the Altos Hornos Zapla. *Source* Ministerio de Defensa (1948)

Fig. 15.3 Engineer Victorio Angelelli receiving the platinum Konex award in 1983. *Source* A.G.A. archives

This region of the Zapla anticline was known from the studies of Otto Schlagintweit (1937), then YPF exploration geologist. Schlagintweit, a great expert on the Subandean System, presents the first study "*On the stratigraphic position of the Zapla iron deposit and the distribution of the Zapla glacial horizon in Argentina and Bolivia*", a fundamental work that describes the importance of its areal extension (Schlagintweit 1943). The initial studies of Angelelli (1943), based on the stratigraphy of Schlagintweit and the work carried out in the central sector with the partial collaboration of Luciano Catalano, Pascual Sgrosso, and occasionally

Eduard Trümpy, led to his classic contribution on the "*Mina 9 de Octubre the main iron deposit of Zapla*" (Angelelli 1946).

The large reserves of this mine prompted Savio to create the *Altos Hornos Zapla* in Palpalá, which produced its first cast of pig iron shortly after in October 1945 "*...with the help of the geologist Victorio Angelelli, the main gallery of the mine was excavated in Zapla, baptized "October 9" in homage to the date of the founding of the DGFM. Built at two ends over a length of 500 m, that is, starting from its ends, trying to join in its middle part. An unusual method, contrary to all universal practices, adopted because the drilling teams could not advance more than one meter per day, while the deadline set by Savio required another pace*" (Brodtkorb and Shalamuk 1991).

Savio finally hired engineer Victorio Angelelli (1908–1991), who worked in the Direction of Mines, to supervise these mining activities in 1946 (Angelelli 1948). Angelelli headed the Department of Geology and Mines until 1956 and led a notable group of mining geologists. Among them stood out Hernán B. Navarro, Rómulo R. Romani, Eduardo M. González Stegemann, Enrique C. Rayces, Pedro J. Quiroga, Luis A. Barrionuevo, Horacio R. González Laguinge, Augusto Nieniewski, Esteban Wleklinski, and Alejo Brodtkorb, among others (Méndez 2004a).

Exploratory activities extended to the Unchimé deposit in Salta (Fig. 15.4), which had been described by another YPF geologist Enrique Mauri in 1943, who indicated these iron layers in his regional studies. Unchimé was exploited starting in 1956 to provide iron to Zapla, and later in 1958 to launch Altos Hornos Güemes, a private venture in Salta.

Both Zapla, Sitio Viejo, and Unchimé had the problem of a relatively low iron grade (Boso and Monnaldi 1999), which did not allow them to be competitive with other ores at an international level, with the first two interrupting their activity when they were privatized in 1989. However, the iron was used for other purposes such as the manufacture of cement, ochers, and paints. Currently, Altos Hornos Zapla uses minerals from Chile and Brazil. The facilities of the Sitio Viejo mining site were transferred to the National University of Jujuy for the training of students as the first "*Mine School*" in the region.

15.2 The Sierra Grande Iron-Ore Basin

The discovery of iron in the surroundings of Sierra Grande motivated a brief visit by Pascual Sgrosso of the Direction of Mines in 1946 and the intervention of the Argentine Industrial Credit Bank that materialized with the work of Amílcar Herrera in 1947 (Fig. 15.5). This study was the basis of his doctoral thesis presented at the University of Buenos Aires (Herrera 1950, see Ramos 2022). The General Direction of Fabricaciones Militares, based on Angelelli's initial reconnaissance in 1946 and 1948, was interested in its mining potential. The National Direction of Mining was in charge of the deposits between 1949 and 1956, carrying out detailed studies (Valvano

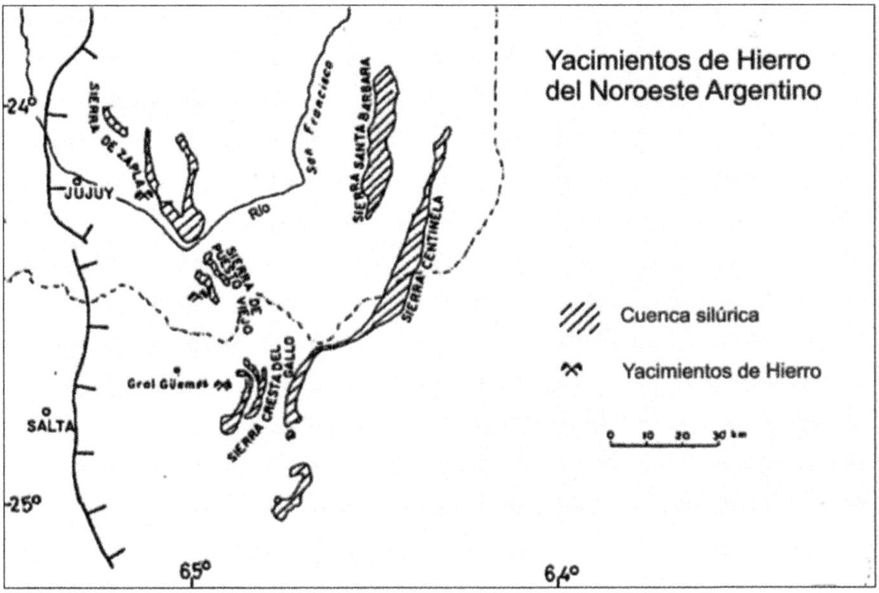

Fig. 15.4 Location of the iron deposits of Zapla, Sitio Viejo, and Unchimé in the provinces of Jujuy and Salta. *Source* Boso and Monnaldi (1999)

1949, 1954; Zollner 1951; and others) and regional studies by Enrique de Alba between 1950 and 1951 for the survey of Sheet 41g Sierra Grande.

The exploitation of the ferriferous horizons was carried out by Fabricaciones Militares between 1958 and 1960. In those years, the research by Navarro (1960) stands out, studying the deposits to intervene in the opposing interpretation options of Valvano and Zollner concerning their structure. The studies had the support of United Nations experts for the magnetometric survey of the ore.

Since 1961 *Minera Siderúrgica Patagónica Sierra Grande S.A.* (Misipa D.G.S.A.) has taken charge of the deposits, hiring a German company to carry out detailed studies. Years later, this company withdrew from exploration, and through National Law 1777/68, *Fabricaciones Militares* took over the exploration work at the end of 1968. In 1969, driven by *Fabricaciones Militares*, Hipasam was created, a state company that started the exploitation and produced about 15 million tons of ore with an average grade of 55% iron with the production of pellets of 68% until 1991 when the operation was closed (Menoyo and Herrmann 2004). During that period the *Fabricaciones Militares* continued to carry out geological profiles, regional geology, aeromagnetic and terrestrial prospecting in the different deposits (Fig. 15.6) with reconnaissance throughout the iron basin (see details in Zanettini 1981, 1999).

The activities were stopped in 2006 when a Chinese company acquired the Sierra Grande Mining and restarted exploitation, which was interrupted in 2012, due to environmental impact conflicts and lack of profitability in mineral prices. *Fabricaciones Militares* was involved for decades in the exploration and evaluation of the site

Fig. 15.5 Jorge Valvano (1922–1960) of the Direction of Mines and Amílcar Herrera (1920–1995) of the Industrial Credit Bank, two geologists, with doctorates from the University of Buenos Aires, made the first studies of the Sierra Grande deposits. *Source* A.G.A. archives

until it passed into private hands and exploitation was finally halted. The 96 km of galleries and tunnels and a central shaft more than 400 m deep remained as witnesses to this activity.

15.3 Exploration of Strategic Minerals

Since its creation, *Fabricaciones Militares* had the mission of evaluating other minerals throughout the country in addition to iron from the ferriferous basins. At that time, stand out the exploration and exploitation of the sulfur deposits of the Julia Mine and the La Casualidad benefit plant in Salta along the border with Chile. They were located in the Cenozoic volcanic rocks of the Western Cordillera at more than 5,000 m high. Production was active between 1941 and 1978.

A national mining evaluation was also carried out, an action that began under the direction of engineer Angelelli and was developed over almost two decades. The known mineral deposits and those discovered later were studied and evaluated, analyzing 500 potential deposits. Non-metal deposits important for the steel industry such as limestone and dolomite were also studied. The studies carried out in districts such as Capillitas in Catamarca, La Mejicana in La Rioja, Paramillos de Uspallata in Mendoza, among others, where new trenches and drilling were carried out stand out (Méndez 2004b, p. 244). These studies concluded that, with a few exceptions such

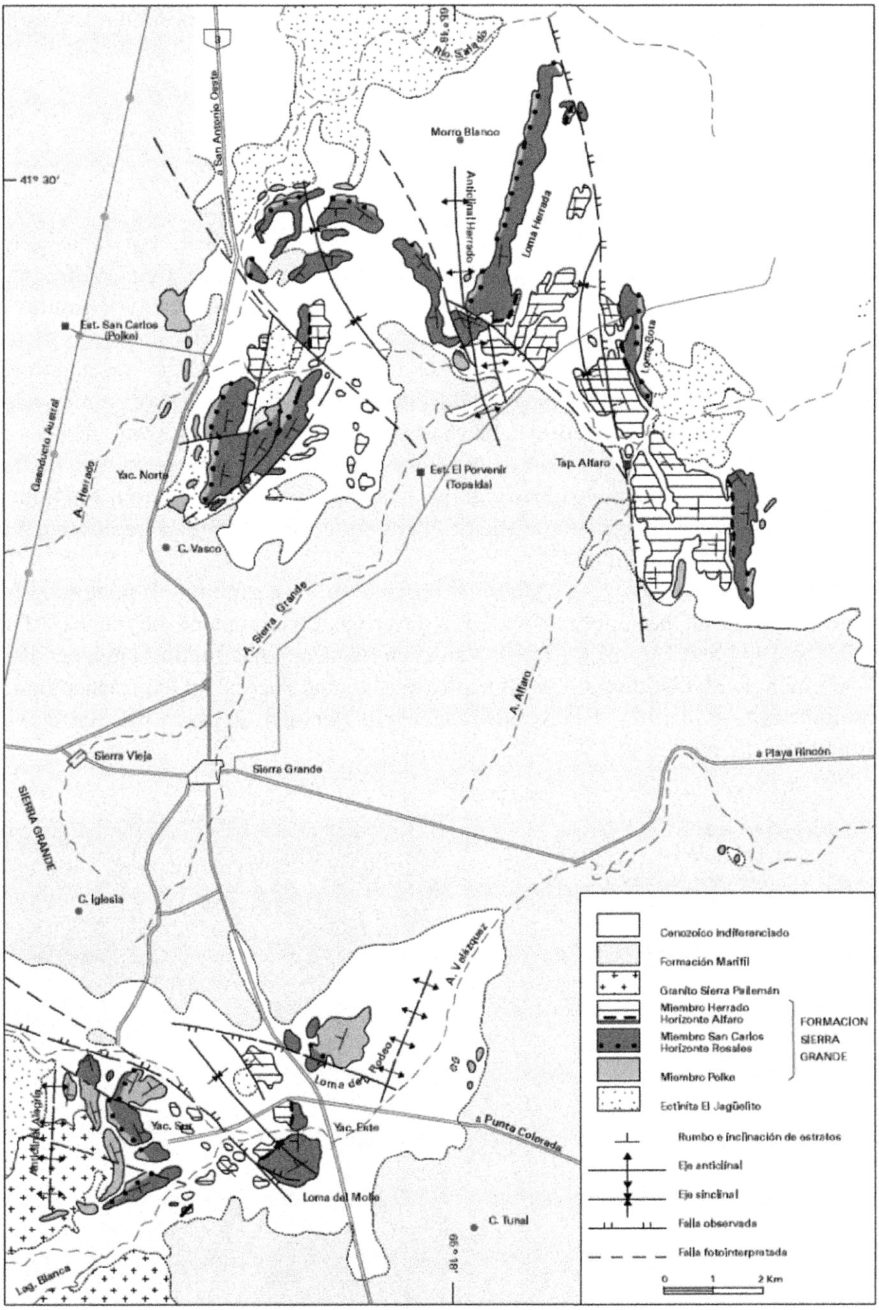

Fig. 15.6 The North and South deposits of the Sierra Grande iron district. *Source* Zanettini (1999)

as the Aguilar District in Jujuy, the vein deposits did not have important reserves for economic exploitation.

15.4 A New Stage with Large Mining

The negative results obtained in those years in the evaluation of vein deposits led to the search for new prospecting horizons. In 1963, Edgardo Menoyo (1927–2021) joined the corps of geologists at *Fabricaciones Militares*, who had received training in Canada from the Industrial Bank where he previously worked (Fig. 15.7). His participation was key to changing the direction of exploration. In 1965, he was appointed head of the Department of Geology and Mines and under his direction, a new stage was opened. The observation of what had been happening for several decades in Chile and Peru, as well as in North America, led him to modify the programs that were being executed to try to identify large-scale deposits, with disseminated mineralization and open pit exploitation with operating costs lower than traditional vein mining (Méndez 2004b, p. 245).

This led to the conceiving of the first regional mining exploration plan, the Plan Cordillerano, with the support of the United Nations Development Program (UNDP) between 1964 and 1968 (Fig. 15.8). This plan had Canadian mining engineer John S. Carman as its first director, who had extensive and successful experience in our country starting in 1965. The Swedish geologist Pontus Ljunggren was the second director of the plan.

Fig. 15.7 As head of exploration, Edgardo Menoyo boosted the activity of *Fabricaciones Militares* for several years. *Source* Courtesy Norma Pezzutti

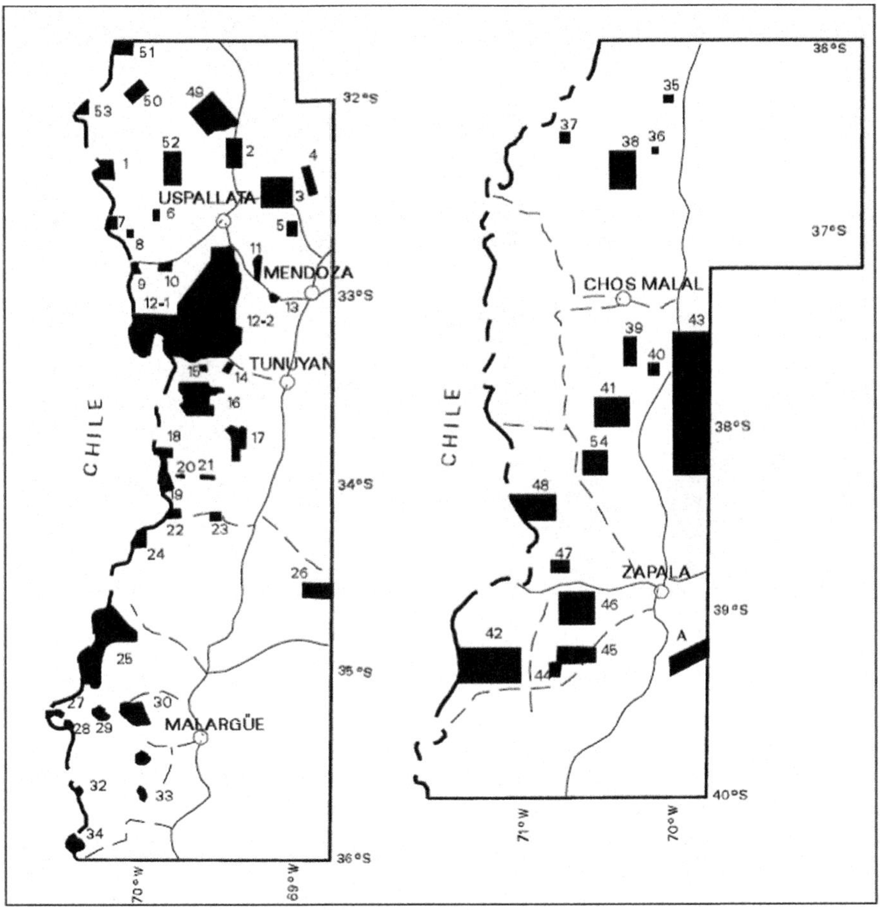

Fig. 15.8 Reserve areas of the Cordillerano plan in the provinces of San Juan. Mendoza and Neuquén. *Source* Méndez (1999)

This plan was responsible for the expeditious survey of the mountain areas of Mendoza and Neuquén and covered an area of 130,000 square kilometers, later extending to the province of San Juan. To survey this area, aerial photographs were obtained, mosaics were made at a scale of 1:50,000 and a photogeological map of the mountain region was made. Dr. Osvaldo Bracaccini was in charge of this survey supported by a small group of geologists between 1963 and 1964, presenting the results in a fundamental work on "*Organization and summary of the geological background related to the mountain area*" and the "*Structural geology of the "cordilleran zone" of the provinces of Mendoza and Neuquén*" accompanied by photogeological maps at a scale of 1: 50,000 and 1: 200,000, which served as the basis for the selection of the 54 reserve areas of the Plan Cordillerano (Bracaccini 1964a, b).

Fig. 15.9 Rómulo R. Romani and Alejo Brodtkorb, two geologists with extensive experience in *Fabricaciones Militares*. *Source* A.G.A. archives

The geologists of Fabricaciones Militares Hernán Navarro, Remo R. Romani, Horacio González Laguinge, Rodolfo González Amorín, and Adrián Soto, carried out an internship in porphyry copper deposits in the United States that subsequently contributed to the selection and study of the reserve areas (Fig. 15.9). Alejo Brodtkorb was in charge of the analytic laboratories in Mendoza. As a result of these studies, drilling and subsequent analysis, the first two disseminated copper porphyries were discovered in 1963, Paramillos Sur and Paramillos Norte, where geologists Hernán Navarro and Rómulo R. Romani played a leading role (Navarro 1972).

15.5 *Fabricaciones Militares* and YMAD

The participation of *Fabricaciones Militares* had begun in 1953 in the Agua de Dionisio region in Catamarca to carry out exploration work in the Los Viscos area through an agreement with the National Direction of Mining. *Yacimientos Mineros Agua del Dionisio* (YMAD), a company created in 1958, confirmed after a series of surveys and evaluations the existence of disseminated mineralization in Bajo de La Alumbrera in 1969 (Figs. 15.10 and 15.11).

From a geological point of view, the first survey of Bajo de la Alumbrera by Adolfo Mezzetti in 1963 and the studies on the volcanic complex by Llambías (1970) stand out. It is up to the geologist Rómulo R. Romani based on his geological and

Fig. 15.10 First photograph of Bajo de la Alumbrera is from 1963 when Adolfo Mezzetti carried out the first detailed geological survey. *Source* Mezzetti and Llambías (1963)

Fig. 15.11 View of the open pit of Bajo de la Alumbrera at the end of exploitation. *Source* Segemar archives

geochemical reconnaissance to establish in 1969 the existence of a copper porphyry disseminated with gold, confirmed by subsequent studies in 1973 by experts such as Richard Sillitoe (see Alderete 2004).

At the end of 1974, given Bajo de la Alumbrera's importance as a disseminated copper deposit for the country, an agreement was signed between *Fabricaciones Militares* and YMAD, which was an important contribution to the geological exploration plan. On the recommendation of John Gilbert, three deep drillings were carried out in 1975 to expand the reserve potential. Between 1976 and 1979 *Fabricaciones Militares*, which was trying to agree with YMAD to carry out a pre-feasibility study, managed to hire a company to carry it out, before the international bidding. The Malvinas War in 1982 definitively distanced *Fabricaciones Militares* from the destiny of Bajo de la Alumbrera, which became the first large-scale mine to be developed in Argentina (Méndez 2004b).

Minera La Alumbrera began exploiting the deposit between 1994 and 1995, starting production in 1997. Years later, in 2015, another copper porphyry, Bajo del Durazno, was incorporated into exploitation.

15.6 Cordillera Norte Plan and Military Fabrications

In 1969, the Cordillera Norte Plan began with the support of the United Nations. The National Institute of Geology and Mining was in charge of the provinces of Catamarca, Tucumán, and Santiago de Estero, while *Fabricaciones Militares* is concentrated in Salta and Jujuy (Fig. 15.12). Operations began in Salta in March 1969 with the geological photointerpretation of photographs of the entire northwest region obtained the previous year by Spartan Air Services at a scale of 1:50,000. The photogeology carried out and the field surveys produce a great advance in the geological knowledge of this sector, substantially improving the 1:200,000 scale maps available.

The mining exploration project in the northwest region led by *Fabricaciones Militares* ended in 1976 with the selection of 34 areas of mining interest in the provinces of Salta and Jujuy, which identified several projects that later became a reality.

In the geological surveys carried out at that time by *Fabricaciones Militares*, the contributions of Vicente Méndez (Fig. 15.13) on the structure of Salta and Jujuy and the identification of the Eastern Puna Eruptive Belt stand out (Méndez 1975; Méndez et al. 1973), the distribution of the Upper Cretaceous marine ingression in the northern Puna (Navarini et al. 1973), and the discovery of fossiliferous Ordovician in the western Puna (Ramos 1972). The surveys carried out by the author during his stay at *Fabricaciones Militares* in 1969 were the basis of his doctoral thesis on "*The structure of the first buttresses of the Puna between San Antonio de los Cobres and El Moreno, provinces of Salta and Jujuy*" presented in the University of Buenos Aires (Ramos 1970, published in Ramos 1973). In turn, Méndez's work on "*Structures of the provinces of Salta and Jujuy from the meridian 65°30′ west to the limit with the Republics of Bolivia and Chile*" was presented as a doctoral thesis at the National University of La Plata (Méndez et al. 1973, published in Méndez 1975). The two theses were directed by Juan C.M. Turner at both universities. These contributions were complemented by the works of Juan Carlos M. Zanettini in the region of Abra Pampa and Tres Cruces (Zanettini 1974) and those of Méndez in the geology of the region of Mina Aguilar and surroundings (Méndez 1974). These contributions were

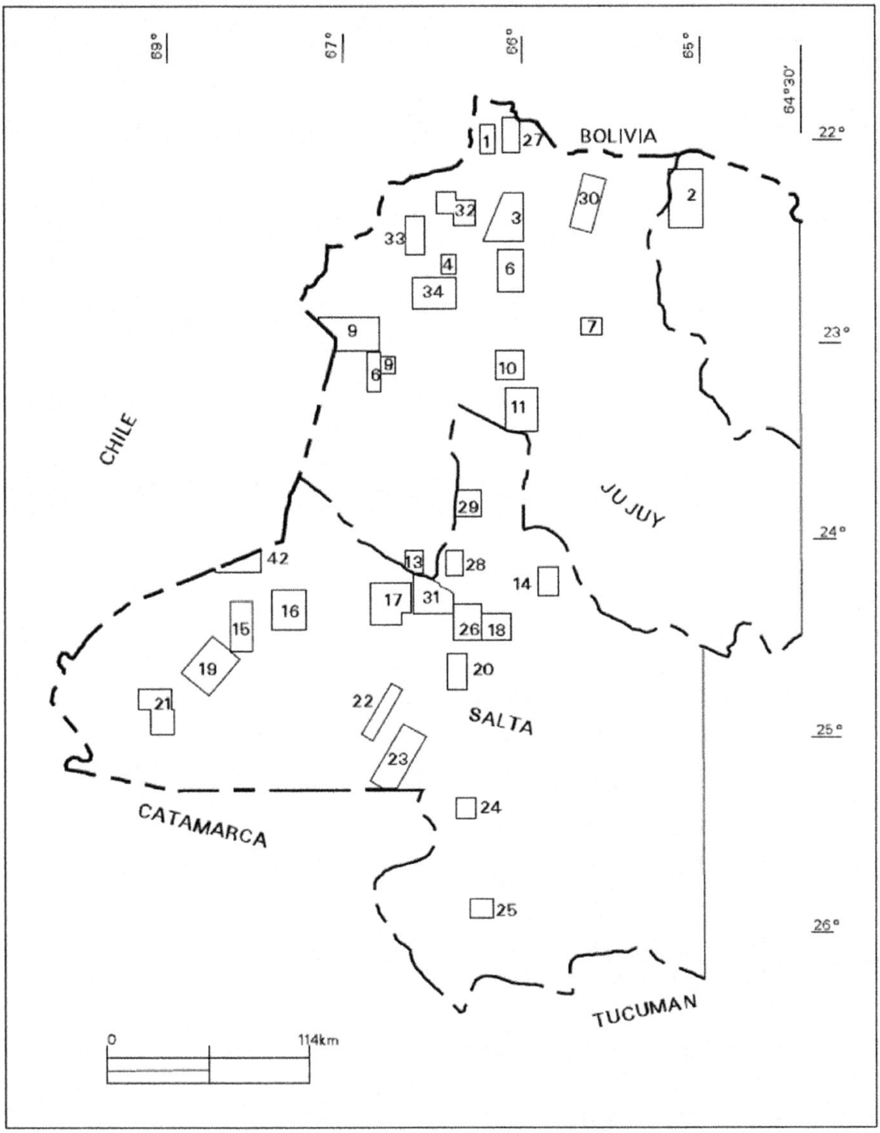

Fig. 15.12 Reserve areas of the Cordillerano Norte plan in the provinces of Salta and Jujuy. *Source* Méndez (1999)

integrated years later in the second Argentine Regional Geology symposium on the characterization of the Puna by Turner and Méndez (1979).

Another geologist of *Fabricaciones Militares* we shall remember is Luis Conti. He graduated from the University of Buenos Aires in 1950 and did postgraduate studies

Fig. 15.13 Vicente Méndez, a referent of Fabricaciones Militares

at the Colorado School of Mines. Based on his experience at the National Development Bank and *Fabricaciones Militares*, he was appointed in the early seventies as Undersecretary of Mining, where he was in charge at the beginning of the NOA Plan for Catamarca, Tucumán and La Rioja and of the supervision of the exploration of disseminated porphyry copper deposits. Unfortunately, in January 1976, working again at *Fabricaciones Militares*, while trying to reach the Pachón porphyry copper prospect, the helicopter lost lift and crashed in the foothills of the Frontal Cordillera (Fig. 15.14). The pilot and Conti died on that occasion, with Héctor Péndola, a geologist from the National Development Bank, miraculously surviving.

15.7 Final Considerations

Fabricaciones Militares in the history of our Geology had a leading role in promoting new concepts regarding our mineral resources. For almost two decades under the direction of engineer Victorio Angelelli, traditional vein deposits were evaluated, or attempts were made to promote the development of two ferriferous basins, which due to their intrinsic characteristics yielded negative results. However, thanks to the tenacity of General Savio, a steel industry was developed in the country. The main credit of *Fabricaciones Militares*, a product of the initiative of Edgardo Menoyo, is that he managed to demonstrate that on the eastern side of the Andes Mountain there were disseminated porphyry copper-type deposits, unknown until that time. The first two identified, Paramillos Sur and Norte, have not been developed yet, but thanks to the methodology used and the experience acquired it was possible to identify, develop and put into exploitation the copper and gold porphyry of Bajo de La Alumbrera.

Fig. 15.14 Luis Conti (1919–1976) died dramatically in the line of duty. *Source* Courtesy of Carlos Conti

This opened the way for new developments currently in advanced exploration that predict an interesting future for these deposits.

Although at the beginning it was theorized that if they existed in Chile, they should also exist on this side of the mountain range, today we know that they exist due to the development of horizontal subduction zones that have caused the magmatic arcs to migrate from the axis of the mountain range towards the foreland (Ramos 1995).

As a by-product of these mining exploration studies, *Fabricaciones Militares* bequeathed an important advance in the geological knowledge of the Main Cordillera of Mendoza and Neuquén, as well as the sectors of the Puna and Eastern Cordillera of Salta and Jujuy (Méndez et al. 1979). The studies carried out contributed in the following years to form a database that supported subsequent studies in a significant way.

References

Alderete MC (2004) Distrito Yacimientos Mineros Agua de Dionisio (YMAD). In: Lavandaio E, Catalano E (eds) Historia de la Minería Argentina, Tomo 2. Anales del Instituto de Geología y Recursos Minerales, SEGEMAR 40:21–42, Buenos Aires

Angelelli V (1943) El Yacimiento ferrífero de Zapla. Su exploración en el tramo comprendido entre el Abra de los Tomates y la proximidad del Arroyo Pantanillos. Estudio Minero-Geológico. Sierra de Zapla, provincia de Jujuy. Dirección de Minas y Geología, unpublished report, Buenos Aires

Angelelli V (1946) La geología y génesis del yacimiento ferrífero de Zapla, Mina "9 de Octubre" (Departamento de la Capital, Provincia de Jujuy). Revista de la Sociedad Geológica Argentina 1(2):117–147

Angelelli V (1948) La cuenca ferrífera de Sierra Grande. Dirección General de Fabricaciones Militares, Unpublished report, Buenos Aires

Boso MA, Monnaldi CR (1999) La cuenca ferrífera del norte argentino, Jujuy y Salta. In: Zappettini EO (ed) Recursos Minerales de la República Argentina, Anales del Instituto de Geología y Recursos Minerales, Segemar 35:529–544, Buenos Aires

Bracaccini IO (1964a) Ordenación y resumen de los antecedentes geológicos relativos a la zona cordillerana de las provincias de Mendoza y Neuquén, República Argentina. Dirección General de Fabricaciones Militares, Contrato 1103, informe No 1, 295 p, Buenos Aires

Bracaccini IO (1964b) Geología estructural de la zona cordillerana de Mendoza y Neuquén, República Argentina. Dirección General de Fabricaciones Militares, Contrato 1103, informe No 4, 220 p, Buenos Aires

de Brodtkorb MK, Schlamuk I (1991) Ingeniero Victorio Angelelli, (1908–1991), Nota necrológica. Revista de la Asociación Geológica Argentina 56(3–4):358–360

Catalano LR (1941) El hierro argentino de Zapa, provincia de jujuy. Dirección General de Fabricaciones Militares, unpublished report, 110 p, Buenos Aires

Catalano LR (1942) Los yacimientos de calizas de León y San Juancito (Jujuy) como base para proveer fundente a la planta siderúrgica de Palpalá. Geología Estructural de los yacimientos Sur y Este de Sierra Grande. Sierra Grande, Provincia de Río Negro. Dirección General de Fabricaciones Militares, unpublished report, 32 p, Buenos Aires

Catalano LR (1965) El Hierro Argentino de Zapla (Provincia de Jujuy)—Calizas de León y San Juancito (Provincia de Jujuy), Estudio preliminar de geología económica. Ministerio de Economía de la Nación, Secretaría de Industria y Minería, Subsecretaría de Minería, Serie Argentina 5:1–143, Buenos Aires

Herrera AO (1947) La cuenca ferrífera de Sierra Grande, Territorio de Río Negro. Banco Industrial de la República Argentina, unpublished report, Buenos Aires

Herrera AO (1950) La cuenca ferrífera de Sierra Grande. Facultad de Ciencias Exactas y Naturales, Universidad de Buenos Aires, unpublished PhD Thesis, Buenos Aires

Larra R (1980) Savio: el argentino que forjó el acero. Anfora, 185 p, Buenos Aires

Llambías EJ (1970) Geología de. los Yacimientos Mineros Agua de Dionisio, Provincia de Catamarca, República Argentina. Revista de la Asociación Argentina de Mineralogía, Petrología y Sedimentología 1:2–32, Buenos Aires

Méndez V (1974) Geología de la comarca de Mina Aguilar y alrededores (Provincias de Salta y Jujuy). 5° Congreso Geológico Argentino, Actas 3:467–470

Méndez V (1975) Estructuras de las provincias de Salta y Jujuy a partir del meridiano 65°30′ oeste, hasta el límite con las Repúblicas de Bolivia y Chile. Revista de la Asociación Geológica Argentina 29(4):391–424

Méndez V (1999) Historia del desarrollo minero argentino. In: Zappettini E (ed) Recursos minerales de la República Argentina, Servicio Geológico Minero Argentino, Anales del Instituto de Geología y Recursos Minerales, Segemar 35(1):3–27, Buenos Aires

Méndez V (2004a) La Dirección General de Fabricaciones Militares. In: Lavandaio E, Catalano E (eds) Historia de la Minería Argentina, Tomo 1, Anales del Instituto de Geología y Recursos Minerales, Segemar 40:243–251, Buenos Aires

Méndez V (2004b) Los pórfidos cupríferos. In: Lavandaio E, Catalano E (eds) Historia de la Minería Argentina, Tomo 2, Anales del Instituto de Geología y Recursos Minerales, Segemar 40:379–392, Buenos Aires

Méndez V, Turner JCM, Navarini A, Amengual R, Viera V (1979) Geología de la región noroeste, provincias de Salta y Jujuy. Dirección General de Fabricaciones Militares-PNUD, 118 p, Buenos Aires

Méndez V, Navarini A, Plaza D, Viera O (1973) Faja Eruptiva de la Puna oriental. 5° Congreso Geológico Argentino, Actas 4:89–100, Córdoba

Menoyo E, Herrmann C (2004) Hierro y siderurgia en la Argentina. In: Lavandaio E, Catalano E (eds) Historia de la Minería Argentina, Tomo 2, Anales del Instituto de Geología y Recursos Minerales, Segemar 40:351–359, Buenos Aires

Mezzetti AM, Llambias E (1963) Informe sobre las áreas de alteración hidrotermal en Agua de Dionisio Distrito Haulfin. Departamento Belén. Provincia de Catamarca. 44 pp. Buenos Aires, Dirección Nacional de Geología y Minería. Secretaría de Estado de Industria y Minería, Unpublished report, 44 p, Buenos Aires

Ministerio de Defensa (1948) En memoria del general de división D. Manuel N. Savio. Dirección General de Fabricaciones Militares, Editorial Peuser, 126 p, Buenos Aires

Navarini A, Méndez V, Zanettini JCM (1973) Sobre la presencia de la Formación Yacoraite en el anticlinal del cerro Chaupiorco, Departamento Santa Catalina (Prov. de Jujuy). Revista de la Asociación Geológica Argentina 28(1):88–90

Navarro H (1960) Geología Estructural de los yacimientos Sur y Este de Sierra Grande. Sierra Grande, Provincia de Río Negro. Dirección General de Fabricaciones Militares, unpublished report, 36 p, Buenos Aires

Navarro H (1972) Área de reserva Nº 3 zona "Paramillo Norte" provincia de Mendoza. 4° Jornadas Geológicas Argentinas (Mendoza, 1969), Actas 3:105–125, Buenos Aires

Ramos VA (1970) Geología de los primeros contrafuertes de la Puna saltojujeña entre San Antonio de los Cobres y el Moreno. Universidad de Buenos Aires, Facultad de Ciencias Exactas y Naturales, PhD Thesis 94 p, Buenos Aires

Ramos VA (1973) Estructura de los primeros contrafuertes de la Puna Saltojujeña y sus manifestaciones volcánicas asociadas. 5° Congreso Geológico Argentino, Actas 4:159–202, Córdoba

Ramos VA (1972) El Ordovícico fosilífero de La Sierra de Lina, Provincia de Jujuy. Revista de la Asociación Geológica Argentina 27(1):84–94

Ramos VA (1995) Evolución tectónica del segmento de subducción horizontal de los Andes Centrales (27–34°S) y su control en las manifestaciones auríferas epitermales. 5° Congreso Nacional de Geología Económica (San Juan), Simposio de Metalogenia Andina de Sudamérica: 82–94

Ramos VA (2022) El doctor Amílcar Herrera, mucho más que un geólogo. Fundación Azara, Universidad Maimónides, Revista de Historia Natural 12(1):139–163

Schlagintweit O (1937) Observaciones estratigráficas del Norte Argentino. Boletín de Informaciones Petroleras 14(156):1–49, Buenos Aires

Schlagintweit O (1943) Posición estratigráfica del yacimiento de hierro de Zapla y la difusión del horizonte glacial de Zapla en la Argentina y en Bolivia. Revista Minera 13(4):115–127, Buenos Aires

Turner JCM, Méndez V (1979) Puna. In: Turner JCM (coord) Geología Regional Argentina. Academia Nacional de Ciencias, Publicación Especial, pp 13–56, Córdoba

Valvano JA (1949) Depósitos ferríferos de Sierra Grande. Territorio de Río Negro. Dirección General de Industria Minera, informe preliminar, 26 p, Buenos Aires

Valvano JA (1954) Génesis de los Yacimientos de hierro de Sierra Grande, Territorio Nacional de Tierra del Fuego. Revista de la Asociación Geológica Argentina 9(4):193–209

Zanettini JCM (1974) Geología de la comarca de Abra Pampa y Tres Cruces (provincia de Jujuy). Revista de la Asociación Geológica Argentina 28(3):331–340

Zanettini JCM (1981) La Formación Sierra Grande (provincia de Río Negro). Revista de la Asociación Geológica Argentina 16(2):160–179

Zanettini JCM (1999) Los depósitos ferríferos de Sierra Grande, Río Negro. In: Zappettini E (ed) Recursos minerales de la República Argentina, Servicio Geológico Minero Argentino, Anales del Instituto de geología y Recursos Minerales, Segemar 35(1):745–765, Buenos Aires

Zollner W (1951) Informe geológico económico del yacimiento ferrífero de Sierra Grande. Dirección Nacional de Minería, unpublished report, Buenos Aires

Chapter 16
The Foundation of the Argentine Geological Society

Abstract The creation of the Argentine Geological Society was one of the main events of the mid-twentieth century for the community. The foundation of the society allowed the members to gather with a common goal, to have a press medium where to publish the geological contributions and paleontological findings of our country and neighboring regions. It was founded in 1945 under the direction of Carlos Storni and a group of young enthusiast geologists in the Aula Aguirre of the School of Geology of the University of Buenos Aires. It was established during the first years in the historic Manzana de las Luces where the University of Buenos Aires had been founded in 1821, and then with the effort of the members the *House of the Geologist* was bought, which is the current headquarters of the Geological Association. Throughout these almost eighty years it has uninterruptedly edited the Journal of the Geological Association, has organized the Argentine Geological Congresses, has published numerous texts for the study of different disciplines, and has maintained a series of books edited at each congress that present the geology of the region where it is held. This activity is shared with updated courses and social events held at the *House of the Geologist.*

Keywords Carlos Storni · Geological journal · Congresses · House of the geologist · Special publications

One of the main events that marked Argentine Geology in the mid-twentieth century is linked to the founding of the Argentine Geological Society. In those years, the geological community had grown steadily with dozens of graduates from the national universities of La Plata, Buenos Aires and Córdoba, most of them working in the oil industry, in institutions such as the National Direction of Geology and Mining or teaching in the universities.

The need to have an institution that brought together Argentine geologists dates back to 1929 when Erwin Kittl (1890–1983) founded the Argentine Society of Mining and Geology and created the Mining Magazine. Kittl had arrived in the country in 1920 to work for mining companies and except for two years in which he directed the School of Mining Engineers of Oruro in Bolivia, he was working in our country. His

V. A. Ramos, *History of the Geological Sciences in Argentina*, The Latin American Studies Book Series, https://doi.org/10.1007/978-3-031-83139-3_16

Mining Magazine, the first Argentine geological publication, appeared uninterruptedly from 1929 until 1980, although the society for various reasons did not prosper and had a small number of partners.

16.1 Dr. Carlos Storni and the First Years of the Society

Years later, a group of young people were gathered and encouraged by Don Carlos D. Storni (1884–1977) to found a society that would bring together geologists and whose objective was to promote better knowledge of our discipline in society (Fig. 16.1). Carlos Storni was already a leading figure in the natural sciences. This man from Tucuman with a Swiss father and Creole mother had two stages in his life. In the first, he graduated as a teacher in Tucumán and later studied agriculture in Casilda, Córdoba. His excellent performance as a student led him to be awarded a scholarship to pursue higher studies at the *Istituto Sperimentale per la Viticoltura ed Enologia of Conegliano*, one of Italy's pioneering schools in Veneto, where he graduated with honors in 1912. He returned to the country and he carried out various advisory and management tasks, founded various schools of Agriculture, received numerous awards, and retired in 1940 as Director General of the Ministry of Agriculture (Bracaccini 1977).

Its second stage begins with a fortuitous event in 1918. Traveling with troops through the south of Neuquén, he found a motor vehicle stuck in a rocky field. A German gentleman, his wife, and his little son were traveling in it. Once the necessary aid has been provided, their paths separate. However, being Director of the School

Fig. 16.1 Carlos Storni as president of the Argentine Geological Association. *Source* A.G.A. archives

of Agriculture in Córdoba in 1926, he was visited by the Teutonic knight, who was none other than Anselmo Windhausen. This meeting motivated him to take doctoral courses in Natural Sciences with "*the fundamental purpose of always finding myself qualified to serve the country as best as possible*" (Bracaccini 1977) at the National University of Córdoba, where he graduated at the beginning of 1932. In those years he held management positions in the Argentine Society of Natural Sciences and in the Argentine Society of Geographic Studies GAEA.

Already retired in Buenos Aires, he returns to teaching, his first vocation, as a Professor of Mineralogy at the *Colegio Alemán Incorporado Burmeister* in Belgrano. His enthusiasm for teaching awakened several vocations for the study of Geology (see Böckman de Cianciarulo 2019).

For a long time, there has been in the geological environment the purpose of creating an association that would bring together all the geologists in the country. Storni, with his tireless management spirit, led a group of young geologists, after several preliminary meetings, to hold an organizational meeting that took place on June 30, 1945, in the Aula Aguirre of the Manzana de las Luces.

There was the site of the Faculty of Exact, Physical and Natural Sciences, (Fig. 16.2) and that classroom was the main venue for geology classes on the second floor of Perú 222, named in honor of the first Argentine professor of this discipline in the country. In addition to Storni, Victorio Angelelli (1908–1991), Félix González Bonorino (1918–1998), Christian Petersen (1912–1962), Osvaldo Bracaccini (1913–1979), Armando Leanza (1919–1975) and Horacio J. Harrington (1910–1973), group of young people who had an average age of 30 years. The latter was appointed President of the Organizing Commission. One of the main purposes was to publish a Journal of the Argentine Geological Society, to constitute "*an adequate dissemination body to publish the results of local geological and paleontological studies*" (Stipanicic and Camacho 1995). The Aguirre classroom was established as the Society's headquarters, with the corresponding permission of the dean, Eng. Pedro Mendiondo authorized placing a plaque on its door, identifying the society.

At this meeting, it was decided to name the first two honorary members, a distinction that went to Juan Keidel and Pablo Groeber, two old professors with legendary activity for their geological studies in our country. Keidel was already retired, but with his characteristic generosity he donated 300 pesos to collaborate with the society's fund.

In turn, they decided to appoint Dr. Arnold Heim as the first Corresponding Member (Fig. 16.3) at that meeting, motivated by his departure from the country and the important contributions he made during those years (Ramos 2021a). He was invited to give a lecture in the Aula Magna of the Faculty of Exact, Physical, and Natural Sciences of the University of Buenos Aires, and a dinner was then served as a tribute to his appointment. Heim expressed years later that "*I have excellent memories of the collaboration and friendship with my Argentine colleagues, especially Professors Harrington and Leanza and Miss Dr. Verena Kull, from the Direction of Mines. The newly founded Argentine Geological Society gave me proof of its recognition, naming me as its first corresponding member, during a festive farewell*" (Heim 1960).

Fig. 16.2 Manzana de las Luces where the Faculty of Exact, Physical and Natural Sciences of the University of Buenos Aires operated. The door on the right was the entrance to Perú 222 through which you accessed the Aula Magna of the Faculty where the assemblies were held and the Aula Aguirre of the Department of Geology where the Argentine Geological Society was founded and was its headquarters from 1946 to 1948. *Source* A.G.N. archives

Fig. 16.3 Arnold Heim (1882–1965) in Zurich upon his return from Argentina. *Source* ETH archives

In its first assembly held the following year, the Society met in the YPF Assembly Hall, provided by this company to also hold the Communications Meetings. In this assembly, the Statutes of the Society prepared by doctors Armando Leanza and Carlos Storni were analyzed and the new authorities were elected, with Carlos Storni as president, supported by Osvaldo Bracaccini.

The first issue of the Magazine appeared in January 1946 and expressed in its pages the enthusiasm with which the brand-new Society had been received, which the year after its foundation had more than 200 members and was beginning to receive numerous subscription requests.

The first Communication Conferences were held in June 1946, directed by Horacio Harrington and with scientific presentations by the geologists Osvaldo Bracaccini, Christian Petersen, Raul Dessanti, Carmelo De Ferrariis, Luis Lambert and Armando Leanza. These communication meetings continued until the First Argentine Geological Congress in 1960.

The Society founded by these young people had to change its name in April 1948 to the Argentine Geological Association, due to the prevailing regulations to obtain legal status, a name that persists to this day. In the first years, it operated at the headquarters at Calle Perú 222 until the first months of 1948. The Faculty of Exact, Physical and Natural Sciences of the University of Buenos Aires created an Institute of Geology at the end of 1944, whose first director was Horacio Harrington. He moved the Paleontology part and the collections to a new building at Ituzaingó 1060. In 1948, when Armando Leanza joined as a professor, he was appointed director of the Institute, a position he held until 1952. Leanza in 1948, who together with Storni, had been in charge of the magazine publication, enabled the Association to have a room for meetings and a small warehouse to store the files and published magazines. From that moment on, the headquarters officially moved to Ituzaingó 1060, where the Geology Institute was working. The Association remained in this location until 1954 when the Faculty of Engineering finally moved from Perú 222 to its new building. The space liberated on that occasion allowed the facilities and teachers of the Institute of Geology to be moved back to the Manzana de Las Luces, leaving the Association without a place to operate (see details in Ramos 2018).

Dr. Storni always said that in 1954 he had to move with all the belongings and the magazine archives to the Casa Coni printing press at Perú 684, where they gave him a section of the warehouse, but he did not have the necessary facilities for the Association. Although he recognized the generosity of the company that printed the magazine, the situation became very difficult. Harrington had gone to the United States, Leanza stopped being a professor at the Faculty and had to wait for González Bonorino, in July 1957, to return from the United States and, as vice-dean of the Faculty, once again offer him a small office and warehouse on the third floor of the Department of Geology in the Manzana de Las Luces (Fig. 16.4) where the association operated until 1966.

The Association operated in this location and that was where I met Dr. Storni, who in the early sixties was the factotum of the institution. He registered the members, collected the fees, received the works for the magazine, corrected them, took them to the printer, and brought the publications back on successive trips for distribution.

Fig. 16.4 The interior patio of the Manzana de las Luces where the small office occupied by Doctor Storni for many years operated on the third floor, where the official headquarters of the Argentine Geological Association was located from 1956 until its move to the new House of the Geologist. *Source* Biblioteca Digital F.C.E.N

The little room was not comfortable, a very hot tin roof, where he spent hours writing letters to the authors, to the benefactor members, arranging the finances of the Association, etc. Despite his long 76 years, he was full of new projects, and he told us about his proposal to create an Academy of Geology, how he had chosen the names for his forty seats among the pioneers who developed Argentine Geology and he read us the draft decree for its creation. In 1960 he received a retroactive retirement benefit on July 9 Bonds, which he donated to the Association. With these funds, the Storni Prize was created and in 1970 a gold medal and diploma began to be awarded for the best annual work published in the magazine for young people under 35 years of age, an award that continues to the present.

Leanza took charge of editing the works for the magazine from the founding of the Society until 1956 with the support of Storni. His departure as a professor for political reasons separated him from these duties.

16.2 The Association's Own Headquarters

Since 1967, the Association had been struggling to acquire its own headquarters in conjunction with the Superior Professional Council of Geology, the Argentine Paleontological Association, and the Argentine Center of Geologists. A pro-own headquarters commission was created, chaired by Pedro N. Stipanicic. This commission, with a strong contribution from its members plus the contribution of numerous associates, and donations from benevolent companies, acquired the property at Maipú 645, first floor, where it currently carries out its activities. This joint effort between the four institutions was difficult, but it came to a successful conclusion in 1970 (Fig. 16.5), when, on the occasion of the 25th anniversary of the Association, the headquarters were handed over in a festive event shared by authorities and different institutions from various parts of the country.

In its new headquarters, the activities of the Association were expanded, providing facilities for the meetings of the different commissions, thus generating greater participation of members in the different disciplines. Among them, the Argentine Stratigraphy Committee and the Tectonics Commission, with their distinct subcommittees stand out, among several others.

In addition, the headquarters itself has been a meeting place to celebrate different events such as Geologist's Day on June 9 of each year (Figs. 16.6 and 16.7). This date was born from an initiative of the Argentine Center of Geologists, a union entity that brought together professionals in this discipline and was formally made official in an assembly held at the Argentine Museum of Natural Sciences on that date in 1948.

The House of the Geologist is currently maintained by the Argentine Geological Association, the Argentine Paleontological Association, and the Professional Council of Geology, since the Argentine Center of Geologists ceases its functions.

Fig. 16.5 The president of the pro-sede commission Pedro N. Stipanicic hands over the House of the Geologist to the four institutions in the pleased presence of Carlos Storni and representatives of other organizations. *Source* Revista de la Asociación Geológica Argentina 23(4): 504

Fig. 16.6 Pedro Stipanicic, Horacio Harrington and Roberto V. Tezón, and the latter talking with Carlos Storni, at one of the evenings at the House of the Geologist in 1972. *Source* A.G.A. archives

Fig. 16.7 The Secretary of Mining of the Nation Fernando Puca Prota, Dr. Margarita Reynoso, Lic. Nello Duranti, and Dr. Mario Sosic at the House of the Geologist. Doctors Emilio González Díaz, Juan Carlos Gamerro, Sergio Archangelsky, and Alwine M. Bertels at one of the evenings in 1972, shortly after its official inauguration. *Source* A.G.A. archives

16.3 Presidents of the Argentine Geological Society (Association)

Beginning with the presidency of the Organizing Commission by Horacio Harrington, a host of illustrious geologists followed one another who were in charge of the Association's destinies. In the first 50 years, we have had 18 presidents (Figs. 16.8, 16.9, 16.10, 16.11, 16.12, 16.13 and 16.14).

In these almost eighty years the Association has had only four women presidents, Norma E. Pezzutti, Maisa Tinik, Cecilia M. Spagnuolo and Silvia I. Carrasquero.

Fig. 16.8 Carlos Storni (1946–1949; 1963–1971); Horacio Harrington (1950–1952) and Félix González Bonorino (1952–1954). *Source* A.G.A. archives

Fig. 16.9 Amando F. Leanza (1954–1956); Raúl N. Dessanti (1956–1957) and Christian Petersen (2957–1958). *Source* A.G.A. archives

Fig. 16.10 Tomás Suero (1958–1960); Horacio H. Camacho (1961–1963; 1974–1975) and Pedro N. Stipanicic (1971–1974). *Source* A.G.A. archives

Fig. 16.11 Roberto L. Caminos (1975–1977). Norberto Malumián (1977–1979) and Victor A. Ramos (1979–1981). *Source* A.G.A. archives

Fig. 16.12 Gualter Chebli (1981–1983); Francisco E. Nullo (1983–1985) and Carlos A. Cingolani (1985–1987). *Source* A.G.A. archives

Fig. 16.13 Héctor A. Leanza (1989–1991); Jorge O. Codignotto (1989–1991), and Alberto C. Riccardi (1991–1999). *Source* A.G.A. archives

Fig. 16.14 Norma Pezzutti (1999–2003), Maisa Tunik (2013–2015), Cecilia M. Spagnuolo (2015–2017), and Silvia I. Carrasquero (2021–2023). *Source* A.G.A. archives

16.4 The Journal of the Argentine Geological Association

Since its first issue edited in 1946, the journal has had an uninterrupted edition until today, publishing eighty consecutive volumes (Fig. 16.15). It is the dissemination organ of the Association that has presented and continues to transmit the advances in knowledge of Argentine Geology in its different disciplines. It has become one of the first Earth science journals on a continental level and in its pages, different special issues analyze priority topics and show specific proposals and contributions in areas of interest to the geological community.

Starting in 1975, the position of Director of the Journal was formalized independently of the authorities, a position that fell for the first time to Roberto L. Caminos. Previously and starting in 1967, it was explicitly stated that the President of the Argentine Geological Association was formally the Director of the Journal, as required by government provisions to allow the Registry of the Intellectual Property of our publication (1975, *Revista de la Asociación Geológica Argentina* 12(4): 280).

The first ten years of the Journal were in charge of a subcommittee led by Armando Leanza who served as ad hoc editor until 1956. Storni was a member of that subcommittee, and when Leanza left, different geologists passed through it, but the responsibility for its publication remained in his hands. Storni worked from correcting the text to establishing the size of the bronze plates with which the figures were printed. Clichés were one of the biggest costs of printing, so figures tended to be small unless the authors covered the cost. He was in charge of the journal until 1971 when he was replaced by Stipanicic during his presidency until 1974.

With the creation of the position of Director of the Journal, other fundamental changes were introduced in 1975, such as submitting works to arbitration by two reviewers. It was up to Caminos to initiate these processes, and I am reminded of the angry complaints of some authors about the criticism they received. Some were offended because strangers to the work requested modifications or justification for what was written.

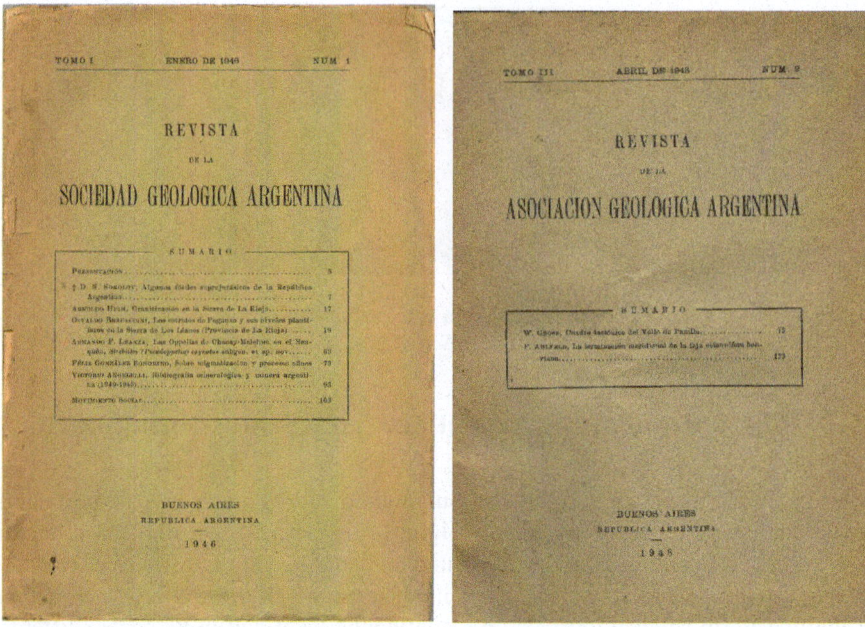

Fig. 16.15 The first issue of the Journal of the Argentine Geological Society of January 1946 and the first issue of the Association of April 1948. *Source* A.G.A. Revista de la Asociación Geológica Argentina archives

The direction of Roberto Caminos (1975–1979) of the journal was followed by Francisco E. Nullo (1980–1983, 1987–1991); Victor A. Ramos (1983–1987, 2004–2009); Susana E. Damborenea (1991–1997); Roberto A. Scasso (1998–1999) and Milka K. de Brodtkorb (1999–2001). In 2010, the figure of Editor was created, with Pablo Pazos (2010–2015) occupying that position; Augusto Rapalini (2016–2019), and Diego A. Kietzmann (2019–2023).

It was the function of the director, and it is nowadays the editor of the journal, to also ensure the editing and printing of other publications such as Series A, Monographs; Series B, Didactic and Complementary; Series C, Reprints; Series D Special Publications and Series E, Field Guides.

Together they make up about 30 publications, several of which have been fundamental contributions to Argentine Geology (Figs. 16.16 and 16.17).

16.5 The Geological Congresses

Another fundamental contribution of the Association to the geological community has been and is the organization of the Argentine Geological Congresses. The first of them corresponds to the First Argentine Geological Conference that was held in

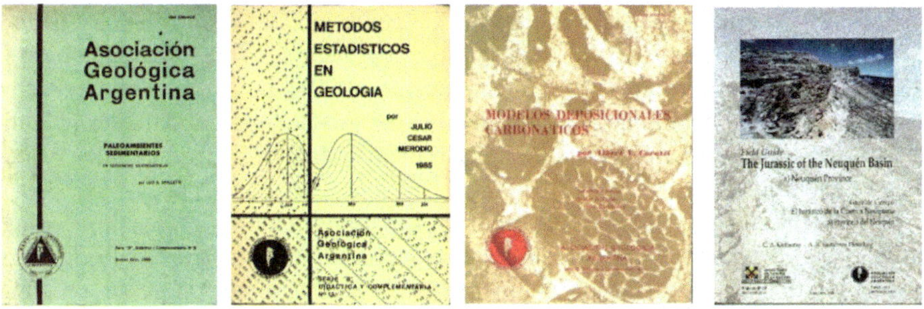

Fig. 16.16 Some of the publications such as *Sedimentary Environments* by Luis Spalletti, *Statistical Methods in Geology* by Julio Merodio, *Carbonate Depositional Models* by Albert Carozzi, and *Field Guides* by Carlos Gulisano, among several classic works from the different series of the Association. *Source* A.G.A. Asociación Geológica Argentina archives

Fig. 16.17 Some more recent special publications from the Argentine Geological Association. *Source* A.G.A. Asociación Geológica Argentina archives

San Juan in 1960, which had Emiliano Aparicio as president and Mr. Pablo Groeber as honorary president. They were followed by the Second Conference in Salta in 1963 under the presidency of Guillermo Furque, the Third Conference in Comodoro Rivadavia chaired by Eduardo Methol, and the Fourth Conference in 1966 organized by Pedro Criado Roque in Mendoza with the honorary presidency of Carlos Storni.

From those days they began to be called Argentine Geological Congresses, the fifth being organized in Villa Carlos Paz in 1972 by Pedro Stipanicic; the sixth in 1975 in Bahía Blanca by Horacio Camacho; the seventh in 1978 in Neuquén by Edgardo Rolleri; the eighth in 1981 in San Luis by Marcelo Yrigoyen; the ninth in 1984 in Bariloche by Victor A. Ramos, the tenth in 1987 in Tucumán by Florencio G. Aceñolaza; the eleventh in 1990 in San Juan by Bruno Baldis; the twelfth in 1993 in Mendoza by Gualter Chebli; the thirteenth in 1996 in Buenos Aires by Enrique Mainardi; the fourteenth in 1999 in Salta by Gustavo González Bonorino; the fifteenth in 2002 in Calafate by Miguel Haller; the sixteenth in 2005 in La Plata by Isidoro Schalamuk; the seventeenth in 2008 in Jujuy by Beatriz Coira; the eighteenth in 2011 in Neuquén by Jorge Vallés; the nineteenth in 2014 in Córdoba by Guillermo Albanesi; the twentieth in 2017 in Tucumán by Sergio Georgieff and the twenty-first in 2022 (delayed by the Covid pandemic) in Puerto Madryn by José Paredes, which was held virtually.

These conferences published different presentations in multi-volume proceedings (Fig. 16.18), which together contain important contributions to Argentine Geology. Starting with the Bahía Blanca Congress in 1975, each of these congresses prepared a report that presents a synthesis of the geology and natural resources of the province where the meeting took place. These books written by recognized specialists in different topics have become important reference works updated before congress. In 1996, when the congress was held in the city of Buenos Aires, the report consisted of the geology and natural resources of the Argentine continental shelf, in which the academy and the different companies that explored the offshore geology of Argentina participated.

Another important contribution that the Association has had is the implementation of courses on different topics and disciplines. The first of them given by Luis Spalletti on "*Sedimentary paleoenvironments in silicicoclastic sequences*" in 1980, which later became a classic text published in the special series, marked the beginning of other successful courses such as that of Albert Carozzi on "*Carbonate depositional model*" from 1983 or that of Mario Mazzoni on "*Pyroclastic processes and deposits*" from 1986. In those years the courses organized at the House of the Geologist on "*Fundamentals of Plate Tectonics*" that we started with Eng. Daniel Valencio in 1982, had a great attendance and impact and was repeated in Mendoza and other places in the country.

Fig. 16.18 Examples of the latest reports from the Argentine geological congresses of Jujuy, Córdoba, and Chubut. *Source* A.G.A. Asociación Geológica Argentina archives

As the years went by, these updated courses given on very different topics selflessly by the teachers were at the time a strong financial aid to continue funding the publication of the journal.

16.6 Distinctions and Awards

Since the first meetings of the Argentine Geological Society, the recognition of pioneering geologists who have contributed significantly to the progress of Argentine Geology has been kept in mind. On the occasion of the 50th anniversary of its foundation, the Organizing Committee of the events decided to pay a fair tribute to the "*Great Masters of Argentine Geology*" whose portraits adorn the House of the Geologist and the reasons for their selection were presented by Yrigoyen (1995). Those teachers that we have referred to individually in the previous chapters were unanimously selected by the commission of notables that had the responsibility of choosing them. These twenty teachers of Argentine geology were listed by the date of the beginning of their work are:

1871. Alfredo Stelzner, Academia Nacional de Ciencias

1874. Luis Brackebusch, Academia Nacional de Ciencias

1875. Florentino Ameghino, Academia Nacional de Ciencias

1885. Guillermo Bodenbender, Academia Nacional de Ciencias

1885. Federico Kurtz, Academia Nacional de Ciencias

1897. Carl Burckhardt, Museo de La Plata

1905. Walther Schiller, Museo de La Plata

1906. Juan Keidel, División de Minas y Geología

1909. Ricardo Stappenbeck, División de Minas y Geología

1909. Anselmo Windhausen, División de Minas y Geología

1910. Enrique Gerth, División de Minas y Geología

1910. Franco Pastore, División de Minas y Geología

1911. Pablo Groeber, División de Minas y Geología

1913. Roberto Beder, División de Minas y Geología

1913. Joaquín Frenguelli, Universidad Nacional de La Plata

1917. Ricardo Wichmann, División de Minas y Geología

1920. Erwin Kittl, Universidad de Buenos Aires

1925. Egidio Feruglio, Yacimiento Petrolíferos Fiscales

1933. Horacio Harrington, División de Minas y Geología

1935. Osvaldo Bracaccini, Yacimiento Petrolíferos Fiscales.

It is interesting to note that among these professors there are only three born in the country, Franco Pastore, Horacio Harrington and Osvaldo Bracaccini. Pastore, the first geologist received in the country, dedicated himself to lecturing at the University of Buenos Aires for almost fifty years, the first twenty of which were done on an honorary basis, training numerous generations of geologists in Mineralogy and Petrography, in addition to his daily work in the Direction of Mines. Harrington was one of the most brilliant geologists of the first half of the twentieth century with a strong international projection. Bracaccini, who in addition to having made important contributions at YPF from assistant geologist to general manager, later had important contributions to applied geology (Ramos 1979). Although Florentino Ameghino was born in Italy, he came to Argentina when he was very young, where he studied and became one of the most prominent naturalists of the nineteenth century. We should also consider him as Argentine, leaving aside the discussions and controversies that his nationality aroused at the time.

At the constitutive meeting of the Society, Keidel and Groeber had been named Honorary Members, followed by Pastore and Frenguelli. Many of the great masters died before the founding of the Society. This distinction was awarded to outstanding Argentine and foreign geologists who, according to the members of the Board of Directors, have had outstanding performance in our geology. Over the years, numerous geologists have been appointed, currently reaching about 75 members, of which almost 60 have passed away. The list of honorary members and their date of appointment can be seen in Riccardi (1999).

Furthermore, in compliance with the duly established statutes, outstanding personalities in geology who have worked in the country or have made significant contributions have been distinguished as Corresponding Members of the Argentine Geological Association. That list begins with Arnold Heim, but is completed with numerous South American, European and North American geologists who have provided important services to our geology (see Ramos 2021b).

In addition to appointments of Honorary and Corresponding Members, the Geological Association has created a series of awards to recognize outstanding colleagues. We have already mentioned the Storni Award, a stimulus award for young people under 35 years of age that has been awarded since 1970. In addition, awards for outstanding career are awarded to geologists over 55 years of age, which bear the name "*Franco Pastore Award to scientific research*". They began in 1984 and are offered every three years on the occasion of the Argentine Geological Congresses. Its first winners were Alberto Mingramm, Félix González Bonorino, Sergio Archangelsky, Milka K. de Brodtkorb, among others, reaching up to the present day. In addition to the previous one, the "*Juan José Nágera Prize: for the dissemination of geology*" is awarded, which has been received since 1984 by Horacio Camacho, Edgardo Rolleri, Telasco García Castellanos, Pedro Stipanicic, among several others.

Since 1995, the *Argentine Geological Association Award* has been approved, which is given annually in different disciplines that alternate and whose jury is made up of active *Honorary Members.* Since 1997, the *Institutional Merit Award* has been granted to those who have contributed in a personal and outstanding way to the activities of the Argentine Geological Association.

The regulations and recipients of these awards are described in the *Societary Vademecum* prepared by Alberto Riccardi in 1999.

16.7 Societies Linked to the Argentine Geological Association

Over the years the Association has been linked to different international societies. Firstly, the members of the Board of Directors have been recognized since 1979 as the "*IUGS National Committee*" of Argentina, given that the Association is an Adhering Member of the International Union of Geological Sciences. The presidency of this institution fell between 2008 and 2012 on Alberto Riccardi, by election of the IUGS—Council carried out by all the countries participating in the International Congress of Oslo in 2008.

In turn, since 1997 the Association was incorporated into the American Association of Petroleum Geologists as an affiliated entity. In 2000 the Association was designated an Allied Society by the Council of the Geological Society of America and as such is considered a strategic partner. Similar relationships are maintained with the Cuban Geology, Chilean Geological, Uruguayan Geological, Brazilian Geological societies and with the Official College of Geology of Spain.

In addition, it maintains close ties with the Argentine Association of Sedimentology, the Argentine Mineralogical Association, the Argentine Paleontological Association and other institutions in different disciplines.

16.8 Final Considerations

The intention of writing this chapter is to highlight the importance that the creation of the Argentine Geological Society in 1945 has had on the activities and progress of both the academic and industrial community. It provided a means of communication for geological news, new studies and findings that were disseminated through the pages of its journal. The journal taught through professors such as Groeber, González Bonorino, Teruggi, Leanza among many others, who presented excellent reviews of the studies carried out. Pablo Groeber's observations along the 70th meridian, from northern Argentina to Patagonia, were a milestone in understanding the stratigraphy, mainly volcanic, and the structure of the Andes with its longitudinal variations. González Bonorino's descriptions and analysis of the Sierras Pampeanas, not only of its crystalline basement but also of the Tertiary structure and tectonic behavior in depth, allowed us to share these novel concepts with students and colleagues, leading to a better knowledge of our geology. Dozens of works published in the journal, such as those by Cesar Vilela on the Sierras Subandinas, or those by Carmelo Ferrariis introducing the Huincul High and its structure, were for many years the only way to study Argentine Geology.

Since 1960, the Association sponsored and promoted the Argentine Geological Congresses, which have been and are a great opportunity to exchange experiences, to listen and participate in tough debates and vivid discussions, with the sole objective of improving the understanding of our knowledge, with a strong didactic component, which enriched several generations.

In current times with instant access to knowledge and carrying more information than the Encyclopedia Britannica on a small cell phone, it is difficult to understand the rudimentary way in which the characteristics of a region or a geological process were transmitted and learned. First, the journal and later the voluminous volumes of the congresses, made it possible to disseminate the stratigraphy, the petrological processes, the structure, and tectonic evolution of our country. Taking into account the parochial characteristics of that knowledge, they were the appropriate means for this transfer and federalizing that knowledge to all regions of the country.

In addition to improving geological discernment through its courses, the Association and the House of the Geologist were for many years a place of contact and exchange that allowed numerous geologists to bring together and communicate. In those years when months were spent in fieldwork, these events provided the opportunity to update oneself, share experiences, and establish or recover old friendships.

The entire community should be grateful to the Argentine Geological Association for the opportunities offered, which should be extended to its authorities and management bodies that have provided and still provide their work generously and selflessly to promote the improvement of the community.

I was able to complement this brief history of the Argentine Geological Association with the synthesis carried out by Stipanicic and Camacho (1995), the data presented by Riccardi (1999), in addition to the valuable details uploaded by Norma Pezzutti in 2016 to the Association's page. To all of them our greatest recognition.

References

Böckman de Cianciarulo S (2019) Una vida dedicada a la economía minera y a la geología ambiental. Ciencia e Investigación, Reseñas 7(1):6–22

Bracaccini IO (1977) Sr. Dr. Carlos David Storni (1884–1977), Necrología. Revista de la Asociación Geológica Argentina 32(2):156–160

Heim A (1960) América del Sur. La vida y la naturaleza en Chile, Argentina y Bolivia. Primera Edición. Editorial Labor S.A, 277 p, Buenos Aires

Ramos VA (1979) Dr. Inocencio Osvaldo Bracaccini (1913–1979). Revista de la Asociación Geológica Argentina 34(3):249–252

Ramos VA (2018) El desarrollo de la Geología en la Universidad de Buenos Aires. Revista Facultad de Ciencias Exactas, Físicas y Naturales, Nueva Serie 5(Suplem 1):23–32
Ramos VA (2021a) Arnold Heim (1882–1965). In: Ramos VA (ed) Los Miembros Correspondientes de la Asociación Geológica Argentina, Serie D, Publicación Especial 15:12–17
Ramos VA (2021b) Editor. Los Miembros Correspondientes de la Asociación Geológica Argentina. Asociación Geológica Argentina, Serie D: Publicación Especial 15, 152 p
Riccardi AC (1999) Vademecum Societario. Editor. Asociación Geológica Argentina, Serie B: Didáctica y Complementaria 24, 64 p
Stipanicic PN, Camacho HH (1995) Historia de la Asociación Geológica Argentina. Revista de la Asociación Geológica Argentina 50(1–4):19–20
Yrigoyen MR (1995) Los grandes maestros de la Geología Argentina. Revista de la Asociación Geológica Argentina 50(1–4):13–18

Chapter 17
The Geology of Radioactive Minerals

Abstract Although radioactive minerals have been known since the late nineteenth century, the first economic manifestations were identified in 1937. The creation of the National Atomic Energy Commission in 1952 during the presidency of General Perón allowed the beginning of a systematic search for radioactive mineral deposits and their exploitation. Once again, the initial operations were led by Eng. Victorio Angelelli, who was in charge of the exploration program. This included aero-radiometric flights and important field surveys, mineralogical studies, and drilling in the more promising areas. These studies allowed the discovery of important sedimentary deposits in the Salta Group basin in the north, in upper Cretaceous deposits in the Neuquén basin, and in the San Jorge Gulf basin in Patagonia. Although there were exploitations in the basement of the Sierras Pampeanas such as the Schlagintweit mine, the sedimentary deposits were the most important. Among them, the Don Otto, Huemules and Sierra Pintada mines stand out, which allowed the country to meet its needs. In more recent years, the sedimentary deposits found in the Upper Cretaceous Chubut Group in Patagonia, where the most important reserves are located, have gained importance.

Keywords Atomic energy · State institution · Exploration · Discoveries · Large reserves

The Direction of Mines and Geology, when it still depended on the Ministry of Agriculture under the direction of engineer Tomás Ezcurra, made the first discovery in the country of uranium mineralization in 1937, as part of the geological-mining studies undertaken by the division.

These discoveries in Calamuchita had an important significance, which were followed by those uranium mineralizations in La Rioja and San Luis. These findings open the way to the most outstanding ores corresponding to the Doctor Baulíes mines in the Sierra Pintada district in Mendoza, Cerro Solo in the Pichiñán district in Chubut, and Don Otto, in Salta.

The successes obtained years later in the exploration of radioactive minerals correspond to the geologists of the National Atomic Energy Commission (Fig. 17.1). This is a scientific and technological organization that was created in 1952 during the

V. A. Ramos, *History of the Geological Sciences in Argentina*, The Latin American Studies Book Series, https://doi.org/10.1007/978-3-031-83139-3_17

Fig. 17.1 Headquarters of the National Atomic Energy Commission in the city of Buenos Aires

presidency of Juan Domingo Perón, as part of the country's industrialization plan dedicated to the exploration and development of nuclear energy and its applications. At the request of the Presidency of the Nation, engineer Victorio Angelelli was appointed, first as advisor to this institution, then as head of the Mining Service, and later as director of the Department of Geology and Mining. Later, in 1966, he was appointed member of the Board of Directors, retiring in 1969. In those years he had a leading role in directing the first studies of our uranium mineral deposits.

17.1 Historical Background

The first uranium minerals were recognized in the province of San Luis in 1874 by Germán Avé-Lallemant (Fig. 17.2), who was a pioneer in the description of the geology of the Cuyo region. In his work, he states that he has found uranium ocher, zippeite, on pitchblende, in veins of gold-bearing quartz in Las Peñas, Saladillo, in the Quinto River, and also several times chalcolite (torbernite), cupric uranium phosphate, in the sericitic schists of the copper mines of Rincón, in the San Luis range (Avé-Lallemant 1874, 1892).

The uranium minerals in the Sierra de Comechingones constituted the first known deposit of radioactive minerals in the country. They were studied by Rigal (1938), who established the presence of gummite, autunite, and torbenite, as secondary products of pitchblende or uraninite, and uranothorite. These minerals appeared associated with Ordovician pegmatites with tantalite-columbite in the metamorphic igneous

Fig. 17.2 Germán Avé-Lallemant (1835–1910) was a pioneer of the geology of San Luis and a nodule rich in uranium minerals described by Remigio Rigal. *Source* Ferrari (1993) and Rigal (1938)

basement (Rinaldi 1968). Years later, these deposits gave rise to the first uranium exploitation in the country, extracting the first tons of this mineral in the Sierra de Comechingones until 1990 (Candiani et al. 2016). For more details about this district, refer to Galiski (1999).

17.2 The Systematic Exploration

The systematic prospecting, exploration, and exploitation of uranium in Argentina began in 1952 with the consolidation of the Atomic Energy Commission under the initial supervision of engineer Angelelli. Ground and aerial geophysical and geochemical methods were used which resulted in the discovery of new deposits. In that decade, the geology of various uranium districts in the country was studied. The results were presented by the main geologists (Fig. 17.3) at the First Argentine Geological Congress held in San Juan in 1960, with important advances in geological knowledge and the explorations carried out. In a volume dedicated to Mineral Deposits, they present 18 works on the geological characteristics of the different ores and recognized prospects that range from northern Argentina to Patagonia, showing a promising future for the exploitation of these minerals. Among these works stand out, the syntheses carried out of the vein deposits by Angelelli (1962) and those in sedimentary rocks of Stipanicic et al. (1960).

Fig. 17.3 Oscar L. Baulíes (1923–1968), Héctor Achen (1920–1981), and Armando Ortega Furlotti (1923–1993), were some of the first geologists to begin the exploration and development of uranium deposits in the region of Cuyo in the 1950s. *Source* Stipanicic (1969, 1993) and Rodrigo and Ortega Furlotti (1982)

17.3 Deposits in Sedimentary Rocks

Uraniferous manifestations are known in different environments, the main ones being those of the Salta Group basin in northern Argentina, the Precordillera, the Cretaceous basin of the Neuquén-Mendoza foreland, the Chubut Group in Patagonia and the Paleogene deposits of the Sierras Pampeanas.

17.3.1 The Salta Group Basin in Northern Argentina

At the beginning of 1959, it was decided to stop exploring the vein deposits to concentrate on the sedimentary deposits, given that various manifestations had been found in the Salta Group that motivated the implementation of a regional prospecting plan associated with the development of the Yacoraite Formation and the basal levels of the Santa Bárbara Formation of Maastrichtian-Paleocene age. The aero-radiometric survey carried out with small planes equipped with high-sensitivity instruments developed by the CNEA, had highly positive results, recognizing various prospects in the Salta Group, which resulted in the Don Otto, Pedro Nicolás, Martín M. de Güemes, Los Berthos, and Emmy mines, and numerous demonstrations (Romano 1999).

The uranium mineralization is contained mainly in the coastal facies transitioning to fluvial environments of the Yacoraite Formation (Fig. 17.4) and is housed in the limbs of broad synclines developed in the sedimentary deposits of the Salta Group.

A uranium concentration plant was inaugurated in 1970, which produced about 202 tons from the Don Otto Mine between 1963 and 1981.

Fig. 17.4 Entrance and camp of the Don Otto mine in outcrops of the Salta Group. *Source* Bianchi (2022)

17.3.2 The Neuquén Group Basin in the South of Mendoza

Again in 1952, the presence of yellow uranium minerals associated with copper mineralization was detected in the deposits of the Diamante Formation, correlatable with the Neuquén Group in the southern region of Mendoza. Of the various manifestations identified, Cerro Huemul was the most prominent (Fig. 17.5), making the uranium district of Malargüe the most important in the country at that time in terms of its reserves and potential (Rojas 1999a).

The Huemul ore was studied by Brodtkorb (1966) and its isotopic characteristics by Linares (1966) (Fig. 17.6). It is found in various levels of conglomeratic sandstones and medium to coarse-grained sandstones associated with bituminous levels produced by the migration of oil from the underlying units, which generated the reducing conditions for the precipitation of uranium associated with copper and vanadium (Gorustovich et al. 1992; Crosta and Belén Palacio 2015).

Fig. 17.5 Distribution of the main occurrences associated with the Neuquén Group in the uranium district of Malargüe in the south of Mendoza, where the Huemul Mine stands out. The entrance of the Huemul mine developed in these sandstones. *Source* Rojas (1999a)

Fig. 17.6 Pedro N. Stipanicic (1921–2008), Milka Kronegold de Brodtkorb (1932–2019), and Enrique Linares (1928–2014), three leaders of the Raw Materials Management of the National Atomic Energy Commission for their contributions to the exploration and development, metallogeny and geochronology. *Source* A.G.A. archives

The Huemul mine was run by Héctor Achen (1920–1981) and Armando Ortega Furlotti (1923–1993) in various periods, and its exploitation extended from 1952 to 1976, producing 208 tons of uranium.

Other manifestations of uranium were found on the Dorso de los Chihuidos in Neuquén associated with the continental deposits of the Rayoso Formation and the Neuquén Group in manifestations of the Cu-V-U association (Ramos 1975; Rojas 1999b). Mineralization in sandstones is apparently controlled by the presence of remains of bituminous material, a product of the migration of oil in sandstones of these units.

17.3.3 The San Jorge Gulf Basin

The discovery of radioactive minerals in this sector of Patagonia in 1952 was due to María Teresa Carri de Riggi (1906–1989), head of the National Museum of Geology and Mining (Fig. 17.7). Dr. Carri de Riggi was the second woman to graduate from the University of Buenos Aires and registered number 5 in the Professional Council of Geology, just after her husband who was the number 3 in the register. She examined some samples of asphalt sandstone collected by Anselmo Windhausen from the upper valley of the Chico River and found that they were radioactive (Bianchi 2022; Ramos 2023). Dr. Riggi was hired by the Atomic Energy Commission where she served as Head of the Department of Geology and Mining, and for several years she collaborated in the prospecting and determination of radioactive minerals. These data provided by Dr. Riggi, currently preserved in a secret file (File No. 277/52 "Secret" in C.N.E.A.) were complemented by verifying that the oil drilling carried out in this sector of the basin had in their gamma profiles radioactive anomalies. Stipanicic verified that the anomalies were in the Río Chico Formation and the Cerro Barcino

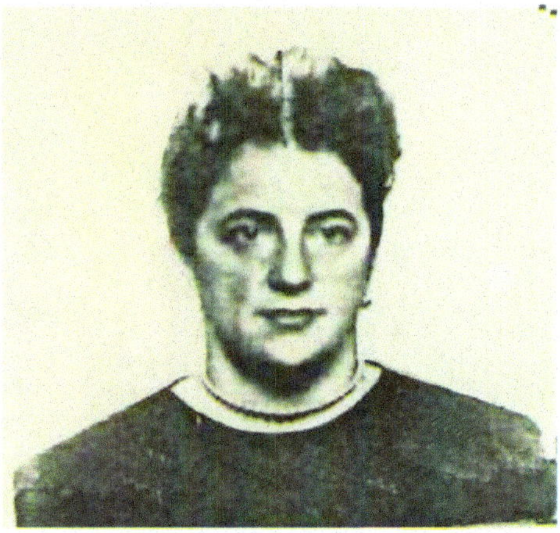

Fig. 17.7 María Teresa Carri de Riggi (1906–1989). *Source* Segemar archives

Formation of the Chubut Group. Years later, in 1955, through ground surveys, the occurrences could be confirmed and in 1958, through new aerial surveys, the surface extension was recorded that allowed the Los Adobes Mine to be discovered in the same basal section of the Chubut Group. This mine went into production between 1974 and 1976, obtaining 120 tons of uranium concentrates with an average grade of 0.141 (Bianchi 2022).

New aero-radiometric explorations made shortly after in 1971, identifies the Cerro Condor anomaly in the vicinity of Los Adobes, again in the lower section of the Chubut Group. In this unit, uranium mineralization is hosted in conglomeratic-sandy facies with lentiform morphology (Fuente and Gayone 1999). This deposit has produced about 56 tons of uranium concentrates.

In this region, the most important uranium mineralization appears in the area of influence of the Los Adobes Mine in the northern sector of the Central Plateau. It is the Cerro Solo deposit (Fig. 17.8) located in a stratigraphic unit similar to the previous ones. The mineralization is uranium and molybdenum and is found at a level between 50 and 130 m depth.

The drilling carried out has discovered reserves of 4,960 t of uranium and 3,236 inferred tons. The closure of the Los Adobes and Cerro Cóndor mines in 1976 has left Cerro Solo as one of the most important uranium mineral reserves in the country. There are also other manifestations in the region of less potential.

17.3.4 The Permo-Triassic Basin of Sierra Pintada

The first discoveries of uranium mineralization were made by the National Atomic Energy Commission between 1956 and 1958 in the Sierra Pintada region (Fig. 17.9), located west of San Rafael, Mendoza. Although the first manifestations were of little

Fig. 17.8 Cerro Solo site camp where you can observe the position of the sub-horizontal strata in the middle of the central plateau of Chubut. *Source* Bianchi (2022)

potential, aerial exploration carried out in 1968 revealed important anomalies associated with the Permian-age sandstones. Terrestrial studies allowed the Dr. Baulíes and Los Reyunos sites to be located in the huge Tigre anticline, among other manifestations (Salvarredi 1999). This anticline is an extensive basement structure flanked by direct faults that have inverted and folded the deposits during the Andean orogeny.

The mineralization is located in the sandy sequences of the middle member of the Los Reyunos Formation of Permian age. The mineralization has been dated to 265 ± 57 Ma and is associated with the weathering processes of the ignimbrites of the upper member of this formation. The region has experienced a series of subsequent tectonic events during the Cenozoic that have reactivated faults and fractures in the

Fig. 17.9 Geological map of the Tigre anticline with the main manifestations of the Sierra Pintada district, Mendoza and view of the open pit exploitation of the Dr. Baulíes mine in Sierra Pintada. *Source* after Salvarredi (1999) and Bianchi (2022)

country rock and allowed the mobilization and concentration of uranium. Organic matter would be the main precipitant of uranium.

The beginning of exploitation was carried out in the Dr. Baulíes deposit in 1975, which continued until 1995 and produced about 1,600 tons of uranium. All these deposits are paralyzed in a remediation process.

17.3.5 Foreland Deposits of the Sierras de Córdoba

These uranium deposits were discovered in the western foothills of the Sierra Chica de Córdoba in the vicinity of the city of Cosquín. The first reconnaissance works were carried out in 1956. Drilling and galleries confirmed the epigenetic mineralization hosted in the Cosquín Formation of the Eocene age. Its structure is associated with reverse faults that repeat the basement rocks above this unit (Timoneri and Linares 1960). Although conspicuous reserves were discovered, it was never put into operation and activities were abandoned in the mid-seventies.

17.4 Vetiform Deposits

It is interesting to note that the first deposits of nuclear minerals that were explored were those of a vein-like nature. In his first synthesis of these manifestations, Angelelli (1962) mentions, in addition to the pegmatites of the Sierra de Comechingones discovered by Rigal in 1937, another series of high-grade but small vein deposits.

The first discovery of uranium in veins associated with sedimentary deposits dates back to 1946 in the Soberanía mine of bentonite and gypsum and was made by Demetrio Ortega and José Cabrera. It is located near Estancia San Isidro, in the foothills, west of the city of Mendoza. Between 1947 and 1948, the Fabricaciones Militares carried out detailed studies in which the National University of Cuyo also participated in 1951. In the vicinity of that mine, the Papagayos deposit was found. This was studied by Alberto Belluco (1956) in his doctoral thesis, concluding that it had been formed by the leaching of granites such as that of Cacheuta, which produced the subsequent concentration in the Triassic sediments. Between 1952 and 1957, around 1,900 t of ore with 0.25% uranyl were extracted from the deposit (Belluco et al. 1974). It was the first uranium deposit in production in the country and with it, the National Atomic Energy Commission was practically born in 1954.

Among these first vein deposits, the San Santiago mine in La Rioja stands out. This mine known since the late nineteenth century for its nickel mineralization is in the western Sierras Pampeanas. It was initially explored in 1950 by Fabricaciones Militares who found uranium mineralization and later in 1952 by the Atomic Energy

Commission. It is characterized by a short northeast vein where uranium mineralization is associated with nickel. A pocket of 15 by 17 m was exploited where 76 tons of ore were extracted with a grade of 18% Ni and 1.1% $^3U^8O$ (Brodtkorb 1999).

In his synthesis, Angelelli (1962) mentions other manifestations in the Puna, the Sierra de Famatina, and the Frontal Cordillera. None of them were of economic interest and at the end of the 1950s, exploration turned to the exploration of uranium manifestations associated with sedimentary deposits. Perhaps the most important exceptions were the vein deposits in granites, where prospecting was concentrated in the 60s.

17.4.1 Vetiform Deposits in the Achala Batholith

The occurrence of uranium in granitic rocks of this batholith was found by Alberto Belluco and Andrés Lencinas in 1965. They made the first manifestation of the discovery, which was the Schlagintweit mine (Fig. 17.10).

The mineralization was concentrated in highly tectonized sectors linked to important faults. The mine is located on the eastern slope of the Sierra Grande de Córdoba and is entirely associated with the early Paleozoic granites of the Sierras Pampeanas. It was exploited in the open pit, extracting 207 tons of uranium with an average grade of 0.0152% (Blasón 1999a). It has currently been abandoned since 1989 and provincial laws prohibit open-air exploitation.

Fig. 17.10 Open pit exploitation at the Schlagintweit uranium mine is currently deactivated. *Source* Bianchi (2022)

17.4.2 *Vetiform Deposits in the Cerro Áspero-Alpa Corral Batholith*

On the western slope of the Sierra de Comechingones in San Luis, uranium minerals were identified in 1956. This led the National Atomic Energy Commission to explore the region, finding several manifestations that became the Estela and Bella Vista mines. Both are located in the early Paleozoic granitic rocks of the Cero Áspero-Alpa Corral batholith. Its control is structural, and the mineralization is of the epithermal type similar to that of the Schlagintweit mine. 54.5 tons of uranium were extracted from the La Estela mine (Blasón 1999b).

17.5 Final Considerations

It has been seen how the systematic exploration of uranium began in the 1950s through the National Atomic Energy Commission as part of the country's industrialization plan, motivated by the interest aroused in World War II by nuclear minerals. Geological exploration allowed the discovery of important uranium deposits in the provinces of Mendoza, Salta, Chubut, and Río Negro. In the 1970s, there was an increase in demand for uranium for nuclear power generation, leading to an increase in uranium prospecting and production in Argentina. The country became one of the leading producers of uranium in Latin America, with production peaking in the 1980s. Concern over the environmental effects of uranium mining led to increased regulation and opposition to mining activity in some areas, which ended up paralyzing this activity.

In those years, research and development of nuclear technology allowed the construction of the first two nuclear reactors in Latin America, both located at the Constituyentes Atomic Center in Buenos Aires. In the 1960s, the Atucha I Nuclear Power Plant was built, located in the province of Buenos Aires, the first nuclear power plant in Argentina. In the following years, the Embalse Nuclear Power Plant was built in the province of Córdoba, and the Atucha II Nuclear Power Plant in Buenos Aires. Financial and environmental problems in the 1990s affected its ability to conduct research and develop technologies.

In all those years, the support that the exploration and exploitation of geological resources provided to make the nuclear fuel cycle a reality. In this action, in the beginning, the engineer Victorio Angelelli had a prominent role, leading the exploration, later giving way to Dr. Pedro N. Stipanicic, who had a leading role in guiding the exploration, evaluating the identified manifestations, and mobilizing geological resources in favor of the exploration of the country's needs (Riccardi 2009). The Commission in turn developed the necessary technology to process uranium concentrates and produce the fuel elements necessary for natural uranium nuclear power plants, achieving complete self-sufficiency.

The history of the geology of radioactive minerals shows once again that when there are decisions and well-defined work plans, the proposed objectives can be achieved. The development of geology in our country made it possible to mobilize human resources between 1950 and 1975 to comply with these exploration plans on time.

References

Angelelli V (1960) Características de nuestras manifestaciones y depósitos uraníferos vetiformes. Primeras Jornadas Geológicas Argentinas, Anales 3(1962):1–15, Buenos Aires

Avé-Lallemant G (1874) Mineralogische Mittheilungen aus der argentinischen Provinz San Luis. IV-Uranhaltende Mineralien, Monatschrift 10:129–178

Avé-Lallemant G (1892) Observaciones sobre el mapa del Departamento de Las Heras. Provincia de Mendoza. Anales del Museo de La Plata, Sección Geológica y Mineralógica 1:5–20

Belluco AE (1956) Las vetas de cuarzo uranífero del Yacimiento "Presidente Perón", Mendoza. United Nations, Conferencia Internacional sobre utilización de la Energía Atómica con Fines Pacíficos, Actas 6:96–104, Ginebra

Belluco AE, Diez JD, Antonietti CE, Achen H, Valerdi CI (1974) Los depósitos uraníferos de las provincias de Mendoza y Neuquén. 5 Congreso Geológico Argentino, Actas 2:35–53, Buenos Aires

Bianchi R (2022) Origen de la exploración y producción de uranio en Argentina. ENULa, Energía Nuclear Latinoamericana. https://enula.org/2022/01/origen-de-la-exploracion-y-produccion-de-uranio-en-argentina

Blasón R (1999a) Yacimiento Schlagintweit, distrito uranífero Batolito de Achala, Córdoba. In: Zappettini EO (ed) Recursos Minerales de la República Argentina, Anales del Instituto de Geología y Recursos Minerales, Segemar 35:609–612, Buenos Aires

Blasón R (1999b) Yacimiento La Estela, distrito uranífero Comechingones, San Luis. In: Zappettini EO (ed) Recursos Minerales de la República Argentina, Anales del Instituto de Geología y Recursos Minerales, Segemar 35:621–624, Buenos Aires

de Brodtkorb MK (1966) Mineralogía y consideraciones genéticas del yacimiento Huemul, provincia de Mendoza. Revista de la Asociación Geológica Argentina 21(3):165–179

de Brodtkorb MK (1999) El yacimiento de níquel y uranio San Santiago, La Rioja. In: Zappettini EO (ed) Recursos Minerales de la República Argentina, Anales del Instituto de Geología y Recursos Minerales, Segemar 35:631–632, Buenos Aires

Candiani JC, Ulacco H, Ojeda G (2016) Hoja Geológica 3366 II Villa de Merlo, provincias de San Luis y Córdoba. Instituto de Geología y Recursos Minerales, Boletín del Servicio Geológico Minero Argentino 415:1–127, Buenos Aires

Crosta S, de Belén Palacio M (2015) Carta minero-metalogénica 3569-III, Malargüe, província de Mendoza. Programa Nacional de Cartas Geológicas de la República Argentina, Boletín del Servicio Geológico Minero Argentino 407:1–89, Buenos Aires

Ferrari R (1993) Germán Avé-Lallemant, Introducción a la obra científica y técnica de Germán Avé-Lallemant en la República Argentina (ca. 1869–1910). Instituto Científico y Cultural "El Diario", 171 p, San Luis

Fuente A, Gayone MR (1999) Distrito uranífero Pichiñán: Los Adobes, Cerro Cóndor y Cerro Solo, Chubut. In: Zappettini EO (ed) Recursos Minerales de la República Argentina, Anales del Instituto de Geología y Recursos Minerales, Segemar 35:1255–1260, Buenos Aires

Galiski M (1999) Distrito pegmatítico Comechingones, Córdoba. In: Zappettini EO (ed) Recursos Minerales de la República Argentina, Anales del Instituto de Geología y Recursos Minerales, Segemar 35:361–364, Buenos Aires

Gorustovich S, de Brodtkorb MK, Benítez A, Maloberti A (1992) Los yacimientos de la paragénesis U-V-Cu del Cretácico de la República Argentina. Recursos Minerales y Energéticos del Cretácico de América Latina, Actas: 25–90
Linares E (1966) Geología isotópica del yacimiento Huemul, provincia de Mendoza. Revista de la Asociación Geológica Argentina 21(3):181–189
Ramos VA (1975) Los ciclos sedimentarios y la biorhexistasia en el control de las manifestaciones cupríferas del Neuquén extra andino. II° Congreso Iberoamericano de Geología Económica (Buenos Aires), Actas 5(4):373–394
Ramos VA (2023) El expediente secreto de la CNEA: María Teresa Carri de Riggi, descubridora del uranio en la Patagonia. 6° Congreso Argentino y 1° Congreso Latinoamericano de Historia de la Geología, Resúmenes: 82–86. Tucumán
Riccardi AC (2009) Necrológica, Prof. Dr. Pedro Nicolás Stipanicic (1921–2008). Revista de la Asociación Geológica Argentina 64(2):352–356
Rinaldi C (1968) Estudio de las pegmatitas uraníferas de la Sierra de Comechingones, Provincia de Córdoba. Revista de la Asociación Geológica Argentina 23(3):161–195
Rigal R (1938). Las minas de columbita y tantalita y el descubrimiento de minerales de uranio radiactivos en la Cañada de Álvarez, departamento Calamuchita, Córdoba. Boletín de la Dirección Nacional de Minería 45:1–45, Buenos Aires
Rodrigo F, Ortega Furlotti A (1982) Dr. Hector Achen (1920–1981). Revista de la Asociación Geológica Argentina 37(1):144–145, Buenos Aires
Rojas G (1999a) Distrito uranífero Pampa Amarilla, Mendoza. In: Zappettini EO (ed) Recursos Minerales de la República Argentina, Anales del Instituto de Geología y Recursos Minerales, Segemar 35:1135–1140, Buenos Aires
Rojas G (1999b) Distrito uranífero Los Chihuidos-Las Cárceles, Neuquén. In: Zappettini EO (ed) Anales del Instituto de Geología y Recursos Minerales, Segemar 35:1143–1146, Buenos Aires
Romano HI (1999) Distrito uranífero Tonco-Amblayo, Salta. In: Zappettini EO (ed) Recursos Minerales de la República Argentina, Anales del Instituto de Geología y Recursos Minerales, Segemar 35:959–970, Buenos Aires
Salvarredi J (1999) Yacimiento Doctor Baulíes y otros depósitos del distrito uranífero Sierra Pintada, Mendoza. In: Zappettini EO (ed) Recursos Minerales de la República Argentina, Anales del Instituto de Geología y Recursos Minerales, Segemar 35:895–908, Buenos Aires
Stipanicic PN, Baulies OL, Rodriguez F, Martinez CG (1960) Los depósitos uraníferos en rocas sedimentarias de la República Argentina. Primeras Jornadas Geológicas Argentinas, Anales 3(1962):351–384, Buenos Aires
Stipanicic PN (1969) Dr. Oscar Leopoldo Baulíes (1923–1968). Revista de la Asociación Geológica Argentina 24(l):77–78, Buenos Aires
Stipanicic PN (1993) Nota necrológica: Dr. Armando Ortega Furlotti (25-02-93). Revista de la Asociación Geológica Argentina 48(2):187–188, Buenos Aires
Timoneri AJ, Linares E (1960) Informe preliminar geológico-minero del yacimiento Rodolfo. Comisión Nacional de Energía Atómica, unpublished report, Buenos Aires

Chapter 18
New Institutions in Northern Argentina

Abstract The University of Tucumán, founded in 1913, was the first one in northwestern Argentina, but it was not until the creation of the Miguel Lillo Institute that research and teaching of Natural Sciences began. This university created the Institute of Geology and Mining of Jujuy in 1946, which for many years concentrated geological and mining studies in the provinces of northern Argentina. This institute hosted numerous European geologists who emigrated to the country after World War II, especially from Italy and to a lesser extent from Germany. Among them, Giovanni Cecioni and Federico Ahlfeld stood out for their important contributions. The director of the Institute was Abel Peirano, who is credited with having discovered the Agua del Dionisio district that became some years later the Bajo de la Alumbrera copper and gold mine. Peirano returned to Tucumán in 1951 and developed geological studies at that university. An excellent teaching staff was formed and the first students in geology graduated in 1960. Natural Sciences began in Salta in 1952, dependent on the University of Tucumán. In 1962 the first geologists graduated and in 1972 the creation of the University of Salta allowed the consolidation of geological studies and research.

Keywords Italian geologists · Mining exploration · Natural sciences · New schools · Graduates

18.1 Introduction

The University of Tucumán was created by a provincial law in 1913. In 1924, it was incorporated as a National University, the fourth after the universities of Buenos Aires, Córdoba and La Plata. However, in its beginnings, it did not contemplate the teaching of Geology (Aceñolaza 2018).

The natural sciences entered Tucumán with the help of Miguel Lillo (1862–1931), a self-taught scholar and botanist who studied the flora and fauna of Tucumán for decades. His formal studies were carried out at the National College of Tucumán,

V. A. Ramos, *History of the Geological Sciences in Argentina*, The Latin American Studies Book Series, https://doi.org/10.1007/978-3-031-83139-3_18

and he had figures such as the German scholar Federico Schickendantz (1837–1896) as a teacher at the secondary school (Peña de Bascary 2014) (Fig. 18.1).

Lillo was not able to pursue university studies, but his work from 1885 as an assistant in the Municipal Chemical Office under the direction of Schickendantz allowed the continuity of his studies in Natural Sciences in Tucumán. In a few years he was in charge of the Chemical Office and, encouraged by his teacher, he toured France, Germany and England, where he visited scientific centers. Upon his return in 1891, he told his teacher that he had found his vocation in the Natural Sciences and among them in botany. In those years he made fundamental publications on the flora of Tucumán and its birds, being recognized by the National University of La Plata in 1914, which awarded him the Doctor Honoris Causa. That same year he began teaching Chemistry at the brand-new University of Tucumán and dedicated himself fully to research. He compiles an important herbarium of thousands of specimens, along with various materials from different places in the province. In 1918 he retired from teaching but maintained his honorary position as director of the University's Natural History Museum, a position he held until his death. His collections, as well as one hectare of his property in the city and cash, were donated to the University in 1931, where the Miguel Lillo Institute of Natural Sciences, formally inaugurated in 1933, currently operates.

One of his main disciples was Abel Peirano (1896–1969). He was born in Buenos Aires, but in 1904 his parents settled in Tucumán where he completed his primary and secondary studies. At the Colegio Nacional, he had Miguel Lillo as a teacher, who instilled in him his passion for nature. Between 1915 and 1919, he returned to Buenos Aires and studied Pharmacy and Biochemistry at the University of Buenos Aires from where he graduated with an honors diploma (Aceñolaza 2020). He returns to Tucumán where he sets up a pharmacy but interested in natural sciences he makes various trips to the mountains of northwest Argentina. After a short period of residence in Santa María, Catamarca, he received a call from Miguel Lillo, who summoned him to return to Tucumán in 1930 and collaborate at the Museum of

Fig. 18.1 The German chemist Federico Schickendantz, his disciple the botanist Miguel Lillo and Abel Peirano, were the three pioneers who contributed to the development of Natural Sciences in Tucumán. *Source* A.G.N. archives

Natural History. His mission was to study the rocks, minerals, and fossils collected by Lillo and Schickendantz that were in the museum. This relationship with the museum marks Peirano's beginning as a geologist (Aceñolaza 2013a).

In 1936 he was appointed head of Mineralogy at the museum and two years later he was appointed Director of the Institute of Mineralogy and Geology. In those years he intensified his studies on the Agua del Dionisio mining district with field work and chemical analyses that began to show its potential. His first contributions are made known in his Notebooks of Mineralogy and Geology which he publishes through the University of Tucumán. In the early 1940s, he continued studying the district, published several publications, and began to request mining permits. He later donated these mining belongings to the National University of Tucumán.

His university companion in Buenos Aires, Horacio R. Descole (1910–1984), was appointed rector of the University in 1946. He created the Institute of Geology and Mining in Jujuy (Fig. 18.2) as a unit of the Tucumán University and appointed Abel Peirano as director. He gives him a significant budget to hire internationally renowned scientists.

Fig. 18.2 Current facilities of the institute of geology and mining of Jujuy, belonging to the national university of Jujuy since 1976

18.2 The Institute of Geology and Mining of Jujuy (1946–1976)

A decree of the national executive branch created the institute dependent on the National University of Tucumán in May 1946. The great war had ended, and Europe was plunged into misery. Many scholars had no horizon for their activities and that is why they were invited to work in Jujuy. At that time, mining in this province had a great boost with the studies of the iron deposits of the Sierra de Zapla, the development of the Aguilar lead and silver mine, the tin and silver of the Pirquitas mine, the zinc district and lead from Pumahuasi, the Pan de Azúcar lead and silver mine; the gold alluvium of Coyahuaima and Olaroz, among other important occurrences (Alcalde and Chayle 2014; Alonso 2021a).

In its first years, it had a budget that allowed it to hire a series of Italian scientists and technicians who carried out intense work in northwest Argentina.

Among these scientists was a plethora of Italian geologists and engineers who, after the Second World War, sought new horizons. Among them was Giovanni Cecioni (1914–1987), born in Tuscany, who arrived in the country with good international experience (Figs. 18.3 and 18.4). He was only two years at the institute, but he carried out a study on fossil nautiloids from the early Paleozoic of northwestern Argentina that is still valid (Cecioni 1953). In 1950 he went to work in Chile, first in oil exploration and later as a professor at the University of Chile where he had a fruitful teaching career. As a researcher, he published important contributions such as his Outline of Chilean Paleogeography, a book in which he poured out his experience from those years in Chile (Cecioni 1970). He directed numerous theses and undergraduate reports and was a pioneer in the teaching of Stratigraphy and Paleontology at that university (Riccardi 1990; Charrier et al. 2018).

Fig. 18.3 Giovanni Cecioni (1914–1987) and Federico Ahlfeld (1882–1982), two illustrious members of the institute of geology and mining who made notable contributions to the Northern Argentine and its director Abel Peirano. *Source* Jujuy I.G.M. archives

Fig. 18.4 Giovanni Cecioni gave a class to a team of Argentine geologists from the institute of geology and mining led by Abel Peirano. First row seated from left to right, Juan C. Porto, Gregorio de la Puente, Enrique Alabí and Raúl Chomnales, second row Luis de la Fuente, Carlos Córdoba, Wilfredo Lyons, José S. Iglesias and Rogelio Bellmann. *Source* Aceñolaza (2013a, b b)

Another geologist was Renato Loss, a Genoese paleontologist who studied Ordovician graptolites and their stratigraphic levels from the Quiaca in Jujuy to Cerro San Bernardo in Salta, making the first publications on these fossils in Argentina. He had to return to Italy in 1952 due to family problems where he continued researching. In his short stay at the Institute, he made pioneering contributions to the stratigraphy and graptolites of the Eastern Cordillera of Salta and Jujuy (Loss 1948, 1949a, b, c a, b, c, 1951, and 1953).

Among them were mining engineers such as Andrés Giordana, who studied the concentration processes by flotation of iron ore from Santa Bárbara and Unchimé and those from Farallón Negro, collaborating extensively with Peirano in Agua del Dionisio. When the brand-new National University of Jujuy was created, he was incorporated as a professor in mining engineering (Aceñolaza 2013b).

Other professionals were also hired at the institute, such as Antonio S. Bonino, a Piedmontese engineer who in the early years collaborated in the construction of the Agua Tapada mine and who later settled in Tucumán; Jorge Melchiori, born in Perugia and doctor in Physics who was in charge of the geophysics of the institute; Martín Giachetto, Piedmontese mechanical technician who directed the installation and operation of the machinery in Agua Tapada; Guido Donnini, native of Tuscany, in charge of the machinery at the Agua del Dionisio mine; Orlando Colafigli, born in Abruzzo who worked as a mining expert, and the chemist, and geologist Giorgio

Ghigi, a Roman who was head of the Mineralogy and Petrography section until 1952. A large part of these Italians who came to work at the Jujuy Institute collaborated with Peirano in the Agua del Dionisio projects. However, after a couple of years and due to the growth in the cost of living in the early years of the 1950s, when their salary demands were not met, they left the institute and some, citing family reasons, returned to their native Italy.

Among these professionals, the German mining engineer Federico Ahlfeld (1892–1982) stood out, born in Marburg and graduated from Clausthal University with postgraduate degrees in Geology and Mining. He arrived in Bolivia in 1923 at the height of tin mining, where he stayed six years before returning to Germany to take over the chair of Mineralogy, Petrography, and Economic Geology at the University of Magdeburg in Saxony (Toselli 2007). In 1935 he returned to Bolivia and was appointed head of the Geology section of the Direction of Mines and Petroleum. In those years he published the first edition of the *Geology of Bolivia*, which would be followed by two new editions, a classic and fundamental text of that country (Ahlfeld 1946; Ahlfeld and Branisa 1960 and 1972). In 1946 Peirano hired him to work at the institute as head of the Mining Department. During his stay of two years, he made fundamental works such as geological studies and mineral deposits of the province of Jujuy where he established the southern continuity of the Bolivian tin belt in northern Argentina and described 65 deposits in Salta and Jujuy (Ahlfeld 1948; Ahlfeld and Angelelli 1948). Furthermore, he described in detail the main mines and metallogeny of Jujuy in a series of international publications. Ahlfeld is recognized as the most outstanding and meritorious geologist and mining engineer of Bolivia who studied its geology and its mines and was a professor of the first generation of geologists at the Universidad Mayor de San Andrés.

Federico Ahlfeld's passage through Jujuy was recorded in dozens of publications made in that short period where he addressed the study of gold alluvium, antimony deposits, deposits of silver, lead, and zinc, alluvial tin and veins, the geology of the Chañi mountains, deposits of onyx, borates, salt, etc., as well as a mining map of the province (Alonso 2021a, b).

The institute also had a series of Argentine geologists such as Juan José de Benedetti, who graduated from Stanford in 1939, and joined the institute in 1947. He was in charge of convening other Argentines who trained with foreign geologists such as Rogelio Bellmann, Carlos Córdoba, Gregorio de La Puente, Celestino Danieli, Enrique Alabí, and Raúl Chomnales, young graduates of the University of Córdoba (Aceñolaza 2013a, b b). Wilfredo Lyons and José S. Iglesias, both graduates of La Plata, also stood out, along with Juan Carlos Porto, a graduate of Buenos Aires and a doctorate in Tucumán.

Wilfredo Lyons (Fig. 18.5), from Jujuy by birth, barely returned to his homeland in 1948 after graduating from university and worked at the institute, where he participated in numerous studies of the Jujuy sites under the tutelage of Federico Ahlfeld. He evaluated various mines in Bolivia and Peru, worked in Colombia and Venezuela, and traveled from Patagonia to the most remote places in the country for more than 50 years in private activity. He carried out studies of deposits in Afghanistan, and supervised explorations and exploitations in different districts at the foot of the Andes

Fig. 18.5 Wilfredo Lyons (1923–2005) worked in the first years at the institute and Beatriz L. Coira and Waldo Chayle participated in its integration with the National University of Jujuy. *Source* Jujuy I.G.M. archives

from Venezuela to Argentina, meeting death in full activity at the age of 82 (Méndez 2005).

Raúl Chomnales (1921–1986) graduated from the National University of Córdoba in 1948 and began professionally at the Institute where he worked from assistant geologist to director starting in 1973. He stood out for his mining studies, especially in opaque mineral analyses, having studied the main mining districts of the province of Jujuy (Chayle 1986).

When Peirano left in 1951 to return to Tucumán, only Rogelio Bellmann, Raúl Chomnales, José Saturnino Iglesias and engineer Andrés Giordana remained in Jujuy, although budget problems and their lack of institutional recognition caused their activity to decline significantly, as Enrique César Rayces highlights, one of its directors in the 50 s. In 1971, the institute joined two other institutions, the Higher Institute of Economic Sciences and the Institute of Altitude Biology which depended on Tucumán, to create the Provincial University of Jujuy, which 2 years later was nationalized. In 1976 the institute was fully integrated into the university, promoting research and teaching in geology, which reached full development with the participation of Beatriz Coira, José Alcalde, María Camacho, and Waldo Chayle.

It should be noted that in 1975 the institute once again edited the Magazine of the Institute of Geological Sciences, where dozens of works by researchers from the universities of northwest Argentina were published, some of which have become classic studies of the region. In this way, the first publications of the Institute of Geology and Mining of Jujuy carried out in the 1940s by Renato Loss and Federico Ahlfeld continue.

18.3 The Miguel Lillo Institute and the Geology Degree

The Miguel Lillo Institute was born from the legacy of Miguel Lillo upon his death in 1931, organizing itself as such from 1933. Years later, the director of the Lillo Institute and rector of the university Horacio Descole requested to create the Institute of Geognosia in 1950. It depended on the Faculty of Exact Sciences and Technology (Fig. 18.6) and put Peirano at the head of the institute.

While Abel Peirano was director of the institute in Jujuy, he did not interrupt his studies and analyses in the Agua del Dionisio district, which years later gave rise to the Farallón Negro and La Alumbrera deposits. On the contrary, many of the Italians hired by that institution worked actively in various aspects of the exploration and exploitation of that district. When in 1951 he returned to Tucumán he was accompanied by two of his disciples Celestino Danieli (1920–1987) and Juan C. Porto who continued to actively collaborate with his projects.

In 1953, the Institute of Geognosia came under the jurisdiction of the Lillo Institute and a few years later the University School of Natural Sciences was created (Fig. 18.7), which began the teaching of the Geology degree (Figs. 18.8 and 18.9). In addition to the teaching of Geology, the Fossil Vertebrate Laboratory was founded at the Institute in 1958 by Osvaldo Reig, which quickly acquired international recognition with his and José Bonaparte's studies. These two self-taught paleontologists, one with his paleobiological studies and the other with his research on dinosaurs, produced a great change in these disciplines in our country (Fig. 18.10), developing topics at the frontier of knowledge in those years, as can be seen in their trajectories presented by Quintana (2012) and Apesteguía et al. (2022).

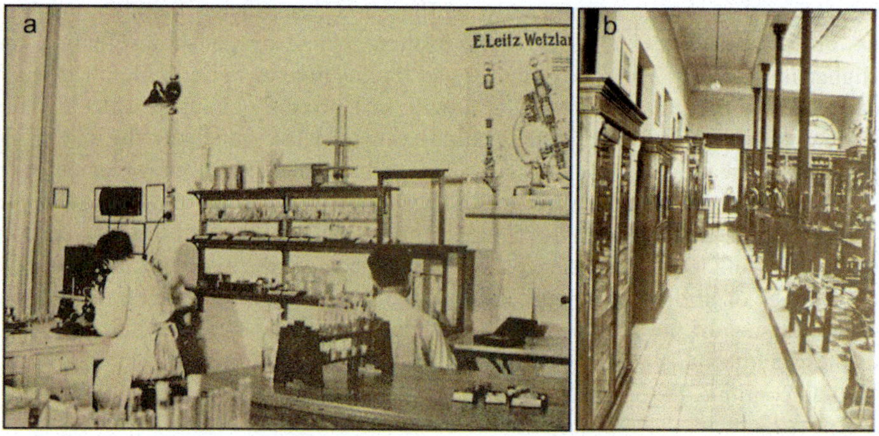

Fig. 18.6 Petrography laboratory of the national university of Tucumán in 1938. On her back analyzing samples under the microscope is Dr. Zulema Chiesa, one of the first geologists to graduate from the University of Buenos Aires who worked for several years as a petrographer at the University of Tucumán. *Source* Aguirre Urreta and Ramos (2021)

Fig. 18.7 Faculty of natural sciences and Miguel Lillo institute of the national university of Tucumán and the museum of natural sciences of the Miguel Lillo foundation

Fig. 18.8 Abel Peirano (1896–1969), Celestino Danieli (1920–1987), and Osvaldo Reig (1929–1992) were the first professors of the brand-new Geology degree. *Source* Instituto Lillo archives

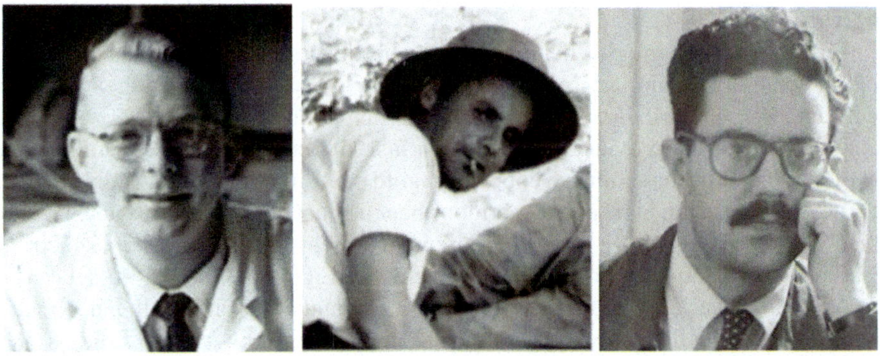

Fig. 18.9 Sergio Archangelsky (1931–2022), José F. Bonaparte (1928–2020), and Rafael R. González (1924–1990) professors at the Lillo Institute in the first decades. *Source* Instituto Lillo archives

Fig. 18.10 Fieldwork team from the Lillo Institute in 1959 that carried out the first expeditions to Ischigualasto led by Osvaldo Reig (standing with a white shirt), José Bonaparte (with a white beret), and numerous *baquianos* and technicians. *Source* Apesteguía et al. (2022)

Among the teaching staff, Sergio Archangelsky (1931–2022) stood out at that time, among others, who graduated from the University of Buenos Aires, and did his thesis under the direction of Alberto Castellanos (1896–1968). He arrived at Lillo in 1955 and worked during those years on his thesis on the Glossopteris flora in Patagonia. His five-year stay in Tucumán, apart from teaching subjects related to his discipline, allowed him to direct his first doctoral thesis corresponding to Rafael Herbst, who was also oriented towards paleobotany, especially paleopalynology (Archangelsky 2014).

Peirano's studies in Agua del Dionisio and his donation to the National University of Tucumán of the mining rights obtained from his searches provided the institution with important contributions due to his co-participation in the profits. Based on the agreement with the province of Catamarca in 1958 and from the beginning of the exploitation of the disseminated copper and gold deposit of Bajo de la Alumbrera, the University of Tucumán received enormous benefits that continue to this day. Peirano died in 1969, subsisting on a meager retirement until his last days (Aceñolaza 2015).

Starting in 1960, the first graduated in Geology (Fig. 18.11), where Rafael Herbst stood out. He was a palynologist who made important contributions to paleobotany and continued the Paleontology courses when Archangelsky went to London on a scholarship (Ottone 2018). Another of the first graduates was Gerardo Bossi, who carried out numerous works on clay sedimentology and dedicated his entire academic life to teaching and research at the Lillo Institute (Georgieff 2015; Muruaga and Gavriloff 2015).

Fig. 18.11 Rafael Herbst (1936–2017) and Gerardo Bossi (1939–2015), the first graduates of the geology degree from the national university of Tucumán. *Source* A.G.A. archives

The School of Natural Sciences of the Lillo Institute became the Faculty of Natural Sciences and Lillo Institute of the National University of Tucumán in 197,3, where prestigious geologists held the highest management positions (Fig. 18.12).

At the end of the 60 s, Gilberto Aceñolaza joined as professor of Paleozoology (Fig. 18.12) and shortly after Alejandro Toselli in Petrography, graduates of the National University of Córdoba. With them begins an important period of geological and paleontological studies in northwest Argentina, which allows the consolidation of teaching and research and the training of new graduates. Among his numerous contributions, the studies of Oldhamia ichnofossils stand out, which allowed Aceñolaza to recognize a Lower Cambrian age to the upper part of the Puncoviscana Formation, ending a controversy about its age of several years (see Aceñolaza 2015). Also in those years, Ricardo Mon (1940–2022), a graduate of Buenos Aires and a doctorate at the Sorbonne University in Paris, joined as a professor of Structural Geology who made important contributions to the structure of the basement (Mon 2022; Adler et al. 2022).

Fig. 18.12 The deans Gilberto Aceñolaza (1983–1991), Alejandro Toselli (1991–1998), and Ricardo Mon (2006–2014), who in addition to their management activities stood out for their important geological contributions in their different disciplines. *Source* Instituto Lillo and A.G.A. archives

In the 1970s, intense international collaboration developed, highlighting the participation of Dr. Hubert Miller (1936–2020) through an agreement first with the Münster University, and later with the Ludwig Maximilians University of Munich. In those years, young Germans such as Meinolf Knüver, Ulrich Lottner, Michael Reissinger and Arne Willner, carried out their doctoral theses in different sectors of northwest Argentina, together with local professors and numerous German geologists and geophysicists made important contributions to our geology (Aceñolaza 2022; Aceñolaza and Toselli 2020a, b; Charrier et al. 2020, and Ramos 2021).

In 1984, the Higher Institute of Geological Correlation (INSUGEO) was created, which not only opened important research in northern Argentina but also launched a series of periodic and special publications that in recent decades have become the organ of dissemination and a reference for the advancement of geological knowledge in this region. It is currently a dual dependency institute between the National University of Tucumán and Conicet.

18.4 Teaching Geology in Salta

The teaching of Geology in Salta began in the Provincial Museum of Natural Sciences (Fig. 18.13) where a Higher School of Natural Sciences had been created. The museum and school were run by Professor Amadeo R. Sirolli (1900–1981), who began operating in the early 1950s (Alonso 2018).

Fig. 18.13 Provincial museum of natural sciences in San Martín Park where the teaching of geology began in Salta. *Source* Alonso (2018)

Fig. 18.14 Current headquarters of the faculty of natural sciences of the national university of Salta

The high school was transformed in 1952 into the Faculty of Natural Sciences of Salta dependent on the National University of Tucumán, which 20 years later became part of the brand-new National University of Salta. The history of the faculty headquarters and its successive moves is masterfully told by José Salfity (2017, p. 88–89), until ending up on the current university campus (Fig. 18.14). The first professors who stood out for their vocation to teaching and their geological contributions were doctors Carlos Moreno Espelta (1920–2010) in Introduction to Geology, Marcelo Figueroa Caprini (1923–1995) in Structural Geology, Antonio P. Igarzábal (1925–1997) in Geomorphology, and Domingo Jakúlica (1921–2006) in Petroleum Geology (Alonso 2018).

At that time, two excellent geologists were received, Apolo Ortiz in 1962 and Andrés Lencinas the following year. Both have had an outstanding geological career, one in oil exploration and the second in the mining industry (Salfity 2016; Lencinas 2017).

The creation of the National University of Salta in 1972 gave a strong boost to the Geology career with the incorporation of new professors. At that time Juan Luis Benedetto and Margarita Toro joined in Paleontology; José Viramonte in Volcanology; José Salfity who replaced Félix Celso Reyes (1924–1972) in 1972 (see Russo 1974) in Historical Geology and Stratigraphy; Ricardo Sureda (1946–2020) in Mineralogy (Alonso 2020), Ricardo Omarini (1946–2015) (see Alonso 2016) in Introduction to Geology; Arturo Castaños in Sedimentology; among several others (Figs. 18.15 and 18.16).

In the following years, this group of young professors, most of them graduates of the National University of Córdoba, began to make transcendental contributions

Fig. 18.15 José Viramonte and José Salfity were prominent professors, who consolidated Geology starting in the seventies in the new faculty of natural sciences of the national university of Salta. *Source* Alonso (2018)

Fig. 18.16 Ricardo Sureda and Ricardo Omarini, professors who with their research contributed significantly to the knowledge of the geology of northwest Argentina. *Source* Alonso (2018)

to the geology of northwest Argentina. José Viramonte, after a stay in Nicaragua to study its volcanoes, the subject of a thesis that he defends in Córdoba, settles in Salta and begins his volcanological studies, becoming a national referent in these topics (Viramonte 2017).

José Salfity's contributions to the stratigraphy and geological evolution of the Salta Group marked a sustained development based on the initial paleogeography proposed by Jorge Moreno in 1970, who showed the advance of knowledge obtained by YPF's explorations in northwest Argentina (Moreno 1970). Salfity, in numerous works carried out by him and his doctoral students, made known in local and international media the complex geology of the Upper Cretaceous-Cenozoic of this region (Salfity 2017).

Ricardo Sureda's mineralogical and metallogenic studies of the main ores of the Andean region, in addition to the discovery of new minerals, made a substantial contribution to the History of Mineralogy, not only in our country but also internationally (Sureda 2008).

Ricardo Omarini presented his doctoral thesis in 1982 at the National University of Salta on the Puncoviscana Formation and the Puna Eruptive Belt directed by Bruno Baldis. He also made numerous contributions throughout these years to a better understanding of the geological and tectonic evolution of the Proterozoic basement of northern Argentina and Ordovician magmatic activity. Among his works, his 1999 work on Puncoviscana fold-belt in northwestern Argentina stands out, as one of his latest contributions on the subject (see Alonso 2016).

These works were complemented by the transect program of the International Lithosphere Program (ILP) where geologists from Salta and their German colleagues carried out a complete structural section between the Pacific coast and the Chaco plain at 24° south latitude with wide international repercussion (Omarini et al. 2005).

18.5 Final Considerations

Starting from the initial efforts of the naturalist and famous botanist Miguel Lillo and his disciple Abel Peirano, geological research began in Tucumán in the 1930s at the National University of Tucumán. Only at the beginning of the 1950s did the Geology degree program begin in Tucumán. From this university radiated the Institute of Geology and Mining of Jujuy, dependent on it, which was founded in 1946 under the direction of Peirano. In its first years, it had the active participation of foreign geologists, especially Italians, who did great work in the study of the geology of Jujuy and its mines. Among them, Federico Ahlfeld stood out, a mining engineer who did fundamental works for the knowledge of mining in Jujuy and Bolivia, and bequeathed a classic work, the Geology of Bolivia, which through its different editions continues to be a must-see work. Twenty years later, at the beginning of the seventies, the National University of Jujuy was created, where the teaching of Geology began.

The development of Geology at the National University of Tucumán was linked to the evaluation of the Agua del Dionisio mining district by Abel Peirano, which once in production thanks to Peirano's generosity, is still an important source of income for the university. The teaching activity in those first years was carried out by important figures who temporarily settled in Tucumán, such as Sergio Archangelsky, José Bonaparte, and Osvaldo Reig, and a group of geologists who acted as visiting professors in the first years. The consolidation of teaching and research at the beginning of the seventies was led by young professors such as Aceñolaza, Toselli, Mon and Bossi, among several others who, through the Faculty of Natural Sciences and the Lillo Institute, made important contributions and taught several generations of geologists who graduated from those institutions. The Miguel Lillo Institute is an emblematic institution of science and culture in Tucumán and Argentina and continues to be a reference in the study of northwest Argentina.

The first Faculty of Natural Sciences of Salta in 1953 was an extension of the National University of Tucumán, which twenty years later when the National University of Salta was created, was consolidated with new professors and research that formed the current thriving Department of Geology.

The role that the National University of Tucumán has had was fundamental in generating teaching and research in the provinces of Jujuy and Salta, first directly and later with the creation of the new national universities. The research of these universities together with the oil exploration carried out by YPF and the mountain plans allowed the current knowledge of the geology of northern Argentina.

References

Aceñolaza FG (2013[a]) Farallón Negro: Epopeya de un sueño minero. Instituto Superior De Correlación Geológica, Miscelánea 20, pss 100., Tucumán

Aceñolaza FG (2013b) Los italianos del Instituto de Geología y Minería de Jujuy. 3° Congreso Argentino de Historia de la Geología, p. 13- 28, Salta

Aceñolaza FG (2015) En busca del pasado geológico. Asociación Argentina para el Progreso de las Ciencias, Revista Ciencia e Investigación Reseñas 3(2): 6–19, Buenos Aires

Aceñolaza FG (2018) Reseña histórica de la enseñanza de Geología en la Universidad Nacional de Tucumán. In Ramos VA, Martino RD (eds.) El nacimiento de la Geología en la enseñanza universitaria, Revista Facultad de Ciencias Exactas, Físicas y Naturales 5 (Suplemento 1): 63–70, Córdoba

Aceñolaza FG (2020) Abel Peirano y la Geología del Noroeste Argentino. Revista Facultad de Ciencias Exactas, Físicas y Naturales 7 (Suplemento 1): 15–17, Córdoba

Aceñolaza FG (2022) Los extranjeros del Lillo: científicos y profesores que jerarquizaron la institución. Revista de la Junta de Estudios Históricos de Tucumán 17: 180–201, Tucumán

Aceñolaza FG, Toselli A (2020a) Prof. Dr. Hubert Miller (1936–2020). Revista de la Asociación Geológica Argentina 77(2): 30–334

Aceñolaza FG, Toselli A (2020b) Hubert Miller (1936–2020). J S Am Earth Sci 103:102644

Adler F, Eremchuk J, Hongn F (2022) Ricardo Mon, Nota necrológica. Revista De La Asociación Geológica Argentina 79(2):386–396

Aguirre-Urreta B, Ramos VA (2021) Las cinco primeras geólogas egresadas de la Universidad de Buenos Aires: temprana paridad de género. In Guereschi, A., Martino, R. and Ramos, V.A. (eds.) La Mujer en la Geología, Asociación Geológica Argentina, Publicación Especial 14: 32–43

Ahlfeld F (1946) Geología de Bolivia. Revista Del Museo De La Plata 3(19):5–370

Ahlfeld F (1948) Estudios geológicos y yacimientos minerales de la provincia de Jujuy. Universidad Nacional de Tucumán, Instituto de Geología Minería, p 455., Jujuy

Ahlfeld F, Angelelli V (1948) Las especies minerales de la República Argentina. Universidad Nacional de Tucumán, Instituto de Geología Minería, p 304., Jujuy

Ahlfeld F, Branisa L (1960) Geología de Bolivia. Editorial Don Bosco, p 245., 2da. ed. 1972, Buenos Aires

Alcalde JA, Chayle W (2014) Historia del Instituto de Geología y Minería (1946–2014). San Salvador de Jujuy, Edinju

Alonso RN (2016) Dr. Ricardo H. Omarini, Nota necrológica. Revista de la Asociación Geológica Argentina 73(1): 149–155

Alonso RN (2018) El nacimiento de la Geología en la Universidad Nacional de Salta. In Ramos, V.A. and Martino, R.D. (eds.) El nacimiento de la Geología en la enseñanza universitaria, Revista Facultad de Ciencias Exactas, Físicas y Naturales, Vol 5 (Suplemento 1): 70–79, Córdoba

Alonso RN (2020) Ricardo J Sureda (1946–2020) Revista de la Asociación Geológica Argentina 77(4): 631–839

Alonso RN (2021a) Geología y minería en Jujuy. http://miningpress.com/debate/342873/escribe-alonso-geologia-y-mineria-en-jujuy?fbclid

Alonso RN (2021b) La rica historia de la Geología y Minería en Jujuy. Diario Río Negro, Editorial del Pregón Minero

Apesteguía S, Alvarez SM, Giacchino A (2022) Bonaparte: el amo de los dinosaurios. Fundación de Historia Natural Félix de Azara, p 320., Buenos Aires

Archangelsky S (2014) Paleobotánica y arte: experiencia de una vida. Revista Reseñas, Ciencia e Investigación 2(3): 6–19, Buenos Aires

Cecioni G (1953) Contribución al conocimiento de los nautiloideos eopaleozoicos argentinos. Boletín del Museo Nacional de Historia Natural 26(2): 57–110, Santiago de Chile

Cecioni G (1970) Esquema de paleogeografía chilena. Editorial Universitaria, p 144., Santiago de Chile

Charrier R, Aguirre L, Hervé F, Klohn E, Thiele R (2018) La carrera de Geología en la Universidad de Chile. In Ramos VA, Martino RD (eds.) El nacimiento de la Geología en la enseñanza universitaria, Revista Facultad de Ciencias Exactas, Físicas y Naturales 5 (Suplemento 1): 48–62

Charrier R, Godoy E, Hervé F, Parica C (2020) Prof. Dr. Dr.h.c. em. Hubert Miller (1936–2020) An articulator of international cooperation. Andean Geology 47(2): 466–467

Chayle W (1986) Raúl Chomnales, Nota necrológica. Revista De La Asociación Geológica Argentina 42(3–4):420–421

Georgieff SM (2015). Gerardo E Bossi (1939—2015) Necrológica. Revista de la Asociación Geológica Argentina 72 (3): 438 – 443.

Lago MG (2015) Historia de la Universidad Nacional de Jujuy (1972 - 2015). https://www.unju.edu.ar/historia.html

Lencinas A (2017) Relatos Mineros. 2da. ed. (en español), Sergio Lencinas (ed.), Amazon, p 163

Loss R (1948) Consideraciones preliminares sobre la cronología de las estructuras paleozoicas de los alrededores de La Quiaca (Dto. Yaví, Prov. Jujuy. Instituto de Geología y Minería de la Universidad Nacional de Tucumán, Publicación 456: 23–28, Jujuy

Loss R (1949[a]) Sobre el hallazgo del *Didymograptus climacograptoides* (llolm) en el norte argentino. Instituto de Geología y Minería de la Universidad Nacional de Tucumán, Publicación 504: 29–37, Jujuy

Loss R (1949b) Sobre la presencia de *Phyllograpti* y la estratigrafía del Ordovícico inferior entre La Quiaca y Yaví (Prov. de Jujuy). Instituto de Geología y Minería de la Universidad Nacional de Tucumán, Publicación 505 (II-4): 39–62, Jujuy

Loss R (1949c) Sur le presence de schistes graptoliferes dans l'Ordovicien inférieur de la frontiére Argentine-Bolivie. Compte Rendu des Séances De La Société Géologique De France 8:140–142

Loss R (1951) Contribuciones al conocimiento de las faunas graptolílicas del Norte Argentino. Revista De La Asociación Geológica Argentina 6(1):21–61

Loss R (1953) Sulla presenza del Didymograptus climacograptoides (Holm) del Nord Argentino. Atti della Società Italiana di Scienze Naturali 42:189–197, Milano

Méndez V (2005) Wilfredo Alejandro Lyons (1923–2005), Necrológica. Revista De La Asociación Geológica Argentina 60(3):626–627

Mon R (2022) Cincuenta años dedicados a la Geología Estructural y a la microtectónica, interrumpidos por golpes militares y retomados con el acceso a la democracia. Asociación Argentina para el Progreso de las Ciencias, Revista Ciencia e Investigación Reseñas 10: 40–61, Buenos Aires

Moreno JA (1970) Estratigrafía y Paleogeografía del Cretácico Superior en la Cuenca del Noroeste Argentino, con especial mención de los Subgrupo Balbuena y Santa Bárbara. Revista De La Asociación Geológica Argentina 25(1):9–44

Muruaga CM, Gavriloff IJC (2015) In Memoriam, Nota Necrológica Dr. Gerardo Eugenio Bossi (1939–2015). Latin American Journal of Sedimentology and Basin Analysis 22(1): 79–82

Omarini RH, Götze HJ, Mazzuoli R, Mat teini M, Peralta C (2005) Implication of the New Geochemical and Gravity data in the Lithospheric Crustal Evolution of the Cen tral Andes along the Nazca Plate to Chaco Plain Transect (24° S.L.). International Final Symposium, Deformation Processes in the Andes, 19th Colloquium on Latin-American Geosciences, Terra Nostra 90 p, Potsdam

Ottone EG (2018) Rafael Herbst (1936–2017), Nota Necrológica. Revista De La Asociación Geológica Argentina 75(2):296–302

Peña de Bascary S (2014) Federico Schickendantz, Un científico investigador y vehemente. Revista De La Junta De Estudios Históricos De Tucumán 14:85–124

Quintana CA (2012) Osvaldo Alfredo Reig. Conociendo a nuestros Científicos, Universidad de La Punta, p 41., San Luis

Ramos VA (2021) Hubert Miller (1936–2020). In Ramos, V.A. (ed.) Los Miembros Correspondientes de la Asociación Geológica Argentina. Serie D Publicación Especial 15: 74–80, Buenos Aires

Riccardi AC (1990) Giovanni Cecioni, Nota necrológica. Revista De La Asociación Geológica Argentina 45(3–4):410–411

Russo A (1974) Prof. Dr. Felix Celso Reyes (1924–1972). Revista de la Asociación Geológica Argentina 28(4): 439–440

Salfity J (2016) Apolo Ortiz (1936–2015), Necrológica. Revista De La Asociación Geológica Argentina 73(1):156–159

Salfity J (2017) Una vocación de vida: la geología regional del noroeste argentino. Asociación Argentina para el Progreso de las Ciencias, Revista Ciencia e Investigación Reseñas 5(1): 85–99, Buenos Aires

Sureda RJ (2008) Historia de la mineralogía. InsuGeo, Serie de Correlación Geológica 23, 193 p., San Miguel de Tucumán

Toselli A (2007) Federico Ahlfeld, una figura olvidada de nuestra Universidad. Actas 1° Congreso de Historia sobre la Universidad Nacional de Tucumán. 673–679

Viramonte JG (2017) La ardiente pasión por los volcanes. Asociación Argentina para el Progreso de las Ciencias, Revista Ciencia e Investigación Reseñas 5(2): 55–75, Buenos Aires

Chapter 19
The Geology of Coal and *Yacimientos Carboníferos Fiscales*

Abstract The need for coal supply expanded at the same time as the development of the railway network that operated with coal imported from England. In those years, Francis Rickard's studies established in 1869 that the Marayes coal deposits in San Juan were the most promising. Various studies were carried out for decades in different parts of the country without being able to discover good-quality coal deposits. With the beginning of the Second World War, explorations were restarted in the extra-Andean region of northern Patagonia where there were deposits of a similar age to those being exploited in Chile on the other side of the mountain range at those latitudes. The studies carried out allowed the Pico Quemado mine in Río Negro to be exploited, but no important deposits could be developed. New geological studies, in which Angel Borrello and Mario Hünicken played an important role, allowed the identification of the deposits of the Río Turbio coal district in Southern Patagonia. Under the geological direction of Federico Bergmann, coal exploration and exploitation were concentrated in this region. A state-owned company created in 1958, *Yacimientos Carboníferos Fiscales*, was subsequently in charge of coal development and exploitation.

Keywords Coal needed · Railway fuel · Initial failures · State company · Río Turbio

19.1 Introduction

The need for mineral coal arose at the same time as the development of the railway network whose fuel was imported from England. The rapid growth and expansion of the railway network increased interest in evaluating the possibility of exploiting mineral coal produced in the country. In those years there were a few mentions of the existence of mineral coal. Hóskold (1889, p. 308) in his inventory of the country's mineral resources mentions that before 1856 there was already knowledge about the existence of the Marayes coal basin, in the province of San Juan. A few years later,

V. A. Ramos, *History of the Geological Sciences in Argentina*, The Latin American Studies Book Series, https://doi.org/10.1007/978-3-031-83139-3_19

Victor Martín de Moussy in 1860, mentions his presence in San Javier, Corrientes, currently Misiones (see Chapter 4).

19.2 The Initial Studies

The mining engineer Francis Rickard in the years 1863 and 1868 made a tour of the mineral districts, mines, and existing establishments at the request of the first provincial and later national government, summarizing the results in his report of 1869 (see Chapter 4). The following year, President Domingo F. Sarmiento dictated Law 8198 to promote the exploitation of coal mines. In that law, it was established that *"anyone who discovers a coal mine in the Argentine Republic, in good condition to be exploited with advantages over imported coal for the purposes of commerce and industry, is awarded a prize of 25,000 pesos"*. Foreign dependence was serious, especially with the expansion of the railway network throughout the Argentine territory whose locomotives ran on stone coal.

Rickard in his report had concluded that the largest coal reserves were in the Marayes district, which over the years had sporadic exploitation. It was up to Alfred Stelzner in 1873 to obtain some samples with plant remains that, studied by Bruno Geinitz, allowed establishing a Rhaetian age for the carboniferous strata, which together with those of Cacheuta corresponded to the first discovery of this flora in Argentina. A detailed study of the basin is carried out at the request of the railway company by Juan Rassmuss (1922) who describes in detail the Marayes outcrops. Years later they were evaluated again by Nils Albert Lannefors (1930) with negative economic results. Seth Rosén (1934) presents one of the most complete studies on the stratigraphy and structure of carbonaceous seams. He concludes that although the resources are limited, they can be an important fuel reserve for the metal mining operations of the Sierra de la Huerta. One of the latest studies is presented in the doctoral thesis of Federico Bergmann, who concludes that, despite its lenticular nature and impurities, he expresses a certain optimism regarding the amount of coal obtainable (Bergmann 1947).

In the years 1871 and 1872, Alfredo Stelzner, as part of his research at the National Academy of Sciences, mentioned important carboniferous outcrops in the Tambillos River on the western slope of the Sierra de Famatina in La Rioja. Shortly after, in 1876, Germán Burmeister pointed out the existence of an accumulation of "lignite" in La Rioja, near Guandacol. That same year Ludwig Brackebusch prepared his well-known report on the geological conditions of the Paganzo deposit in this province. As a result of this carbon deposit, history remembers the exaggerated importance given to it by some private entrepreneurs, who invested a certain amount in the execution of various works, improving the conditions of the deposit, and went out to claim the desired prize, with negative results. This is how Hünicken expressed the work of these partners… *"A drilling machine was taken to Paganzo at great expense and Messrs. Irgazabal and Sánchez spent a lot of money drilling there to a depth of 1,000 feet, but no carboniferous seam was found worthy of exploitation"* (Crovara and Hünicken 2004).

Another pioneer of the studies of coal deposits was Guillermo Bodenbender, who from the National Academy of Sciences conducted important studies on various coal deposits and the first synthesis of their geological characteristics in his work *"On the age of some coal formations of the Argentine Republic"* (Bodenbender 1892, 1896 and 1902). He traveled from the north of the country to the central region studying the different manifestations and known coal mines. When it was created in 1902 in the Ministry of Agriculture, the Commission for the Study of Water and Coal Deposits appointed Eng. Enrique Hermitte in charge. In his 1904 synthesis on *"Coal, Oil and Water"*, he makes an inventory of the carbonaceous deposits of various provinces and, based on data from Bodenbender and other authors, presents their main geological and chemical characteristics (Hermitte 1904).

The creation of the General Direction of Mines, Geology, and Hydrology, dependent on the Ministry of Agriculture of the Nation in 1904, again under the direction of Eng. Hermitte, brought together the Division of Mines and Geology and the Commission that studied water and coal deposits. The first head of the Geology Division was Bodenbender who continued to study coal deposits (Bodenbender 1905, 1911).

In those years, numerous ventures in search of coal deposits were carried out in the Precordillera and the western Sierras Pampeanas by individuals stimulated by the offer of compensation. See the details in the review carried out by Janitens and Pezzutti (2004) and by Juan (2004). However, most of them were not economically important.

19.3 New Explorations

With the start of the Second World War in Europe, fearing that it could deprive us of a regular supply of coal, another important cycle of exploration began. On this occasion, attention is concentrated on the northern Patagonian region, where José Román Guiñazú, who made a reconnaissance of the region, makes a synthesis of the stratigraphic position of the known carboniferous deposits (Guiñazú 1940).

José Román Guiñazú (Fig. 19.1), a native of San Francisco del Monte de Oro, a small and historic town in the north of San Luis, after his secondary studies at the Juan Lafinur National School in the capital, moved to Buenos Aires. He studied the first three years of medicine at the University of Buenos Aires, abandoned his studies, and joined the Mines Directorate as a geologist assistant.

He spent several decades working and training as a self-taught geologist during his campaigns with Carl Caldenius in Patagonia. Thus, between 1929 and 1930 he accompanied the glaciological studies of Caldenius, when he found the flora of Pichileufú, in well-stratified Eocene lacustrine clays that contained abundant and beautiful impressions of leaves, branches, and fruits of fossil plants. In this flora, more than 130 species were recognized, which were studied by Edward W. Berry (1875–1945) in his monumental work of 1938, who, apart from thanking the specimens

Fig. 19.1 José Román Guiñazú (1897–1987) and his view of the Pichileufú river valley taken in 1930. *Source* Pérez Gutiérrez (2013) and Berry (1938)

provided and the data on their stratigraphic position, dedicated a kind of laurel to Guiñazú.

In his study of the Pichileufú valley (Fig. 19.2), he correlates the coal deposits with equivalent units in southern Chile that have been exploited over the years (Guiñazú 1940). In this work, he proposes that the Pacific Tertiary seas crossed the mountain range before its uplift and incursion only up to 70° longitude. He disagreed with Windhausen's scheme that brought together the Patagonian Atlantic ingression, after the uplift of the Patagonian Andes, with the Pacific which according to Guiñazú would be younger, as confirmed many years later (Ramos 1982). Furthermore, he associated the Pacific deposits with the coal seam-bearing strata known in northern Patagonia. In his paleogeographic reconstruction, he identifies a series of plant-bearing deposits from the Nahuel Huapi region to Santa Cruz, and Tierra del Fuego, where Hileman (1931) explored the Sloggett Bay coal beds. These deposits correspond in the northern sector to what is known today as the Ñirihuao basin. Although Guiñazú has other later works on the peat bogs of Tierra del Fuego, this study on the strata with charcoal remains from the pre-Andean region of Patagonia can be considered his main contribution to our Geology.

At the end of the 60 s during the survey of Sheet 20f Chepes in La Rioja, I had the opportunity to meet him and analyze together the extent and Pliocene age of the Los Llanos Strata based on his study presented at the First Argentine Geological Congress (Guiñazú 1960). Perhaps his last contribution was the discovery in the Lagarcito Formation of the Lower Cretaceous, together with José Bonaparte in what is now known as Loma del Pterodaustro in the Sierra de las Quijadas National Park, of a pterodactyl that bears his name (Bonaparte 1970). The existence of this national park was a project originally proposed by Guiñazú.

Remigio Rigal commissioned by the Direction of Mines (Fig. 19.3) carried out an inspection of the coal deposits in 1940 located at the foot of the northern Patagonian Andes where he evaluated the Newbery mine between Traful and Bariloche, the Quimey Malal mine north of Ñorquinco also known as Pico Quemado, the La China and Providencia mines in the Cushamen stream and the Epuyén mine in that town. In

Fig. 19.2 *Laurelia guinazui* Berry, named in honor of the discoverer of the Pichileufú flora by Edward W. Berri and the *Pterodaustro Gulnaz* Bonaparte from the Lagarcito Formation in San Luis. *Source* Berry (1938) and Museo de Ciencias Naturales Bernardino Rivadavia

those years, Rigal performed numerous studies of the coal seams of Mendoza, San Juan, and La Rioja, and even of the Chilean sector of Puerto Natales.

These studies did not find coal seams of economic importance and of all these mines the only one that had relatively important activity was that of Pico Quemado in

Fig. 19.3 José María Sobral and Remigio Rigal during fieldwork in northern Patagonia. *Source* Malumián (2011)

Río Negro. In this locality, four coal seams were identified in early Tertiary deposits that had some exploitation starting in 1939, totaling about 4,000 m of galleries and a production of 38,600 tons of raw coal. Although exploitation over the years has been modest, Federico Bergmann recognizes that this coal is one of the few that have coking properties to eventually be used in the San Nicolás blast furnaces (Bergmann, 1960).

19.4 The Creation of *Yacimientos Carboníferos Fiscales*

Given the beginning of the Second World War and the obvious fuel crisis that the country was experiencing, in 1941 *Yacimientos Petrolíferos Fiscales* was commissioned to carry out the necessary studies and searches to determine the possible exploitability of the coal manifestations in our country. The YPF Mineral Coal Division begins work in the Río Turbio and Cancha Carrera region in Santa Cruz.

Ángel V. Borrello (1918–1971) (Fig. 19.4), graduated from the National University of La Plata in 1942, had joined YPF where he later moved as a geologist to the recently founded Mineral Coal Division.

Shortly after, the General Direction of Solid Mineral Fuels was created in 1943, reporting directly to the Ministry of Industry and Energy. The energy situation was worrying for the supply of coal and intense exploration was carried out in different districts of the country, to finally concentrate on the Río Turbio Deposit where the first studies began to open Mine 1 of the deposit.

In those years Borrello made a complete synthesis of the country's coal deposits and manifestations with their respective evaluations. He presented a detailed history

Fig. 19.4 Dr. Ángel Borrello had an outstanding career between 1942 and 1959 in *Yacimientos Carboníferos Fiscales* and his magnum opus Solid Mineral Fuel Resources. *Source* Cingolani (2016)

of the development of exploration in its different stages, which we have followed in general outline in this chapter. He analyzed the different manifestations, dividing them by their stratigraphic position and their different regions, their grades, and energy potential, not only those of mineral coal, but he made a complete review of the asphaltites. Since then, this work has been the most rational evaluation of our coal resources and continues to be an updated reference work due to the validity of its data (Borrello 1956).

In 1946, Mario A. Hünicken (1926–2013) (Fig. 19.5) joined *Yacimientos Carboníferos Fiscales* as a scholarship recipient. After graduating as a geologist at the University of Córdoba, he moved to Río Turbio in 1948. He worked as a geologist, carrying out important studies on stratigraphy and paleontology of the Río Turbio-Cancha Carreras region, which were later used for his doctoral thesis, which he defended in 1951 at that same university. Although after obtaining his doctorate he left the coal exploration activity, he has left us complete studies of the Río Turbio coal basin, its flora, and its fossil fauna (Hünicken, 1955, 1965 and 1967).

The evaluation of the different manifestations distributed throughout the country allowed us to recognize that the Río Turbio deposit is the one that presented the greatest feasibility of exploitation.

The General Direction of Solid Mineral Fuels, which had a modest production of mineral coal in those years, led to the creation in 1958 of the state company *Yacimientos Carboníferos Fiscales*, initiating larger-scale production in Río Turbio.

Fig. 19.5 Mario Hünicken (1926–2013) and the Río Turbio Formation with the Dorotea coal seam, Río Turbio. *Source* Albanesi et al. (2013) and Hünicken (1955)

19.5 Río Turbio Site

Knowledge of the existence of coal manifestations in this sector of the province of Santa Cruz dates back to the news released in 1887 by Lieutenant Agustín del Castillo. During his hydrographic research work on the Santa Cruz River, he discovered carboniferous outcrops in the area of what is now known as Cancha Carrera.

In 1892, the studies carried out by Alcides Mercerat, as part of the explorations carried out by the Museum of La Plata again on the Santa Cruz River, detected large outcrops of mineral coal on its banks up to Río Turbio.

Federico Bergmann had received his doctorate from the National University of La Plata in 1947 with a thesis directed by Pascual Sgrosso on the carbonaceous manifestations of Marayes in the province of San Juan. That same year he began to work at Río Turbio, carrying out a series of detailed geological surveys to understand the structure and geology of the deposit, presenting a series of reports between 1949 and 1950, which allowed production to be channeled. Shortly after, in 1954–1955, he received a Humboldt scholarship to improve his knowledge of deposits and fuel technology at the *Bergakademie Clausthal* (Clausthal Mining Academy). Upon his return, he continued working in Río Turbio where he held the position of head of the Department of Exploration of Fiscal Coal Deposits (Figs. 19.6, 19.7, 19.8 and 19.9), a responsibility he held for many years (Bergmann 1960).

The drilling and surveying carried out have made it possible to expose the continuity of the coal seams of the Rio Turbio Formation of the middle Eocene age from Cancha Carrera in the north to the south of Río Turbio for almost 30 km. It is a sub-bituminous coal deposited in transitional sediments from marine to continental, suitable as industrial fuel and for distillation and gasification processes (Bergmann 1993). The total reserves of the deposit amount to 752 million tons (Carrizo and Valerdi 1990).

An updated synthesis of the geological characteristics of the Río Turbio region is presented by Malumián and Panza (2000), as well as the current state of knowledge of its coal deposits.

Fig. 19.6 First drilling carried out in the Río Turbio Deposit in 1941 and the opening of Mine 1. *Source* Y.C.F. archives

Fig. 19.7 Precarious exploitation conditions in 1947 led to the opening of Mine 2. *Source* Y.C.F. archives

Fig. 19.8 New exploitation systems carried out by the Río Turbio Carboniferous Deposits. *Source* Y.C.F. archives

Fig. 19.9 Thermoelectric plant connected to the national system supplied by coal from Río Turbio. *Source* Y.C.F. archives

In the last decades of the twentieth century, the deposit went through numerous technical, economic, and political problems that first led to its privatization as *Empresa Yacimientos Carboníferos de Río Turbo* (YCRT). In 2004 it was restored to the State and coal is finally being produced to supply a thermoelectric plant interconnected to the national system.

19.6 Final Considerations

It has been seen that for years mineral coal manifestations have been explored from Salta and Jujuy to Tierra del Fuego in outcrops of various ages from the Lower Carboniferous to the Eocene in different geological environments. This exploration was encouraged during periods of global conflagration that put the supply of mineral coal at risk for use in the railway system or industries. After the war periods where there was modest exploitation, production fell until the mines in operation were closed. Of all the manifestations evaluated, the only one that had characteristics for exploitation was the Río Turbio deposit, which was subject to the vagaries of economic policy.

It can be concluded that despite the exploratory effort made, high-quality coal deposits and a large amount of reserves for exploitation have not been revealed .

References

Albanesi GL, Tauber AA, Ortega G, Mazzoni AF (2013) Necrológica Dr. Mario A. Hünicken (1926–2013). Revista de la Asociación Geológica Argentina 70(3): 443–449

Bergmann FA (1947) Contribución al conocimiento de los yacimientos de carbón de Marayes, San Juan. Universidad Nacional de La Plata, unpublished PhD Thesis, La Plata

Bergmann FA (1960) La minería del Carbón Nacional, Problemas Argentinos. Revista de la Universidad 12: 11–24, La Plata

Bergmann FA (1993) Efectos metamórficos del carbono del manto Dorotea del Yacimiento Río Turbio (provincia de Santa Cruz). 12° Congreso Geológico Argentino y 2ª Congreso de Exploración de Hidrcarburos. Actas 5:306–310

Berry EW (1938) Tertiary Flora from Río Pichileufú, Argentina. Geological Society of America, Special Paper 12: 1–157, New York

Bodenbender G (1892) Sobre el carbón y asfalto carbonizado de la provincia de Mendoza. Boletín de la Academia Nacional de Ciencias 13: 151–170, Córdoba

Bodenbender G (1896) Sobre la edad de algunas formaciones carboníferas de la República Argentina. Revista Del Museo De La Plata 7:129–148

Bodenbender G (1902) El carbón Rhético de Las Higueras en la provincia de Mendoza. Boletín de la Academia Nacional de Ciencias 17: 139–160, Córdoba

Bodenbender, G. 1905. Informe sobre una exploración geológica en la región de Oran, Prov. de Salta. Boletín del Ministerio de Agricultura 4: 392–402, Buenos Aires.

Bodenbender G (1911) Constitución geológica de la parte meridional de La Rioja y Regiones limítrofes. Boletín de la Academia Nacional de Ciencias 19: 5–36. Córdoba

Bonaparte JF (1970) *Pterodaustro guiñazui* gen. et. sp. nov. Pterosaurio de la Formación Lagarcito, Provincia de San Luis, Argentina y su significado en la geología regional (Pterodactylidae). Acta Geológica Lilloana 10:207–226

Borrello AV (1956) Recursos Minerales de la República Argentina. III. Combustibles Sólidos Minerales. Revista del Instituto Nacional de Investigaciones de las Ciencias Naturales y Museo Argentino de Ciencias Naturales Bernardino Rivadavia, Ciencias Geológicas 5: 1–665, Buenos Aires

Carrizo R, Valerdi C (1990) Actualización de las reservas del yacimiento Río Turbio, Provincia de Santa Cruz. Actas del 11° Congreso Geológico Argentino 1: 187–190, San Juan

Cingolani CA (2016) La historia de la geocronología en La Plata: el Proyecto Borrello. In Riccardi, A.C. (ed.) La Historia de la Geología en el Bicentenario de la Argentina. Revista del Museo de La Plata, vol. 1, Número Especial: 86–102

Crovara E, Hünicken H (2004) La Rioja hasta los albores del siglo XX. In Lavandaio, E. and Catalano, E. (eds.) Historia de la Minería Argentina, Tomo 2: 157–174, Buenos Aires

Guiñazú JR (1940) El Terciario carbonífero del sur argentino y chileno. Boletín de Informaciones petroleras, Año 17(187): 16–71, Buenos Aires

Guiñazú JR (1960) Los llamados "Estratos de los Llanos" en la provincia de San Luis y su contenido de rocas andesíticas y restos de mamíferos fósiles. Primeras Jornadas Geológicas Argentinas, Actas 2: 87–95, Buenos Aires

Hóskold HD (1889) Memoria General y Especial sobre las minas, metalurgia, leyes de minas, recursos, ventajas, etc. de la explotación de minas en la República Argentina, ilustrada con mapas topográficos, geológicos, mineros, planos, secciones y vistas. 595 p., Imprenta y Estereotipia del Courrier de La Plata, Buenos Aires

Hermitte E (1904) Coal, petroleum and water in Argentine, paper presented in the Saint Louis Exposition. Ministry of Agriculture, p 85., Buenos Aires

Hileman G (1931) Informe Sobre la Comisión en Bahía Slogett, Yacimiento de Carbón, Tierra del Fuego. Dirección de Minas y Geología, unpublished report, p 41., Buenos Aires

Hünicken MA (1955) Depósitos neocretácicos y terciarios del extremo S-SW de Santa Cruz (Cuenca carbonífera de Río Turbio). Revista del Instituto Nacional de Investigación de las Ciencias Naturales, Museo Argentino de Ciencias Naturales Bernardino Rivadavia, Ciencias Geológicas 4(1): 1–164, Buenos Aires

Hünicken MA (1965) Algunos cefalópodos supracretácicos de Río Turbio, Santa Cruz. Revista De La Facultad De Ciencias Exactas, Físicas y Naturales, Universidad Nacional De Córdoba, Serie Ciencias Naturales 52:47–102

Hünicken MA (1967) Flora terciaria de los estratos de Río Turbio, Santa Cruz. Niveles plantíferos del Arroyo Santa Flavia. Revista De La Facultad De Ciencias Exactas, Físicas y Naturales, Universidad Nacional De Córdoba, Serie Ciencias Naturales 56:137–262

Janitens M, y Pezzutti, N (2004) Carbón. In Lavandaio, E. and Catalano, E. (eds.) Historia de la Minería Argentina, Tomo 2: 365–377, Buenos Aires

Juan FJ (2004) Mendoza. In Lavandaio, E. and Catalano, E. (eds.) Historia de la Minería Argentina, Tomo 2: 175–203, Buenos Aires

Lannefors NA (1930) Las Minas de Carbón de Marayes. Dirección General de Minas, Geología e Hidrología, Publicación 73: 1–28, Buenos Aires

Malumián N (2011) Historia de la concepción de un espacio geográfico denominado plataforma continental. Revista De La Asociación Geológica Argentina 68(3):407–414

Malumián N, Panza J (2000) Hoja Geológica 5173-III Yacimiento Río Turbio, Provincia de Santa Cruz. Programa Nacional de Cartas Geológicas de la República Argentina 1:250.000, Boletín 247, 108 p., Buenos Aires

Martín de Moussy V (1860) Description géographique et statistique de la Confédération Argentine. Librairie de Firmin Didot Fréres, Fils et C^{ia}, Tome Premier, p 582., Paris

Pereyra F, Boujon M, Gómez A, Tello N, Tobío MI, Lapido O (2010) Estudio geocientífico aplicado a la evaluación de la aptitud para la urbanización en la Cuenca Carbonífera de Río Turbio, Santa Cruz. Revista De La Asociación Geológica Argentina 66(4):505–519

Pérez Gutiérrez S (2013) Personalidades de San Luis que dejaron huella. Editorial San Luis Libro, p 234. San Luis

Ramos VA (1982) Las ingresiones pacíficas del Terciario en el Norte de la Patagonia. 3° Congreso Geológico Chileno (Concepción) Actas I(A): 262–288

Rassmuss J (1922) La cuenca de Marayes. Ministerio de Agricultura de la Nación. Boletín 32, Serie B, 32: 7–21, Buenos Aires

Rickard FI (1869) Informe sobre los distritos minerales, minas y establecimientos de la República Argentina en 1868–69. Ministerio del Interior, Imprenta, Litografía y Fundición de Tipos a Vapor, p 174., Buenos Aires

Rigal R (1940) Breve reseña sobre algunos Yacimientos de Carbón de la Cordillera Patagónica. Dirección de Minas y Geología, unpublished report, p 7., Buenos Aires

Rosén S (1934) Estudios geológicos del terreno carbonífero de Marayes, provincia de San Juan. Dirección de Minas y Geología, unpublished report, p 121., Buenos Aires

Súnico A (2002) La Minería. In Haller, M.J. (ed.) Geología y Recursos Naturales de Santa Cruz. 15° Congreso Geológico Argentino (El Calafate), Relatorio 4(1): 665–677, Buenos Aires

Chapter 20
The First Syntheses of Argentine Geology

Abstract At the end of the nineteenth century there were attempts to summarize the geology of our country in a single work, but knowledge was still embryonic. It was not until 1931 that Alselm Windhausen produced the first Argentine Geology. It was an extraordinary work due to its comprehensive development and the effort to describe the stratigraphy from the Precambrian to the Quaternary in great detail based on personal observations and an exhaustive analysis of previous works. It was the textbook for several generations of students and a reference work for geologists for decades. A second attempt was made by Armando Leanza in 1958, whose description is approached through the recognition of different morpho-structural units, and the geology is presented by regions. In 1965 Wolfgang Volkheimer tried to make a multi-author Argentine Geology, bringing together the referents and specialists of the different disciplines. However, he was not successful as he was unable to finish the three scheduled volumes. This last attempt motivated the National Academy of Sciences to undertake a multi-author book under the coordination of Armando Leanza as editor. The work was completed and published in 1972 with the participation of twenty-eight authors. The presentation and analysis were carried out with the description of regions proposed by Leanza. The success obtained led to a new update presented in two extensive volumes edited by Juan C. Turner and published in 1979–1980. Sixty-four authors participated in this work, also addressed from a regional point of view. Finally, the National Geological Survey published the latest Argentine Geology presented according to a stratigraphic order. It was edited by Roberto Caminos in 1999 with the participation of 128 authors.

Keywords Early reviews · Windhausen book · Regional symposium · Multiauthor volumes · Updated geology

V. A. Ramos, *History of the Geological Sciences in Argentina*, The Latin American Studies Book Series, https://doi.org/10.1007/978-3-031-83139-3_20

20.1 Introduction

There were some pioneering attempts to carry out a synthesis of Argentine Geology such as those of Juan Valentín in 1895 (Fig. 20.1), Enrique Hermitte in 1914, and Franco Pastore in 1925. However, the state of knowledge of those times did not allow more than to present a fragmentary geological sketch. Among them, Valentín's stands out, not only because it was the first, but because in its 3 years in Argentina, it achieved a pioneering representation of the main structural units, its tectonic style, and the vergences of its deformation. Despite the efforts put into these syntheses, large regions remained uncovered due to a lack of studies and an adequate understanding of our territory could not be obtained.

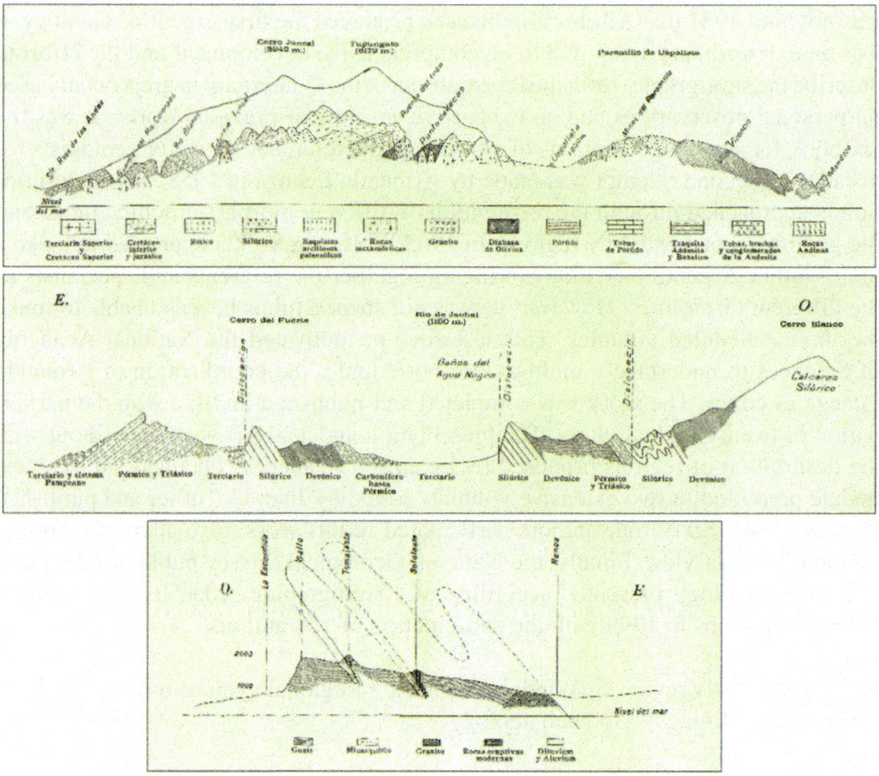

Fig. 20.1 Examples of some of the structural sections that Juan Valentín presented in 1895 to illustrate the style and characteristics of the different geological provinces based on previous work and his observations, where he illustrates the Main and Frontal Cordilleras, the Precordillera, and especially the Sierras Pampeanas where had worked personally. *Source* Valentín (1895)

20.2 The Argentine Geology of Windhausen

The first attempt to make a geological description of the stratigraphy and structure of the entire national territory corresponds to the cyclopean work of Anselmo Windhausen published in 1931 during the period in which he taught at the National University of Córdoba (see Chap. 12). This complete description of Argentine Geology, described in a stratigraphic sense from the Precambrian to the present day, has the merit of being a one-man work where it anticipates original concepts and novel hypotheses, some of which are still valid. Surprising in this work is the exhaustive analysis of the stratigraphy, accompanied by complete geological-chronological tables, detailed structural sections, and numerous paleogeographic sketches, presented with an abundant illustration of the main fossils of each period, with an impression of quality as it already did in those times the Peuser Publishing House (Windhausen 1931). The author says that his book is intended *"for teaching and amateurs"*. However, the topics are treated in depth, which explains why for decades it has been a study and reference book for several generations of geologists (Fig. 20.2).

Years later, in a series of books for secondary schools, several geology books appear, two of which stand out for the treatment they give to Argentine Geology. That of Giambastiani de Peláez (1935 to 1943), which had four editions between 1935 and 1943, and that of Pablo Groeber of 1938.

20.3 The Regional Geology of Leanza

It can be stated that a second attempt to bring together in one work a synthesis of the Regional Geology of our country was presented a few decades later by Armando Leanza in 1958 as part of a larger work. Under the direction of Francisco de Aparicio and Horacio A. Difriery, two famous Argentine geographers, *"La Argentina. Sum of geography"* a complete work presented in eight huge volumes where the first volume highlights the Regional Geology of Armando Leanza (1958). In this synthesis (Fig. 20.3), unlike that of Windhausen, geology is presented by region and its main merit is an identification of the different morphostructural units, characterizing them by their geological history, formed by continental and marine basins, with a specific age of its stratigraphic series. Leanza clarifies in the first lines that the rapid growth of geological knowledge in the last ten years has left previous syntheses outdated. He presents the different units as geological provinces, being one of the first proposals where the country is divided into units that are still in force, with a few exceptions, such as the Western Cordillera. This today is subdivided into its northernmost sector with that name and the central sector as Cordillera Principal.

As a division, it is much more complete than the one presented by Pablo Groeber in his Regional Geology chapter of 1938, which briefly describes the large structural units of Argentina.

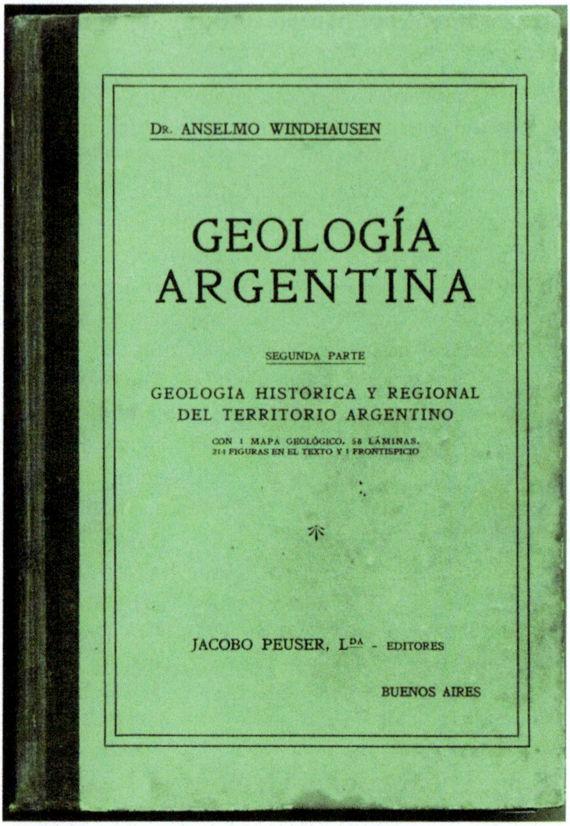

Fig. 20.2 First complete text of Argentine Geology of 645 pages written by Anselmo Windhausen in 1931

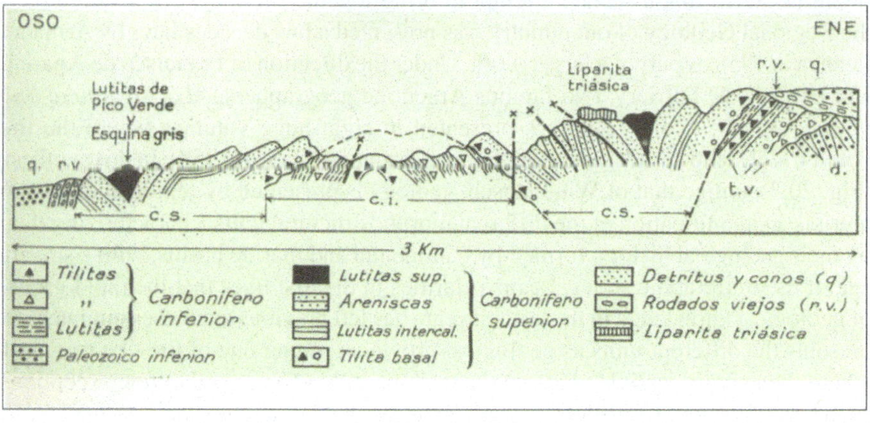

Fig. 20.3 Geological section of the Sierra de Barreal, San Juan, in Leanza (1958) based on Zöllner

In each province, it characterizes its stratigraphy and fossil content, its tectonic characteristics, and the age of its main diastrophisms. For Patagonia, he had the important contribution of Geology of Patagonia by Egidio Feruglio (1949–1950), which in its three volumes presents a complete description of the region, largely based on the studies of the geologists of *Yacimientos Petrolíferos Fiscales* unpublished. It covers the Malvinas Islands and the Argentine Antarctic and culminates with a unified geological history with excellent illustrations, which characterize the Peuser Editorial publications. This work has had a considerable impact on the geographic community and presents a notable bibliographic list with works almost exclusively from recent years.

20.4 Volkheimer's Argentine Geology

In 1965, Wolfgang Volheimer, at that time a professor at the University of Buenos Aires and a geologist at the National Institute of Geology and Mining, aware of the need to update the geological knowledge of our country, decided to produce a multi-author work in four volumes personally inviting the greatest leaders of Argentine Geology to write the different chapters. Its content had been planned to follow the stratigraphic sequence, a volume dedicated to the Precambrian and Paleozoic, another to the Mesozoic, and a final one on the Cenozoic, each of them divided by periods. The work committed to publishing the text in three volumes with Editorial Eudeba, reserving a last volume for a series of geological maps that would cover the entire country on an adequate scale, which Jorge Braidot was going to coordinate. The different texts were advancing rapidly and to standardize their content, the authors were summoned to the assembly hall on the ninth floor of Diagonal Sur 651, where the National Institute of Geology and Mining operated. That day the top leaders of Argentine Geology were gathered in the assembly hall. In his presentation, Volkheimer explained the rules for the final writing of each chapter, which should have a complete description of each period, analyzed by basins, with the different stratigraphic units accompanied by a distribution map of their outcrops, with a good description of their type profiles detailed illustrated, which culminated in a paleogeographic essay.

When Volkheimer finished his presentation, Pedro Stipanicic asked him to go to an intermission room. Stipanicic, Polanski, Rolleri, Cuerda, Yrigoyen, Turner, among several others, met in Guillermo Furque's office, next to mine. They were outraged because a young German who had arrived in the country a few years ago had gathered the will of the leading figures in Argentine Geology to write the book. Stipanicic, assuming the representation of the participants, resumed the intermission room and informed Volkheimer that the authors had decided not to participate in the book and left the meeting. The few of us who were left tried to console Volkheimer, who was so affected by the results of the event that have deprived us of having a well-structured and illustrated book on our geology.

However, that editing effort was not in vain. On the one hand, some of the authors independently published their results. For example, Juan C.M. Turner published shortly afterwards, in the proceedings of one of the Argentine Geological Congresses, three articles, Cambrian, Ordovician and Silurian, which had been written for Volkheimer's book. In turn, Jorge Polanski contacted Eudeba, and his book on Carbonic and Permian of Argentina was published in 1970, part of the frustrated attempt at a new Argentine Geology.

Perhaps the most important thing about this initiative is that Stipanicic interested in the National Academy of Sciences in Córdoba in preparing a new synthesis of our territory. The following year, potential authors were invited to carry it out.

20.5 Argentine Regional Geology Edited by Leanza

At the end of 1969, coordinated by the National Academy of Sciences, the Argentine Regional Geology (Fig. 20.4) was presented in Córdoba on the occasion of the centenary of its foundation. Academic Armando Leanza was in charge of its direction and editing. Before a packed audience, the different geological units by region were presented for a week, the content of which was a turning point in the knowledge of our national territory. The top representatives of the different institutions, including prominent geologists from *Yacimientos Petrolófilos Fiscales* and other companies, shared their experience and updated their understanding of the stratigraphy and structure.

Unlike Windhausen's synthesis, this new proposal, instead of analyzing geological evolution period by period, does so by region, largely following the divisions used by Leanza (1958). Except for the novel *"Mendocino Pampean Mobile Belt"* by Criado Roque, the geological provinces proposed in his previous synthesis were largely respected. Leanza himself, in the prologue of the written work, regrets that units such as the Deseado Massif have not been individualized, referring to it as a part of the *"Extra-Andean Region of Chubut and northern Santa Cruz"*. He also highlights that his proposal of the Patagonian Precordillera as an independent geological unit in the original division planned as editor has been included in the extra-Andean region, despite its unique morphostructural characteristics.

Despite these objections, this first multi-author work that was published by the Academy in 1972, brought together 28 prominent referents to write the different chapters and constituted an important milestone in the knowledge of Regional Geology.

Fig. 20.4 Argentine Regional Geology, written by 28 authors in 869 pages and directed by Armando Leanza published in 1972

20.6 Second Argentine Regional Geology Symposium Edited by Turner

The success obtained in the first symposium led the Academy to organize a Second Regional Geology Symposium (Fig. 20.5), which was coordinated by the academic Juan C.M. Turner in September 1976. When the different units selected by Turner are compared with those of the first symposium, some interesting differences emerge.

Firstly, the Deseado Massif is identified as an independent unit as postulated by Leanza (1958), but it is called Nesocratón del Deseado following the proposal of Harrington (1962). The term nesocraton is practically out of use; had been proposed to designate areas whose stability occurred in the Paleozoic. A similar problem arises with the Somún Cura Massif, also considered a nesocraton, which they identify as the North Patagonian Region. However, it is currently known that it has a basin in its central part and the metamorphic basement has just a peripheral position (see Ramos 2022, p. 1373–1374).

Two well-defined geological provinces such as Ventania and Tandilia, so-called by Nágera (1939 and 1940), are named the Northern and Southern Sierras of the province of Buenos Aires, in the first of which important geological-petrographic contributions are made.

Another novelty that this second edition introduces is the definition of Mesopotamia as an independent unit with specific characteristics. Despite these

Fig. 20.5 Second symposium on Argentine Regional Geology edited by Juan C.M. Turner in 1979–1980 under the auspices of the National Academy of Sciences with the participation of 64 authors in two volumes with 1716 pages

formal problems, it must be recognized that it produces an important advance in geological knowledge both on the surface and in the subsoil and the characterization of the geology of our continental shelf.

20.7 Argentine Geology by Caminos

Twenty years later, Roberto Caminos (1931–1997), at the initiative of the National Geological Survey (Fig. 20.6), planned and edited an Argentine Geology, approach with a rationality typical of his career. Unfortunately, he does not get to see the printed work, which was completed shortly after his death.

This new work addresses the stratigraphic point of view, but with a very completely new historical introduction (Rolleri et al. 1999), a location in the South American geological framework (Yrigoyen, 1999), and an update of the geological provinces of our territory, based on previous proposals (Ramos 1999a).

One of the main characteristics of this work is that in the twenty years since the previous Regional Geology, there has been strong progress in the geochronological knowledge of the metamorphic basement and igneous rocks through the application of new and more precise methodologies. Thus, the different stratigraphic units have

Fig. 20.6 This Argentine Geology edited by Roberto Caminos brought together 128 authors in its almost 800 pages

an adequate chronological analysis while the general knowledge of the environmental characteristics, both of the sedimentary deposits and the plutonic and volcanic rocks, allowed for a general and synthetic integration of the tectonic evolution of Argentina.

20.8 Final Considerations

It has been succinctly presented how geological knowledge of the country has progressed since the first works by a single author, where Windhausen (1931), Giambastiani de Peláez (1935 to 1943), Groeber (1938) and Leanza (1958) stood out. The rapid advance of studies in the second half of the twentieth century made it difficult to undertake successive syntheses with a single author, making it necessary to call on an increasing number of colleagues to carry them out. The first Regional Geology had 28 authors, the second with 64 and the last one in 1999 needed 128, which explains why it is increasingly difficult to update Argentine Geology.

Another important change in these syntheses is that based on the premises of plate tectonics, the division of units has been made more rationally and its understanding enriched the analyses carried out in its last edition in 1999.

However, the rapid advancement of geology in this century due to new technologies and tools for its study has produced an exponential growth of knowledge. This is also reflected in the greater number of local and foreign geologists who contribute to numerous investigations, which has made it a strong challenge to update their knowledge again.

References

Caminos R (1999) Geología Argentina. Servicio Geológico Minero Argentino, Anales del Instituto de Geología y Recursos Minerales, Segemar 29: 1–196, Buenos Aires

Cuerda AJ (1973) Reseña del Ordovícico Argentino. Ameghiniana 10(3): 272–312, Buenos Aires

Giambastiani de Peláez J (1935 a 1943) Nociones de Mineralogía y Geología Argentina. Talleres Jacobo Peuser, p 501, Buenos Aires

Groeber P (1938) Mineralogía y Geología. Espasa-Calpe Argentina, p 492, Buenos Aires

Harrington HJ (1962) Paleogeographic development of South America. Bull Am Assoc Pet Geol 46(10):1773–1814

Harrington HJ (1968) Desarrollo paleogeográfico de Sudamérica. Fundación Instituto Miguel Lillo, Miscelánea 26: 1–74, Tucumán

Hermitte EM (1914) La geología y minería argentina en 1914. Tercer Censo de la República Argentina: 406–494, Buenos Aires

Leanza AF (1958) Geología Regional. In: de Aparicio F, Difriery HA (eds) La Argentina, Suma de Geografía, Tomo I, Capítulo 3. Editorial Peuser, Buenos Aires, pp 217–349

Nágera JJ (1939) Geografía Física de la República Argentina. In Manito O, Nágera JJ (eds.) Geografía Física de las Américas y de la República Argentina, Editorial Kapelusz, p 232, Buenos Aires

Nágera JJ (1940) Tandilia. Historia física de la provincia de Buenos Aires. Universidad Nacional de La Plata, Humanidades I, p 272, Buenos Aires

Pastore F (1925) Nuestra Mineralogía y Geología durante los últimos cincuenta años. Cincuentenario de la Sociedad Científica Argentina (1872–1922), Evolución de las ciencias en la República Argentina 6: 1–47, Buenos Aires

Polanski J (1970) Carbónico y Pérmico de la Argentina. Eudeba, p 216, Buenos Aires

Ramos VA (1999a) Las provincias geológicas del territorio argentino. In: Caminos R (ed) Geología Argentina, Anales del Instituto de Geología y Recursos Minerales, Segemar 29(3). Buenos Aires, pp 41–96

Ramos VA (1999b) Evolución tectónica de la Argentina. In: Caminos R (ed) Geología Argentina, Anales del Instituto de Geología y Recursos Minerales, Segemar 29(24). Buenos Aires, pp 715–784

Rolleri EO, Caballé MF, Tessone MO (1999) Datos para una historia de la Geología Argentina. In: Caminos R (ed) Geología Argentina, Anales del Instituto de Geología y Recursos Minerales, Segemar 29(1). Buenos Aires, pp 1–34

Turner JCM (1972a) Cámbrico. 4tas. Jornadas Geológicas Argentinas (Mendoza), Actas 3: 181–190, Buenos Aires

Turner JCM (1972b) Ordovícico. 4tas. Jornadas Geológicas Argentinas (Mendoza), Actas 3: 191–210, Buenos Aires

Turner JCM (1972c) Silúrico. 4tas. Jornadas Geológicas Argentinas, (Mendoza), Actas 3: 211–224, Buenos Aires

Turner JCM (eds.) (1979–1980) Segundo Simposio de Geología Regional Argentina. Academia Nacional de Ciencias, v. I y II, p 1691, Córdoba

Valentín J (1897) Bosquejo geológico de la Argentina. Gmo. Van Woerden & Cia., p 50, Buenos Aires

Windhausen A (1931) Geología Argentina, segunda parte: Geología Histórica y Regional del Territorio Argentino. Ed. Jacobo Peuser, p 646, Buenos Aires

Yrigoyen M (1999) Situación de la Argentina en el marco geológico de América del Sur. In: Caminos R (ed) Geología Argentina, Anales del Instituto de Geología y Recursos Minerales, Segemar 29(2). Buenos Aires, pp 35–40

Chapter 21
The Conicet Foundation and Its Impact on Geological Sciences

Abstract The creation of a Scientific and Technical Research Council in 1958 allowed the development of exclusive research both in universities and in different institutions. After the first few years when leading figures of our geology such as Félix González Bonorino supported its development together with its first founder and president, Dr. Bernardo Houssay, there were periods full of diverse vicissitudes due to the different military governments. With the arrival of democracy, the institution was consolidated and became the promoter and supporter of the advancement of geology in its various orientations. The creation of regional research centers allowed studies and teaching to be carried out outside the major metropolitan areas. The formation of institutes where geology was developed from Tierra del Fuego and Patagonia to every corner of the country produced sustained progress in the geological knowledge of the different regions. The creation of scholarships spread throughout the country strengthened the learning and training of new geologists in different universities. The existence of a full-time research career allowed for the sustained development of vocations along with the possibility of undertaking postgraduate studies overseas. This academic policy made it possible to reinforce the vacant areas in the different disciplines and to strengthen geological research in different parts of the country.

Keywords Early years · Political problems · Research institutes · Research career · Science support

21.1 Introduction

It has been officially established that Conicet was created through a decree-law on February 5, 1958. However, when its history is analyzed in more detail, it stands out that it has been the culmination of a process that had begun almost a decade ago. In 1950, a National Direction had been created with objectives to promote and create scientific and technical research areas. That institution had not achieved true research development in the years it was in force. Thus, in 1957, when Félix González Bonorino came back from the United States, he was appointed interventor of the

V. A. Ramos, *History of the Geological Sciences in Argentina*, The Latin American Studies Book Series, https://doi.org/10.1007/978-3-031-83139-3_21

Fig. 21.1 Meeting of the national direction of scientific and technical research chaired by González Bonorino where Bernardo Houssay and other prominent scientists of the time participated. *Source* Ramos (2023)

National Direction of Scientific and Technical Research to develop and modernize the country under a state that promotes science and technology (Fig. 21.1). With the collaboration of a group of notable scientists, this geologist presented to the provisional president of the civic-military dictatorship of the moment the project of creating a National Council of Scientific and Technical Research as an autonomous entity dependent on the Executive Branch (Hurtado 2010).

This measure meant the closure of the National Direction and the transfer of its assets and administrative staff to the new organization that was founded in 1958 under the direction and extraordinary drive of Bernardo Houssay along with other prestigious scientists (Fig. 21.2). Once again, González Bonorino was its vice president and member of the board of directors on two occasions between 1958 and 1966. There is no doubt that the brand-new organization gave new directions to scientific research, generating and consolidating new lines of work. Through the creation of the full-time scientific researcher career, for the first time in the country, scientists were able to work on their projects full time.

The following years were marked by the institutional ups and downs of our country. However, over time this institution remained latent and allowed scientific research to develop, where geology played an important role. From the beginning, the presence of Félix González Bonorino allowed not only in the Department of Geological Sciences of the University of Buenos Aires but also in the National University of La Plata, that the development of positions with exclusive dedication together with a scholarship program marked a strong change in research through the first subsidies offered by Conicet. Previously, geological investigations were financed by institutions, preferably from the State, such as the Direction of Mines.

Fig. 21.2 Constitutive session of the national scientific and technical research council, on February 5, 1958. From left to right: Rolando García, Raimundo L. Parodi, Félix González Bonorino, Venancio Deulofeu, Alberto José Zanetta, Pedro E. Aramburu, Bernardo A. Houssay, Ignacio Pirosky, Eduardo Braun Menéndez, Humberto Ciancaglini, Fidel A. Alsina Fuertes, Luis F. Leloir, Eduardo De Robertis; semi-hidden Julio César Gancedo of the Ministry of Education and Justice. Source A.G.N. archives

21.2 The First Decades of Conicet

The beginnings of Conicet were marked by moments of high tension due to the scientific policy to be followed. Bernardo Houssay's essential driving force was basic research of excellence in the fundamental sciences that sought to legitimize itself in the internationalization of knowledge production (Fig. 21.3). A smaller group led by Rolando García was oriented in a direction much more connected to *"national needs"* and maintained that knowledge should be integrated into all sociocultural processes, emphasizing development and industrialization. When the 1962 coup d'état interrupted Arturo Frondizi's government, Rolando García tried to promote a statement against the coup. Not only did it not prosper, but a few days later the National Institute of Microbiology Dr. Carlos Malbrán, directed by Ignacio Pirosky was intervened, who was separated from all his positions, a decision that motivated Cesar Milstein to resign and go into voluntary exile to the exterior. Already in 1965, Rolando García's group had lost weight in the Conicet, which is reflected in the *"noche de los bastones largos"* (night of the long canes) of 1966, where four days

Fig. 21.3 Bernardo Housay (1887–1971), Rolando García (1919–2012), and Félix González Bonorino (1918–1998), three important protagonists of the first years of Conicet. *Source* Foglia (1971) and Biblioteca Digital F.C.E.N

later the institution recognized that it was the Government's responsibility to reform the university legislation to the time that declared "*The violence committed against teachers and students implies a true insult to Argentine culture and affects the good name of the country*", disapproving the forms but not the decisions of the executive branch. Houssay continued to lead the institution until he died in 1971.

From 1971 to 1983, Conicet had more than seven presidents who alternated according to the political changes that occurred. Perhaps the most notable thing is that from those years on, the formation of new institutes was prioritized, which from about ten at the beginning of the seventies rose to 116 in 1983, a process whose dominant characteristic was the separation of these research centers from the universities.

21.3 The Policies Implemented by Conicet Until the Final Crisis

The Conicet area was not immune in those years to the disappearances and persecutions during the last civil-military dictatorship. In the organization, almost a hundred researchers were laid off, and "*news about missing scientists began to circulate in international newspapers and magazines in the sector*". It should be mentioned that "*before the coup of '76, due to the influence of the Argentine Anti-Communist Alliance (Triple A), researchers had already been dismissed under the pretext of National Security*" (Hurtado de Mendoza and Feld 2008; Hurtado 2010). In the field of Earth Sciences, several geologists were laid off, perhaps one of the most charismatic cases being Eduardo J. Llambías and Ricardo Mon (Fig. 21.4) (see Llambías 2013 and Mon 2022). In both cases, local professional jealousy was more important

Fig. 21.4 Eduardo J. Llambías (1937–2018), first from the right, and Ricardo Mon (1940–2022), had to interrupt their academic life to work, one as a miner in San Luis and the other in geology applied to the construction of dams respectively. *Source* Cingolani and Leanza (2018) and Mon (2022)

than their political ideas. These researchers were reinstated, the first in 1980 and the second in 1985, with the advent of democracy.

The arrival of Raúl Alfonsín, the first democratically elected president after a series of dictatorships and civil-military coups, marked a change in Conicet's scientific policies. Carlos Abeledo was appointed to the presidency of Conicet and during his administration not only was the expendability law abolished by which numerous researchers and technicians had been fired from the institution, but a stage of reintegration and recovery of those laid-off scientists began. During this period, the researcher's career, scholarship programs, the implementation of a financing system for research and development projects, effective coordination with universities, along with evaluation processes by the scientific community itself, were strengthened, based on strictly academic criteria (Albornoz and Gordon 2011). However, strong financial restrictions resulted in the freezing of Conicet's budget during the period.

With the arrival of Saúl Menem to the presidency of the Nation in 1989, neurosurgeon Raúl Matera was appointed in charge of the Secretariat of Science and Technology. After Carlos Cavotti's brief tenure at the head of Conicet, geologist Bernabé Quartino was appointed president of the institution. Quartino *"publicly maintained his intention to separate the Conicet from research in the universities, in line with the policy followed during the military dictatorships and in contrast to the policy of rebuilding the link between the Council and the universities carried out during the management of the radical government"* (Albornoz and Gordon 2011).

Quartino stepped away from the Conicet presidency in 1992 after a confrontation with the Secretary of Science and Technology, Raúl Matera, who asked him to resign. He told him that he did not know science or Conicet and that he had accepted his position as hard work, not to be a vassal, and had not managed Conicet as a personal fiefdom. Matera, in disagreement with the policies carried out, took charge of Conicet without resigning from the Secretariat, until his death, being replaced by Domingo Liotta. It is noteworthy that in this period (1989–1994) there was an important dismantling of the policies devised under Alfonsinism.

Liotta was followed as president of Conicet by another geologist, Gilberto Aceñolaza in 1996. Although he had a short period where he tried a series of regionalization measures for Conicet, it was a conflictive moment, so he left the presidency and remained part of the Conicet. directory until 1998 (Aceñolaza 2015).

These were times of important changes in Science and Technology and in the scientific policy of Conicet, which led Juan Carlos del Bello to the Secretariat of Science in 1996, the National Agency for Scientific and Technological Promotion was also created and Enrico Stefani was appointed as Conicet's interventor. In 1998, dissent between the Secretariat and the Council led to Stefani's resignation and his replacement by Armando Bertranou, with the institution subsisting with diminished resources. These resources were increasingly reduced, a trend that had begun with Domingo Cavallo when he sent researchers *"to wash the dishes"* and that gave rise to a robust reaction from the scientific community to the neoliberal policies that had been implemented.

The situation worsened during the Alliance government, which put Dante Caputo in the Secretariat and Pablo Jacovkis in the presidency of Conicet. Neoliberal measures together with a strong reduction in the Conicet budget imposed by the measures of the International Monetary Fund led to a deep crisis of the scientific system.

The economic, social, political, and institutional crises that shook the country lasted during 2001 and 2002 and led scientific institutions to a serious financial situation. A gradual economic recovery that began to be felt after the government of Néstor Kirchner and the appointment of Eduardo Charreau in charge of Conicet led to a reconciliation between the Science and Technology institutions and the national government. The creation of the Ministry of Science and Technology and a gradual increase in allocations from the scientific system led to the strengthening of the scientific base and technological capacity. The expansion of the number of scholarships and the reopening of admission to the research career marked the scientific development of those years.

21.4 The First Geological Research Centers Created by Conicet

The Austral Center for Scientific Research (CADIC) was the first regional center created by Conicet in Ushuaia in 1969 (Fig. 21.5). Its main objectives were to *"promote, support and execute scientific studies and tasks in the National Territory of Tierra del Fuego, Antarctica and the South Atlantic Islands and adjacent areas"*. Throughout its more than 50 years of existence, it covered studies of its geological and soil resources, hydrometeorology, physical oceanography, bio marine resources, ecology and genetics, and subantarctic forest ecology, among several others. The

Fig. 21.5 Austral Center for Scientific Research (CADIC) in Ushuaia, one of the first regional centers of Conicet

second director of the center was the geologist Carlos Schroeder between 1983 and 1986. In the end, another geologist Jorge Rabassa (2023) was appointed, the first director by a national public competition in 1986. He was followed by Eduardo Olivero (2020) in 1992, Gustavo González Bonorino in 2004, and again Rabassa in 2011.

In 2012, the National University of Tierra del Fuego began its activities, which include the Geology degree, with the center's researchers having an important participation. Geological investigations stood out in the study of glaciations and geomorphology of the region with Jorge Rabassa; the mainly Meso-Cenozoic stratigraphy of its sedimentary deposits and its correlation with the Antarctic Peninsula with Eduardo Olivero; coastal geology, among other studies.

The second regional research center created is the National Patagonian Center (CENPAT) in Puerto Madryn in 1970 (Fig. 21.6), as an appendix to the Observatory of Geoheliophysical Studies of San Miguel, created at the initiative of Monsignor Fortunato J. Devoto S.J. The Center was incorporated into Conicet in 1978 and became part of its system of Regional Centers. The first geologist incorporated was Carlos Beltramone, who worked in geomorphology in the Arid Zones areas. In 1984, geologist Carlos Meister, dependent on the provincial Tourism Direction, joined as a seconded member, who worked on the geological survey of the Reserves. In 1986, Miguel J. Haller joined, who will twice serve as director of the center between 1988–1989 and 2001–2002 (Haller 2020).

The first geologists incorporated constituted a new scientific area, which finally constituted the Patagonian Institute of Geology and Paleontology in 2015, which today has numerous researchers, scholarship holders, and research support technicians. His works with the greatest geological impact are related to the study of volcanism and Jurassic pyroclastic deposits that dominate the Patagonian region at these latitudes, headed by Miguel Haller. CENPAT continues to be an institution dependent exclusively on Conicet.

Fig. 21.6 Miguel Haller, twice director of the National Patagonian Center (CENPAT) of Puerto Madryn. *Source* courtesy Miguel Haller

Another example of the policy of unlinking research with the university that dominated the seventies until 1983 was the Geological Resources Research Center (CIRGEO), created in 1974, with Bernabé Quartino (1925–2003) as director (see Rinaldi 2003), with the participation of Horacio Camacho (1922–2015) (see Camacho 2013; Aguirre Urreta 2016) and Sergio Archangelsky (1931–2022) (see Archangelsky 2014; Del Fueyo et al. 2022). In his twenty years of life, the scientific achievements obtained were not commensurate with the conspicuous financial support received. This center was finally closed in 1994, and its collections were transferred to the Bernardino Rivadavia Museum of Natural Sciences.

A center founded in those early years by Conicet was the Institute of Geochronology and Isotopic Geology (INGEIS) in 1969 (Fig. 21.7), created within the Department of Geological Sciences, as a unit shared between the University of Buenos Aires, Conicet and the National Geological Survey. This institute carried out its first K–Ar dating in 1972 and was under the direction of Enrique Linares (1928–2011) from 1969 to 2000 (Ostera 2015).

Fig. 21.7 Enrique Linares, first director and the new facilities of the Institute of Geochronology and Isotopic Geology (INGEIS) in the Campus of the University of Buenos Aires. *Source* Ramos (2018) and courtesy Maximiliano Naipauer

In 1976 it moved to the current facilities in the Campus of the University, inaugurating its new building in 1982, during Linares Conicet's vice presidency. It produced not only the first K–Ar and Rb–Sr dating, but also had extensive development in stable isotope analyses. For a few years now it has been under the direction of Maximiliano Naipauer.

The Argentine Institute of Nivology and Glaciology (Ianigla) was founded in the city of Mendoza in 1972 (Fig. 21.8) at the initiative of Arturo E. Corte (1919–2001) as a center of Conicet, with the support of the National University of Cuyo and the provinces of San Juan and Mendoza. It was the first institute dedicated to the study of periglacial processes and geocryological studies. Arturo Corte, graduated as a doctor in 1947 with a thesis directed by Otto Schlagintweit at the National University of Córdoba, was an expert who had investigated these processes in Scandinavia, Greenland, and Antarctica, being an international referent on these topics and was widely recognized for his studies on rock glaciers and permafrost (Tromotto and Brown 2001). His Geocryology text remains the first on the subject and one of the few in Spanish. The Institute received support from the General Direction of Water and Electric Energy, the National Institute of Economy, Legislation and Water Administration, and years later from Hidronor due to the importance of its research. Wofgang Volkheimer (1928–2018) was director of Ianigla between 1993 and 2001, developing and emphasizing palynological studies of the Paleozoic and Mesozoic, with special emphasis on Mesozoic paleoclimates (Volkheimer 2014). In 1994, Conicet expanded its name by adding *"and Environmental Sciences,"* to explicitly expand the institute's objectives. Ianigla was established in 2007 as a triple dependency executing unit shared by Conicet, the National University of Cuyo, and the Government of the Province of Mendoza.

Ianigla has different groups in Earth Sciences, highlighting Laura Giambiagi and José Mescua in tectonics; Claudia Rubinstein, Mercedes Prámparo and Ana María Zavattieri in palynology and paleobotany; Osvaldo Bordonaro and Matilde Beresi in paleoinvertebrates, among others.

Fig. 21.8 Arturo Corte is one of the founders of Ianigla and its current facilities. *Source* Dimieri and Frisicale (2018)

An important Conicet institution in the field of Geology is the Geological Research Center (CIG) created in 1980 as an institute of the National University of La Plata (Fig. 21.9), which later became a center of double dependence on Conicet, to develop basic and applied scientific research. His first director was Carlos Cingolani, followed for many years by Mario Iñiguez (1937–1996) (see Poiré 1997), who was his director from 1982 until his death. This institute brought together at the time a plethora of young researchers such as Carlos Cingolani (2015a), Mario Mazzoni (1942–1999) (see Matheos 1999), Luis Dalla Salda (1942–2015) (see Cingolani 2015b), Carlos Rapela (2018) and Luis Spalletti (2015), who over the years made important contributions to our geology.

His research stood out for its contributions to sedimentology, the study of the petrology of the metamorphic basement of our country, especially that of the Sierras Pampeanas, and for his contributions to geochronology, despite not having an isotopic dating laboratory in the Institute.

The Higher Institute of Geological Correlation (InsuGeo) was organized at the National University of Tucumán in 1984 (Fig. 21.10) by a group of geologists interested in investigating the geology of northwest Argentina. Starting in 1996, it became a dual-dependent institute of Conicet, covering research in regional geology, petrography, geochemistry, paleontology of invertebrates and vertebrates, sedimentology, ichnology, dynamic stratigraphy and hydrogeology.

Among its first researchers, Gilberto Aceñolaza and Alejandro Toselli stood out, producing important contributions such as the Regional Geology of Northern Argentine, a book awarded by the National Secretariat of Culture. For several decades they worked with Professor Hubert Miller of the University of Munich, producing numerous publications on different aspects of the Famatina System and the Puncoviscana basin.

Fig. 21.9 Carlos Cingolani first director and current building of the geological research center (CIG) of double dependency between Conicet and the national university of La Plata. *Source* courtesy Carlos Cingolani

Fig. 21.10 InsuGeo facilities in Horco Molle, Tucumán, and associated geological park

This institute has been publishing the Geological Correlation series and other publications on different aspects of geology for forty years. Also noteworthy is the publication of *"Norte Geológico"*, a magazine active between the mid-1970s and 1980s, focused on geological news and the history of geology. It is currently made up of around twenty researchers and numerous scholarship recipients, mostly teachers or researchers from the Faculty of Natural Sciences and the Miguel Lillo Institute.

21.5 The New Institutes of Double Dependence with the Universities

At the beginning of the 2000s, Conicet decided to generate new executing units of double dependence with national universities within a policy of rapprochement between research and university teaching. As a result of this policy carried out by the president of Conicet Eduardo Charreau numerous centers were created in different regions of the country, especially geological centers where studies are currently carried out.

A group of researchers at the Department of Geological Sciences of the University of Buenos Aires had been organized to create the Andean Tectonics Laboratory in 1984. Its main objective was to study the Andes Mountains from the most varied disciplines to establish its tectonic evolution. Over the years and with the participation of Ernesto Cristallini, Andrés Folguera, Matías Ghiglione and Daniel Pérez under the supervision of Victor A. Ramos, a comprehensive survey of the Alta Cordillera de San Juan and Mendoza began. The book *Geology of the Aconcagua Region* (Fig. 21.11) was one of the emblematic results that made the laboratory's work known (Ramos et al. 1996). At the same time, the laboratory continued working on the metamorphic basement of the western Sierras Pampeanas through the studies of Graciela Vujovich in Pie de Palo, who focused on investigating the geochemical and geochronological characteristics of its rocks. She extended this activity with her students to the ophiolites of the Western Precordillera, making important contributions to its tectonic evolution.

The unification of the High-Resolution Biostratigraphy Laboratory under the direction of Beatriz Aguirre-Urreta with the Andean Tectonics Laboratory formed the

Fig. 21.11 Some of the books published by IDEAN researchers allowed the institute to be recognized as a reference center for the study of the Andes worldwide

Don Pablo Groeber Institute of Andean Studies (IDEAN), which was recognized as a double dependency institute by Conicet in 2006. In these years Scientific production related to the evolution of the Andes from biostratigraphy, geochronology, structural geology, and tectonics, turned this institute into an international reference center for the study of the Andean Mountains.

During the government of President Saúl Menem, the Regional Center for Scientific Research and Technology Transfer of La Rioja (CRILAR) was created in March 1998 in Anillaco (Fig. 21.12), Menem's hometown. Its first director was the geologist Mario Hünicken (1926–2013) and in his memory, the center's auditorium is named after him (see Albanesi et al. 2013). It depends on Conicet, the province of La Rioja, the national universities of La Rioja and Catamarca, and the Argentine Mining Geological Survey. His research focuses on Mesozoic paleo vertebrates, paleobotany, and the study of Paleozoic magmatism, especially type A from the lower Carboniferous, and is currently under the direction of Mariano Larrovere.

Fig. 21.12 Mario Hünicken, first director and the Regional Center for Scientific Research and Technology Transfer of La Rioja (CRILAR) in Anillaco. *Source* Albanesi and Fiorelli (2013)

Fig. 21.13 Pedro Depetris, first director and the Earth Sciences Research Center (Cicterra) of double dependency between Conicet and the National University of Córdoba. *Source* Depetris (2014)

A group of researchers from the National University of Córdoba who had been carrying out petrographic, mineralogical, geochemical, and Paleozoic invertebrate fauna studies met to form the Earth Sciences Research Center (Cicterra), which in 2007 was recognized as an executing unit of double dependency between the National University of Córdoba and Conicet (Fig. 21.13).

At this institute, Luis Benedetto (2015), Teresa Sánchez, Ricardo Astini, Pedro Depetris, Diego Gaiero, among many others, stood out in their paleo biological studies, basin analysis, environmental geochemistry, and other disciplines. Its first director was Pedro Depetris, a senior researcher at Conicet with extensive experience in studies of the Critical Zone that analyzes the permeable layer near the surface at the intersection of various disciplines such as pedology, hydrogeology, ecology, and biogeochemistry, where geology and biology are a fundamental part of your knowledge (Depetris 2014).

The Geological Institute of the South (Ingeosur) located in Bahía Blanca at the National University of the South was created in the Department of Geology of that university in 1994 and since 2007 has been an executing unit of double dependence on Conicet. Its first director was Ernesto Bjerg (2021). In recent years he has carried out various studies of regional geology, petrology, mineralogy, paleontology and applied geology. His main referents are Luis Dimieri, Daniel Gregori, Martín Turienzo, and Ernesto Bjerg.

In the city of Santa Rosa, the Institute of Earth and Environmental Sciences of La Pampa (Incitap) was founded in 2008, an executing unit of double dependence between the National University of La Pampa and Conicet. Its first director was Marcelo Zárate and his research focused on Quaternary geology, soil science, and petrology, especially the extra-Andean volcanism of Payenia.

The rector of the National University of Río Negro, Juan Carlos del Bello, established technical working groups in the different regions of the province. In the city of General Roca, the Research Institute in Paleobiology and Geology (IIPByG) was created in 2009 (Fig. 21.14), within the framework of the Alto Valle Campus of the university, determining the area of Earth Sciences as a priority. This institute, which is also dependent on Conicet, had Rodolfo Coria as its first director. Among its researchers, Silvio Casadío, Maisa Tunik, Pablo González, among many others, stand out. The research carried out focuses on studies of sedimentary rocks, geophysics applied to oil prospecting, and paleo-biological studies, preferably concentrated in the Neuquén basin.

In northern Argentina, at the initiative of researchers from the Faculty of Natural Sciences of the National University of Salta, the NOA Institute of Bio and Geosciences (IBIGEO) was created in the city of Salta in 2005. Cecilia del Papa, Fernando Hongn, Alicia Kirschbaum, and Ivan Petrinovic participated in its foundation, and since 2009 has been an executing unit with double dependence on Conicet. His main lines of research are the structural and tectonic geology of northwest Argentina, the petrology of the Paleozoic basement, geothermal energy, and volcanism of the Andes.

The Institute of Basic, Applied, and Environmental Geosciences of Buenos Aires (IGEBA) is an Institute of double dependency between the University of Buenos Aires and Conicet created in 2011, being fully operational as of 2013. It includes the Laboratory of Paleomagnetism Ingeniero Daniel Valencio with a long career and a pioneer in these studies, which was under the direction of Juan F. Vilas. His main lines of research include magnetism of environmental and Quaternary rocks, coastal geology, studies of sedimentary basins, and environmental sciences. Its directors were Oscar Limarino and Augusto Rapalini, and it is currently directed by Roberto Scasso.

Fig. 21.14 Headquarters of the Research Institute in Paleobiology and Geology in the city of General Roca

Fig. 21.15 Patricia Alvarado, seismologist and director of the National Institute of Seismic Prevention in San Juan. *Source* U.N.S.J. archives

At the National University of San Juan, the Geosphere and Biosphere Research Center (Cigeobio) was created in 2012 as an institute with double dependence on Conicet. Among its founding members was Patricia Alvarado (1977–2022), an international leader in the field of seismology. She graduated with a degree in Geophysics from the National University of San Juan, completed a Master's degree in Sciences at the University of Chile, and received a doctorate in Geosciences from the University of Arizona. The institute develops three lines of research, seismotectonics, Quaternary geology, and petrology. Dr. Alvarado's early death found her directing the National Institute of Seismic Prevention in San Juan (Fig. 21.15). During these years she made important contributions of high impact in seismology and seismotectonic interpretation that had made her an undisputed leader in our country on these issues.

Finally, the Center for Andean Geological Studies (CEGA), was created by Ricardo Omarini in Salta in 2014 and is made up of researchers from Conicet and the National University of Salta. He is linked to the InsuGeo of Tucumán with his group of researchers from Salta, including Ricardo Sureda, Cristina Moya, Julio Monteros, Ricardo Alonso (2021), Vanina López de Azarevich, Claudia Galli, among others, as well as interns and technicians. His work addresses topics related to the evolution of the Ediacaran-Cambrian, Lower Paleozoic, and Cenozoic basins of the region, along with studies of geology, mineralogy, paleontology and mineral deposits in northwest Argentina.

21.6 Final Considerations

The actions of Conicet with the creation of geological research centers throughout the country have been highlighted given that it allowed the activity to be strengthened in regions that previously did not have institutes in the field of geology. However, perhaps the greatest contribution made since the 1960s has been the financing through subsidies of numerous research projects by academics and professors working in universities. The creation of the Scientific Researcher Career, which began through training scholarships, consolidated a body of researchers with a wide regional distribution, although with a certain concentration in the central region of the country. Although these activities had moments of crisis, such as at the end of the 1990s, there were moments of greater prosperity. Many of the currently active researchers have started through doctoral and postdoctoral scholarships, which have also had periods of boom and restrictions depending on scientific policies and the economic situation of the country. It must be recognized that together and after numerous twists and turns, the research carried out by Conicet has achieved a certain positive recognition in society that also extends to the Earth Sciences. This acceptance by society of the work of Conicet has been strengthened especially in recent years.

In the following chapters will be highlighted, the geological personalities who, through the support received by Conicet, have made significant contributions, some of them of high impact in the advancement of knowledge of Argentine Geology.

References

Aceñolaza FG (2015) En busca del pasado geológico. Asociación Argentina Para El Progreso De Las Ciencias, Reseñas De Ciencia e Investigación 3(2):6–19

Aguirre Urreta MB (2016) Horacio Homero Camacho (1922–2015), Necrológica. Revista De La Asociación Geológica Argentina 7(2):294–299

Albanesi GL, Fiorelli LE (2013) Prof Dr Mario A Hünicken (1926–2013). Ameghiniana 50 (4): 472–473

Albanesi GL, Tauber AA, Ortega G, Mazzoni A (2013) Dr Mario A Hünicken, Necrológica (1926–2013). Revista de la Asociación Geológica Argentina 70(3): 443–449

Albornoz M, Gordon A (2011) La política de ciencia y tecnología en Argentina desde la recuperación de la democracia (1983–2009). In: Albornoz M, Sebastián J (eds) Trayectorias de las políticas científicas y universitarias de Argentina y España, CSIC. Madrid, pp 1–46

Alonso RN (2021) Puna, boratos y salares: bajo el signo de la mente y el martillo. Asociación Argentina Para El Progreso De Las Ciencias, Reseñas De Ciencia e Investigación 9(4):7–24

Archangelsky S (2014) Paleobotánica y arte: experiencia de una vida. Asociación Argentina Para El Progreso De Las Ciencias, Reseñas De Ciencia e Investigación 2(3):6–19

Benedetto JL (2015) El inabarcable mundo de los fósiles. Asociación Argentina Para El Progreso De Las Ciencias, Reseñas De Ciencia e Investigación 3(1):22–39

Bjerg EA (2021) Hallazgos afortunados: minerales metalíferos y rocas del manto terrestre. Asociación Argentina Para El Progreso De Las Ciencias, Reseñas De Ciencia e Investigación 9(1):9–26

Camacho HH (2013) Mis estudios de la geología y la paleontología de nuestra región austral. Asociación Argentina Para El Progreso De Las Ciencias, Reseñas De Ciencia e Investigación 1(2):34–37

Cingolani C (2015a) Medio siglo descubriendo el tiempo en la historia de la Tierra. Asociación Argentina Para El Progreso De Las Ciencias, Reseñas De Ciencia e Investigación 3(1):44–66

Cingolani C (2015b) Prof. Dr. Luis Hugo Dalla Salda (1942–2015). Revista de la Asociación Geológica Argentina 72(4): 578–586

Cingolani CA, Leanza HA (2018) Eduardo J. Llambías (1937–2018). Revista de la Asociación Geológica Argentina 75(3): 626–634

Del Fueyo GM, Cárdenas O, Carrizo MA, Lafuente Diaz MA, Lezama LR (2022) Sergio Archangelsky (1931–2022): reseña biográfica y su contribución al conocimiento de las comunidades vegetales del Cretácico Inferior de Patagonia Austral. Revista del Museo Argentino de Ciencias Naturales, n.s. 24(2): 197–215

Depetris PJ (2014) Medio siglo investigando en la "zona crítica." Asociación Argentina Para El Progreso De Las Ciencias, Reseñas De Ciencia e Investigación 2(3):31–44

Foglia VG (1971) Bernardo Alberto Houssay (1887–1971). Acta Physiologica Latinoamericana 21: 267–285, Buenos Aires

Haller MJ (2020) De Suiza a la Patagonia, 46 años de practicar geología por todo el país. Asociación Argentina Para El Progreso De Las Ciencias, Reseñas De Ciencia e Investigación 8(2):62–76

Hurtado D (2010) La ciencia argentina: un proyecto inconcluso: 1930–2000. Buenos Aires. Editorial Edhasa, 256, Buenos Aires

Hurtado de Mendoza D, Feld A (2008) Los avatares de la ciencia. Nómada 12:2–7

Llambías EJ (2013) 50 años escrutando la naturaleza de las rocas ígneas. Asociación Argentina Para El Progreso De Las Ciencias, Reseñas De Ciencia e Investigación 1(3):67–75

Matheos S (1999) Dr. Mario Martín Mazzoni (1942–1999), Nota Necrológica. Revista de la Asociación Argentina de Sedimentología 6(1–2): 91–95

Mon R (2022) Cincuenta años dedicados a la geología estructural y a la microtectónica, interrumpidos por golpes militares y retomados con el acceso a la democracia. Asociación Argentina Para El Progreso De Las Ciencias, Reseñas De Ciencia e Investigación 10(1):40–61

Olivero EB (2020) Rocas, fósiles y sucesos en espacio y tiempo: armando un rompecabezas antártico-fueguino. Asociación Argentina Para El Progreso De Las Ciencias, Reseñas De Ciencia e Investigación 8(3):74–87

Ostera HA (2015) Enrique Linares (1928–2014), Necrológica. Revista De La Asociación Geológica Argentina 72(2):291–296

Poiré DG (1997) Adrián Mario Iñiguez Rodríguez (1937–1996), Nota Necrológica. Revista De La Asociación Argentina De Sedimentología 4(1):53–57

Rabassa J (2023) Cincuenta años con geoformas y paisajes, glaciares y suelos congelados. Asociación Argentina Para El Progreso De Las Ciencias, Reseñas De Ciencia e Investigación 11(2):67–94

Ramos VA (2018) El desarrollo de la Geología en la Universidad de Buenos Aires. Revista de la Facultad de Ciencias Exactas, Físicas y Naturales, Nueva Serie 5 (Suplem. 1): 23–32, Córdoba

Ramos VA (2023) Félix González Bonorino: el geólogo que cambió la historia. Fundación Azara, Universidad Maimónides, Revista De Historia Natural 13(1):13–43

Ramos VA, Aguirre Urreta MB, Alvarez PP, Cegarra M, Cristallini EO, Kay SM, Lo Forte GL, Pereyra F, Pérez D (1996) Geología de la Región del Aconcagua, Provincias de San Juan y Mendoza. Anales de la Dirección Nacional del Servicio Geológico 24: 1–510, Buenos Aires

Rapela CW (2018) De la geoquímica a la reconstrucción de continentes. Asociación Argentina Para El Progreso De Las Ciencias, Reseñas De Ciencia e Investigación 6(3):63–77

Rinaldi CA (2003) Bernabé J. Quartino (1925–2003), Nota necrológica. Revista de la Asociación Geológica Argentina 58(3): 479

Spalletti LA (2015) Una vida dedicada a desentrañar los secretos que atesoran las rocas sedimentarias. Asociación Argentina Para El Progreso De Las Ciencias, Reseñas De Ciencia e Investigación 3(3):47–60

Trombotto D, Brown J (2001) Arturo Eduardo Corte 1919–2001. Colorado University, Arctic, Antarctic, and Alpine Research 33(3): 378–379, Boulder

Volkheimer W (2014) De los Alpes a los Andes: investigando los microfósiles vegetales y los paleoclimas del Mesozoico y Paleógeno. Asociación Argentina Para El Progreso De Las Ciencias, Reseñas De Ciencia e Investigación 2(2):21–25

Chapter 22
The National Geological Survey and the Integration of Geology

Abstract In previous chapters, the consolidation of the Argentine Geological Survey was analyzed. In the mid-1970s, an ambitious plan was implemented for the systematic survey of the Patagonian Cordillera of the province of Chubut, which remained largely unknown. For almost a decade, a legion of young geologists studied the Andean and extra-Andean regions. The success obtained led to the complete survey of the southern Patagonian Andes, as well as the Fuegian Cordillera, in the following decade. This allowed not only an advance in the geological understanding of the region but with the implementation of the Patagonia Plan for geological mining exploration, important gold deposits were found in the Deseado Massif. These studies, together with the extra-Andean region, permitted the production of geological maps of the provinces of Chubut and Santa Cruz. The research carried out allowed a great advance in the knowledge of the Patagonian and Fuegian Andes. A fundamental milestone of these years has been Law 24.224 on Mining Reorganization of 1993, which established in its first chapter the National Program of Geological and Thematic Maps.

Keywords Patagonian Andes · Extra-Andean region · Systematic surveys · Gold discovery · Thematic mapping

In the previous chapters, we analyzed how the Division of Mines, Geology and Hydrology created in 1904 under the direction of Enrique Hermitte evolved to become a Direction of Mines and Geology, which over the years received different names and hierarchies, with numerous budgetary ups and downs that were modified according to the politics and economic situation of the country, which have been analyzed in detail on the occasion of the institution's centenary by Zappettini (2004).

In 1976, the State Secretariat of Mining separated the technical jobs from the administrative functions of mining, thus creating the National Geological Survey, which under that name had Dr. Oscar Ruiz Huidobro as director between 1977 and 1984. In those years there was a renewed effort in geological surveys that was concentrated in the Patagonian region.

V. A. Ramos, *History of the Geological Sciences in Argentina*, The Latin American Studies Book Series, https://doi.org/10.1007/978-3-031-83139-3_22

22.1 The Patagonian Cordillera

This sector of the Andean mountains had until then only the works of Félix González Bonorino and Christian Petersen (Fig. 22.1) who had been commissioned to carry out sheets 41 b Río Foyel and 42 b Cholila in the province of Chubut, works that they used as a thesis for the doctorate in Natural Sciences at the University of Buenos Aires. Of these two sheets only the first was presented (González Bonorino 1944), while the second, not having a complete topographic base, could not be finished and there is only one work on its petrography (Petersen and González Bonorino 1947).

The 1:200,000 scale survey of the Andean region continued in the seventies covering the entire Chubut region south of Foyel (Fig. 22.2) with the surveys of Antonio Lizuaín of Sheet 42 a-b El Maitén, which was presented as a thesis at the University of Buenos Aires (Lizuaín Fuentes 1983), Eliseo Sepúlveda of Sheet 43 a-b Esquel, Miguel Haller of Sheet 44a-b Trevelín (Haller 1984), which was also presented as a doctoral thesis at the University of Buenos Aires (Haller 1981b, a), Abel Pesce (1945–2023) of Sheet 45a Lago General Vintter (Pesce 1975, 1979a, b), Omar Lapido of Sheet 46 a-b (Lapido 1977) and Victor A. Ramos of Sheet 47a-b La Paz (Pesce 1979a, b). 47 a-b Lago Fontana (Ramos 1981a). These geological surveys permitted a great advance in the knowledge of the age of the volcanic sequences, the plutonic rocks and the tectonic characteristics of the Northern Patagonian Cordillera, which had been independently disclosed (Ramos 1976, 1978a, b; Ploskiewicz and Ramos 1977; González Díaz 1982; Haller et al. 1981; Haller and Lapido 1982). Much of this unpublished information, together with subsequent works, was later published in the 1:250,000 scale geological maps that currently cover the Andean region.

Fig. 22.1 Christian Petersen (1912–1962) and Félix González Bonorino (1918–1998), geologists from the Direction of Mines and Geology began the systematic survey of the Patagonian Cordillera. *Source* Segemar archives

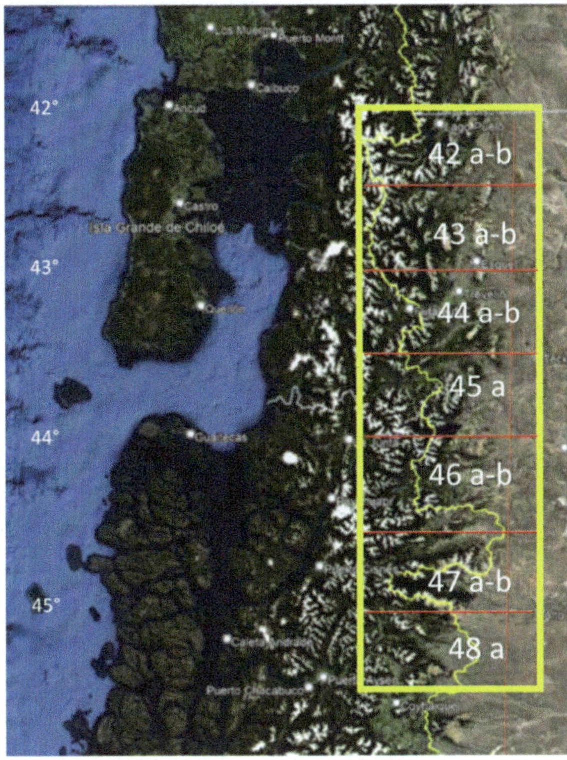

Fig. 22.2 The sector of the Northern Patagonian Cordillera was surveyed in the late 1970s by the National Geological Survey

During the 1980s, attention was drawn to the southern Patagonian Cordillera corresponding to the province of Santa Cruz (Fig. 22.3). In this Andean sector, only Sheet 58 b Lago Argentino had been surveyed, while the sector corresponding to the southern Patagonian Ice remained unsurveyed. This fieldwork had been carried out in 1955 but was published sometime later (Furque 1973).

The 1:200,000 scale survey plan included sheets 52 a-b, Pueyrredón Lake (Franchi 1984); 53 a Monte San Lorenzo and 53 b Meseta Belgrano (Ramos 1980); 54 a-b Río Mayer (Ploskiewicz 1978); 55 a Sierra de Sangra and 55 b Meseta de la Muerte (Ramos 1981b, 1978a); Cerro Fitz Roy (Nullo 1978); 57 b Río Guanaco (Proserpio 1981); 59 a-b Perito Moreno Glacier and Cordón de los Cristales (Nullo 1983a, b). These works allowed a large part of the southern Patagonian Cordillera to be surveyed, leaving only sectors of the southern continental ice sheet unstudied. The effort made by the National Geological Survey through the compilation of these geological sheets did not achieve the necessary dissemination due to the lack of topographical bases at a scale of 1:100,000 where the information could be dumped. Years later, when the Military Geographical Institute produced the necessary topographical sheets, it was decided to implement the millionth sheets according to the recommendations of the World Geological Map, and the maps were compiled with a new format at a scale of 1:250,000, with an update carried out by different authors.

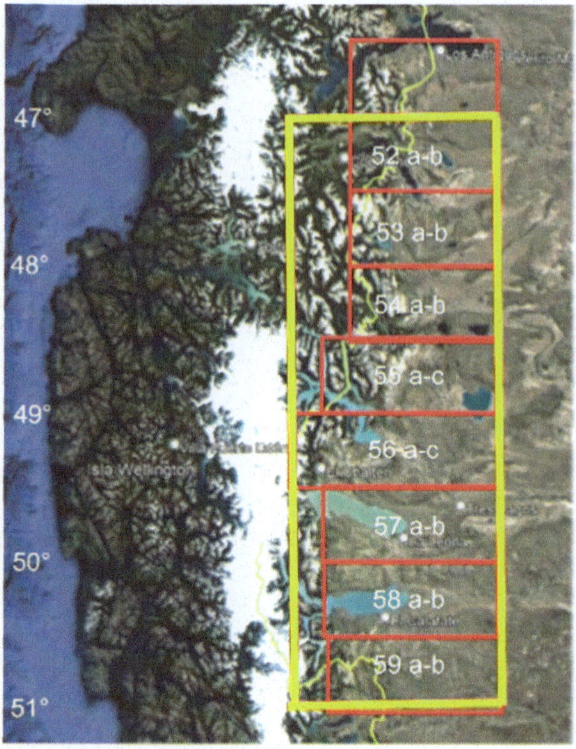

Fig. 22.3 A survey of the southern Patagonian Cordillera was carried out in the late 1970s by the National Geological Survey

The results of these surveys, although not officially published by the institution, produced important contributions that allowed updating the geology of the Patagonian Andes. Among them, the geological-structural studies presented by Ramos (1978a, b, 1979, 1989) and Nullo et al. (1979) stand out, which revealed for the first time the existence of an important fold and thrust belt in the Andean region (Fig. 22.4), which had been traditionally interpreted as a zone where the Meso-Cenozoic rocks inclined towards the east in a normal sequence without major disturbances (Leanza 1972, p. 706). The existence of underthrusts and triangular zones had gone unnoticed (Ramos 1979).

In addition to the advances in the structure of the Patagonian Cordillera from these studies, it was possible to identify the existence of an andesitic series of lower Cretaceous age and characterize the tectonic control of magmatism (Ramos 1978b; Ramos et al. 1982; Ramos and Kay 1992).

These works in the Andean region were carried out by a generation of young geologists who dedicated their efforts and energy to covering the most remote places of the Patagonian Andes with their studies (Fig. 22.5).

In those years, more precisely in 1980, our colleague Osvaldo M. Diez (1947–1980), a geologist who graduated from the Universidad Nacional del Sur, was working in the region of Cerro Diente de Perro on Sheet 41a El Bolsón, going

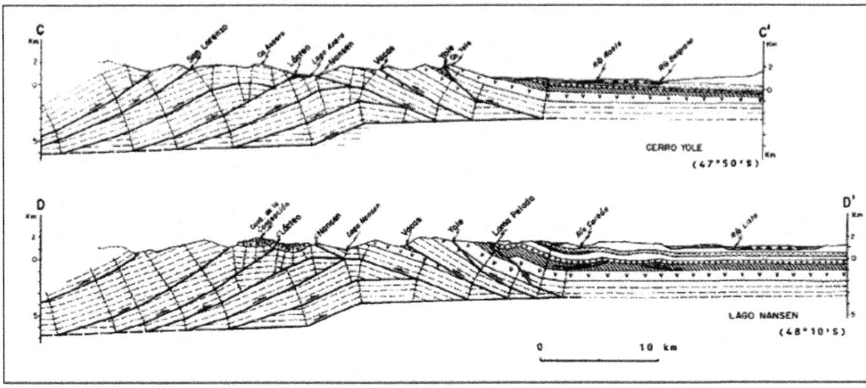

Fig. 22.4 Structural cross-section of Cerro Yole where the imbrication of underthrusts can be observed (Ramos, 1979, 1989)

down to the valley of the Manso River, had an accident. Despite the intense search carried out by his colleagues and the local neighbors, they could not find him. After a year, some local experts found his remains in a cliff that was difficult to access. These few words are a small tribute to the person who gave his life in the fulfillment of his duty (González Díaz 1983).

At that time, led by Roberto Caminos, we carried out a complete field review of the Fuegian Mountain Range, from the border with Chile to the mouth of Lake Fagnano and the Barnevelt and Evout Islands in the South Atlantic (Caminos et al. 1981; Ramos et al. 1986).

22.2 The Extra-Andean Patagonia

A 1973 law enacted an exploration reserve in all Patagonian provinces covering 858,000 km^2, which allowed for regional prospecting in search of mineral deposits in the following years. This Patagonia Plan executed by the National Mining Service allowed the identification of 24 areas of interest, some of which became gold and silver deposits. Among them, the gold veins found through the exploration work of the plan by Adolfo Genini, one of the geologists at the Comodoro Rivadavia headquarters, stand out. These became the important Vanguardia mine, which characterizes a new mining district in the Deseado Massif. Among other deposits, they also identified the Huemules mine, opposite Esquel, an epithermal gold deposit whose exploitation was cut short for environmental reasons.

In parallel to the intense mining exploration, the Geological Survey conducted a survey plan for the entire Patagonian region, including the extra-Andean region (Fig. 22.6). In this sector, there were only two surveyed sheets, in 1950, sheet 49 c Sierra de San Bernardo by Rafael R. González (González 1971) and in 1952, sheet

Fig. 22.5 Young geologists from the National Geological Survey who in the seventies made the first systematic geological mapping of both the Patagonian Cordillera of Chubut and the southern Cordillera in Santa Cruz: Miguel Haller, Antonio Lizuaín, Abel Pesce, Mario Franchi, Omar Lapido, Victor Ploszkiewicz, César Proserpio, Francisco Nullo, and Victor Ramos. *Source* courtesy of those photographed

48 c Ensanche Sarmiento, later published as Lago Musters, by César R. Vilela (Vilela 1971) (see Ramos 2018). Sheet 43 c, Gualjaina also in the province of Chubut was surveyed by Lage (1982).

In the 80 s the following sheets were published: 42 d Gastre (Proserpio 1978); 42 h Puerto Lobos (Cortés 1987); 43 g Bajo de la Tierra Colorada (Page 1987); 43 h Puerto Madryn (Haller 1981b, a); 44 c Tecka (Turner 1982); 44 d Colán Conhué (Turner 1983); 44 e Cañadón Racedo (Proserpio 1987); 44 f Cerro Ponte (Pesce

Fig. 22.6 Rafael R. González (1924–1990), Cesar R. Vilela (1921–2016) and Julio Lage (1936–1995), who carried out the first geological surveys for the National Direction of Mining and Geology in the province of Chubut. *Source* Ramos (2022)

1977); 44 g Cañadón Iglesias (Lapido 1981); 45 c Pampa de Agnia (Nullo 1983a, b); 45 g Monte Triste (Franchi 1983); 47 c Apeleg (Ploskiewicz 1987), among others that have remained unpublished. In the province of Santa Cruz was surveyed Sheet 53 g-h Ría Deseado (Palma 1986). Among all these regional geologists, Francisco Nullo played an important role in the quality and quantity of geological sheets surveyed (Fig. 22.7).

These surveys by the National Geological Survey combined with those previously carried out by YPF in the sectors of the Gulf of San Jorge sedimentary basin allowed

Fig. 22.7 Francisco Nullo, regional geologist who surveyed more than 18 geological sheets in his long career at the National Geological Survey

the completion of the basic knowledge of the region. These were the basis for making the first geological map of the province of Chubut. A preliminary copy of this map, compiled by YPF at a scale of 1:1,000,000, was presented by Chebli et al. (1978) at the 7th Argentine Geological Congress in Neuquén. A corrected and updated version of new surveys was later published by the National Geological Survey (Lizuaín et al. 1995).

During those years, geological sheets of the rest of the country continued to be published, especially of northwestern Argentina, Precordillera, and Sierra de la Ventana by Guillermo Furque, Juan C.M. Turner, among others (see Chap. 13).

22.3 The Creation of Segemar

In the State reforms carried out in 1996, the Argentine Geological and Mining Service (*Segemar*) was created, with Roberto Page as its first geologist president (Fig. 22.8).

Segemar is made up of the Institute of Mining Technology and the National Direction of the Geological Survey, which became the Institute of Geology and Mineral Resources (IGRM), and two other decentralized organizations, the Regional Center for Groundwater (CRAS) and the National Institute for Seismic Prevention (INPRES), based in the province of San Juan. These last two institutions would be separated from Segemar shortly after (Zappettini 2004).

A fundamental milestone of those years has been the Mining Reorganization Law 24.224 of 1993, whose first chapter established the National Program of Geological and Thematic Maps. From those years onwards, the Segemar established a new format for the geological map at a scale of 1:250,000, a millionth format that covers one degree of latitude by one and a half degrees of longitude. The areas of mining interest would be mapped at a scale of 1:100,000. Based on the previous maps at a scale of 1:200,000 and new fieldwork, more than 95% of the mountainous and hilly

Fig. 22.8 Roberto Page was director of the Geological Survey from 1996 to 2016

areas were completed, which are available in the Segemar Digital Repository (see also the Catalogue of Maps and Publications 2019, Segemar, 76 p., Buenos Aires).

In addition to the geological maps, a series of thematic metallogenic maps with multi-element aero-geophysical and geochemical information was started, as well as geoenvironmental maps on different aspects of geological hazards and hydrogeological maps, both at a scale of 1:250,000 and 1:100,000.

In these years, it has continued to produce regional synthesis maps, especially of the different provinces at different scales, thus completing the national territory. Among the contributions made by Segemar in those years, the following stand out: the two volumes of *Mineral Resources of the Argentine Republic* edited by Eduardo Zappettini, the *Argentine Geology*, edited by Roberto Caminos, the two volumes of *History of Argentine Mining* edited by Eddy Lavandaio and Edmundo Catalano, the two volumes of *Sites of Geological Interest*, coordinated by Alberto Ardolino and Hebe Lema and the *Geology of the Aconcagua region* by Victor A. Ramos and collaborators.

In turn, Segemar, in compliance with the recommendations of the *Commission for the Geological Map of the World of the International Union of Geological Sciences* (IUGS), has participated in numerous thematic maps of South America, such as metallogenic, tectonic, metamorphic facies maps, etc.

22.4 Final Considerations

In the different chapters, we have seen the progress of the geological survey of the country starting from the initial surveys carried out predominantly by young German geologists, followed by the systematic survey carried out by Argentine geologists of the economic geological map at a scale of 1:200,000. This map gained special momentum with the creation of the National Geological Survey in 1976. With the creation of the Segemar, the subject matter of geological maps was expanded, not only by changing the scale to 1:250,000 but also with the survey of new geoenvironmental, metallogenic and hydrogeological maps. A characteristic of the new stage is that both the surveys and the texts are available in a publicly accessible digital repository, which has placed the Segemar at the forefront of geological knowledge of the country. This series of maps and reports contribute daily to the development of natural resources, the evaluation of geological risks, as well as their environmental impact.

References

Caminos R, Haller M, Lapido O, Lizuaín A, Page R, Ramos VA (1981) Reconocimiento geológico de los Andes Fueguinos, Territorio Nacional de Tierra del Fuego. VIII° Congreso Geológico Argentino (San Luis), Actas III:759–786

Chebli G, Nakayama C, Sciutto JC (1978) Mapa geológico de la provincia del Chubut, escala 1. 1.000.000. 7° Congreso Geológico Argentino, Neuquén

Cortés JM (1987) Descripción Geológica de la Hoja 42 h, Puerto Lobos, Provincia del Chubut, Carta Geológico- Económica de la República Argentina, Escala 1:200.000. Dirección Nacional de Minería y Geología, Boletín 202, 95 p, Buenos Aires

Franchi MR (1983) Descripción Geológica de la Hoja 45 g, Monte Triste. Carta Geológico-Económica de la República Argentina. Escala 1:200.000. Provincia del Chubut, Boletín del Servicio Geológico Nacional 191:1–62, Buenos Aires

Franchi M (1984) Descripción geológica de la Hoja 52 a-b, Lago Pueyrredón, provincia de Santa Cruz. Servicio Geológico Nacional, unpublished report, Buenos Aires

Furque G (1973) Descripción Geológica de la Hoja 58 b, Lago Argentino, Provincia de Santa Cruz. Carta Geológico-Económica de la República Argentina, Escala 1:200.000. Boletín del Servicio Nacional Minero Geológico 140:1–57, Buenos Aires

González RR (1971) Descripción Geológica de la Hoja 49 c, Sierra San Bernardo. Provincia del Chubut. Carta Geológico-Económica de la República Argentina, Escala 1:200.000, Boletín del Servicio Nacional Minero Geológico 112:1–93, Buenos Aires

González Bonorino F (1944) Descripción Geológica y Petrográfica de la Hoja 41 b, Río Foyel. Boletín de la Dirección General de Minas, Geología e Hidrología 56:1–138, Buenos Aires

González Díaz EF (1982) Chronological zonation of granitic plutonism in the northern Patagonian Andes of Argentina: the migration of intrusive cycles. Earth-Sci Rev 18(3–4):365–394

González Díaz EF (1983) Licenciado Osvaldo M. Diez (1947–1980), Nota necrológica. Revista de la Asociación Geológica Argentina 38(2):286–288

Haller MJ (1981) Descripción Geológica de la Hoja 43 h, Puerto Madryn. Provincia del Chubut. Boletín del Servicio Geológico Nacional 184:1–49, Buenos Aires

Haller MJ (1981) Estructura y mineralización en la Cordillera Patagónica. Facultad de Ciencias Exactas y Naturales, Universidad de Buenos Aires, unpublished PhD thesis, 144 p

Haller MJ (1984) Descripción Geológica de la Hoja 44 ab. Servicio Geológico Nacional, unpublished report, Buenos Aires, Trevelín

Haller M, Lapido O, Lizuaín A, Page RFN (1981) El mar tithoneocomiano en la evolución de la Cordillera Norpatagónica. In: Volkheimer W, Musacchio E (eds) Cuencas sedimentarias del Jurásico y Cretácico de América del Sur 1:221–230, Buenos Aires

Haller M, Lapido O (1982) The Jurassic-Cretaceous volcanism in the Patagonian Septentrional Andes. In: Cordani U, Linares E (eds) Symposium on magmatic evolution of the Andes, vol 18. Earth Science Review, pp 395–410

Lage J (1982) Descripción Geológica de la Hoja 43 c, Gualjaina. Provincia del Chubut. Carta Geológico-Económica de la República Argentina, Escala 1:200.000, Boletín del Servicio Geológico Nacional 189:1–70, Buenos Aires

Lapido O (1977) Descripción geológica de la Hoja 46 a-b Río Pico. del Chubut. Servicio Geológico Nacional, unpublished report, Buenos Aires, Prov

Lapido OR (1981) Descripción Geológica de la Hoja 44g, Cañadón Iglesias. Provincia del Chubut. Carta Geológico-Económica de la República Argentina, Escala 1:200.000, Boletín del Servicio Geológico Nacional 185:1–46, Buenos Aires

Leanza AF (1972) Andes Patagónicos Australes. In: Leanza AF (ed) Geología Regional Argentina, Academia Nacional de Ciencias. Córdoba, pp 689–706

Lizuaín Fuentes A (1983) Geología de la cordillera patagónica entre las localidades de Lago Puelo y Leleque, Provincia del Chubut. Facultad de Ciencias Exactas y Naturales, Universidad de Buenos Aires, unpublished PhD thesis, 135 p

Lizuaín A, Ragona D, Folguera A, Ardolino AA, Franchi M, Panza JLA, Page RN, Ramos VA, Busteros A, Lema H, Caminos RL, Giacosa RE, Nullo FE, Leanza HA, Márquez MJ, Parisi C (1995) Mapa Geológico de la Provincia del Chubut. Servicio Geológico Minero Argentino, Instituto de Geología y Recursos Minerales, Buenos Aires

Nullo F (1978) Descripción Geológica de la Hoja 56 a-b Cerro Fitz Roy, provincia de Santa Cruz. Servicio Geológico Nacional, unpublished report, Buenos Aires

Nullo, F. 1983. Descripción Geológica de la Hoja 59 a-b Glaciar Perito Moreno y Cordón de los Cristales. Servicio Geológico Nacional, unpublished report, Buenos Aires

Nullo F (1983) Descripción Geológica de la Hoja 45 c, Pampa de Agnia, Provincia del Chubut, Carta Geológico-Económica de la República Argentina, Escala 1:200.000. Boletín del Servicio Geológico Nacional 199:1–107, Buenos Aires

Nullo FE, Proserpio CA, Ramos VA (1979) Estratigrafía y tectónica de la vertiente este del Hielo Continental Patagónico, Argentina-Chile. 7° Congreso Geológico Argentino (Neuquén), Actas 1:455–470, Buenos Aires

Page RN (1987) Descripción Geológica de la Hoja 43 g, Bajo de la Tierra Colorada, Provincia del Chubut, Carta Geológico-Económica de la República Argentina, Escala 1:200.000. Boletín de la Dirección Nacional de Minería y Geología 200: 1–77, Buenos Aires

Palma M (1986) Descripción Geológica de la Hoja 53 g-h Ría Deseado, provincia de Santa Cruz. Servicio Geológico Nacional, unpublished report, Buenos Aires

Pesce AH (1975) Informe Preliminar de la Hoja 45 a. Servicio Geológico Nacional, unpublished report, Buenos Aires, Lago General Vintter

Pesce A (1977) Descripción Geológica de la Hoja 44f Cerro Ponte, Provincia del Chubut, Carta Geológico-Económica de la República Argentina, Escala 1:200.000. Dirección Nacional de Minería y Geología, unpublished report, Buenos Aires

Pesce AH (1979a) Estratigrafía da la Cordillera Patagónica entre los de 43°30' y 44° de latitud sur y sus áreas mineralizadas, 7° Congreso Geológico Argentino (Neuquén), Actas 1:257–270, Buenos Aires

Pesce AH (1979b) El plutonismo Cretácico superior y sus relaciones con las zonas de alteración hidrotermal en el tramo medio de la Cordillera Patagónica, República Argentino, 4° Congreso Geológico Latinoamericano, Port of Spain

Petersen CS, González Bonorino F (1947) Observaciones geológicas en el Chubut occidental. Revista De La Sociedad Geológica Argentina 2(3):177–222

Ploskiewicz V (1978) Descripción Geológica de la Hoja 54 a-b Río Mayer, provincia de Santa Cruz. Servicio Geológico Nacional, unpublished report, 95 p, Buenos Aires

Ploszkiewicz JV (1987) Descripción Geológica de la Hoja 47 c, Apeleg. Provincia del Chubut. Carta Geológico-Económica de la República Argentina, Escala 1:200.000. Boletín de la Dirección Nacional de Minería y Geología 204:1–100, Buenos Aires

Ploszkiewicz JV, Ramos VA (1977) Estratigrafía y tectónica de la Sierra de Payaniyeu, provincia del Chubut. Revista De La Asociación Geológica Argentina 32(3):209–226

Proserpio CA (1978) Descripción Geológica de la Hoja 42 d, Gastre. Provincia del Chubut. Carta Geológico-Económica de la República Argentina, Escala 1:200.000, Boletín del Servicio Geológico Nacional 159:1–76, Buenos Aires

Proserpio CA (1981) Descripción Geológica de la Hoja 57 b, Río Guanaco, provincia de Santa Cruz. Servicio Geológico Nacional, unpublished report, Buenos Aires

Proserpio C (1987) Descripción Geológica de la Hoja 44e Valle General Racedo, Provincia del Chubut, Carta Geológico-Económica de la República Argentina, Escala 1:200.000. Boletín de la Dirección Nacional de Minería y Geología 201:1–108, Buenos Aires

Ramos VA (1976) Estratigrafía de los lagos Fontana y La Plata, provincia del Chubut, República Argentina. 1° Congreso Geológico Chileno (Santiago de Chile), Actas I(A):43 64

Ramos VA (1978a) Descripción Geológica de la Hoja 55 b Meseta de la Muerte, provincia de Santa Cruz. Servicio Geológico Nacional, unpublished report, 88 p, Buenos Aires

Ramos VA (1978b) El vulcanismo del Cretácico inferior de la Cordillera Patagónica. 7° Congreso Geológico Argentino (Neuquén). Actas I:423–436

Ramos VA (1979) Tectónica de la región del Río y Lago Belgrano, Cordillera Patagónica Argentina. 2° Congreso Geológico Chileno (Arica), Actas I(B):1–32

Ramos VA (1981a) Descripción Geológica de la Hoja 47 ab, Lago Fontana. Provincia del Chubut. Boletín del Servicio Geológico Nacional 183:1–141, Buenos Aires

Ramos VA (1981b) Descripción Geológica de la Hoja 55 a Sierra de Sangra, provincia de Santa Cruz. Servicio Geológico Nacional, unpublished report, 124 p, Buenos Aires

Ramos VA (1980) Descripción Geológica de la Hoja 53 a Monte San Lorenzo y Hoja 53 b Meseta Belgrano, provincia de Santa Cruz. Servicio Geológico Nacional, unpublished report, 124 p, Buenos Aires

Ramos VA (1989) Foothills structure in Northern Magallanes Basin, Argentina. Bull Am Assoc PetrolGeolog 73(7):887–903

Ramos VA (2018) Cesar Reinaldo Vilela (1921–2016), Necrológica. Revista De La Asociación Geológica Argentina 75(2):309–311

Ramos VA (2022) La evolución del conocimiento geológico en la provincia de Chubut. In Giacosa, R. (ed.) Geología y Recursos Naturales de la Provincia del Chubut, 21° Congreso Geológico Argentino, Relatorio:3–33, Puerto Madryn

Ramos VA, Haller MA, Butrón F (1986) Geología y Evolución tectónica de las Islas Barnevelt: Atlántico Sur. Revista De La Asociación Geológica Argentina 41(1–2):137–154

Ramos VA, Kay SM (1992) Southern Patagonian plateau basalts and deformation: back-arc testimony of ridge collisions. Tectonophysics 205:261–282

Ramos VA, Niemeyer H, Skarmeta J, Muñoz J (1982) Magmatic evolution of the Austral Patagonian Andes. In: Cordani U, Linares E (eds) Symposium on magmatic evolution of the Andes, vol 18. Earth Science Review, pp 411–443

Turner JCM (1982) Descripción Geológica de la Hoja 44 c, Tecka. Provincia del Chubut. Carta Geológico-Económica de la República Argentina, Escala 1:200.000, (Convenio Universidad de Buenos Aires - Secretaría de Estado de Minería), Boletín del Servicio Geológico Nacional 180:1–90, Buenos Aires

Turner JCM (1983) Descripción geológica de la Hoja 44 d, Colan Conhué, Pcia. del Chubut, Boletín del Servicio Geológico Nacional 197:1–78, Buenos Aires

Vilela CR (1971) Descripción Geológica de la Hoja 48 c, Lago Musters. Provincia del Chubut. Escala 1:200.000. Carta Geológico-Económica de la República Argentina. Boletín 113, 85 p, Buenos Aires

Zappettini E (2004) Servicio Geológico Minero Argentino, 1904–2004, 100 Años al servicio del desarrollo nacional. Secretaría de Minería, Segemar, 106 p, Buenos Aires

Chapter 23
Geology in Argentine Antarctica

Abstract Argentina's involvement in Antarctica began with José María Sobral, who between 1901 and 1903 was part of the first Swedish expedition to the Antarctic region. It continued in 1904 with the work carried out on Laurie Island, where Argentina received the meteorological station of the first Scottish expedition to Antarctica. Augusto Tapia, the first Argentine glaciologist, worked on Laurie Island. From the 1940s onwards, the first geological studies of the Peninsula were carried out by Pascual Sgrosso, which preceded the creation of the Argentine Antarctic Institute in 1951, a world pioneer institution in scientific research on the white continent. In the 1950s, numerous expeditions carried out by the institute allowed us to learn about the characteristics of the Argentine Antarctic sector. Among the first studies, we have the works of Isaías Cordini, Juan Olsacher, Néstor Fourcade and Osvaldo Schauer. With the creation of the National Antarctic Direction, regular studies began, in which not only the Institute's staff but also geologists from different universities in the country participated each year. Over the years, geological and paleontological surveys in our Antarctic sector have stood out for the variety and importance of the studies.

Keywords Pioneering expedition · Long-lasting observatory · Scientific research · Antarctic Institute · Antarctic geologists

In the first Swedish expedition to Antarctica (1901–1903) organized by Otto Nordenskjöld with the Norwegian captain Carl A. Larsen on the whaling ship Antarctic, participated Johan G. Andersson, a young geologist, recently graduated from the University of Uppsala. The expedition left in 1901 from Gothenburg and upon arriving in Buenos Aires, Ensign José María Sobral, appointed by the Argentine Navy, was added. His 2 years in Antarctica awakened his geological vocation (Fig. 23.1), turning the young Sobral into the first Argentine geologist (see details of the Swedish expedition in Chap. 8).

The Scottish Antarctic expedition (1902–1903) commanded by William Bruce aboard the Scotia arrived in Buenos Aires a few days after the arrival of the corvette Uruguay with the rescue of the Swedish scientists. They arrived from Laurie Island

V. A. Ramos, *History of the Geological Sciences in Argentina*, The Latin American Studies Book Series, https://doi.org/10.1007/978-3-031-83139-3_23

Fig. 23.1 Members of the Swedish expedition at Snow Hill. Forward Gösta Bodman, meteorologist, and hydrographer; Otto Nordenskjöld, geologist, José M. Sobral, assistant meteorologist; behind Ole Jonassen, sleigh driver; Erik Ekelöf, doctor, and Gustaf Åkerlund, sailor and cook. *Source* Dirección Nacional del Antártico, DNA-IA archives

in South Orkney, where they had founded a meteorological observatory known as Omond House. These facilities were sold by Bruce to the Argentine government who took charge of the observatory, returning the Scotia to put the Argentine scientists in charge on Laurie Island. Bruce donates a series of geological samples to the La Plata Museum with which this institution began its Antarctic collection in 1904 (Carrasquero 2020).

The Argentine Navy has carried out annual expeditions since 1904 to the Laurie Island meteorological observatory to resupply and replace personnel that continue to this day. This observatory is the oldest permanent base in the entire Antarctic territory.

In 1920 Augusto Tapia (Fig. 23.2), who worked in the Direction of Mines and Geology, was invited to be part of the scientific commission to the South Orkney Islands, where he carried out the first Antarctic glaciological studies of our country on Laurie Island (see details in Chap. 10).

Fig. 23.2 *David Livingstone Centenary Medal was* received in 1930 by Sobral, and scientists working at the Laurie Island meteorological observatory in the early twentieth century. *Source* American Geographical Society and Segemar archives

23.1 National Antarctic Commission

In 1940, the National Antarctic Commission was created, whose objectives were to send polar scientific expeditions to the Antarctic Peninsula, given that with the exception of the Nordenskjöld and Tapia studies there had been no Argentine participation in this continent. We must wait until 1942 when the Argentine Navy decides to begin new Antarctic expeditions to once again carry out scientific studies in general and especially geological ones. In the first, known as the Oddera expedition, named after its leader, there were no geologists, but rather specialists in fisheries and climate issues. The following year, in February 1943, the Harriague expedition was organized, named after the commander of the expedition by the Argentine Navy, whose scientific group was led by the geologist Pascual Sgrosso.

When these Antarctic expeditions began, José María Sobral, who had returned to the country, offered his services to the Navy. He had had important recognitions abroad with distinctions such as *Knight of the Order of the North Star and the Order of Swords* by the King of Sweden, being a foreign member of the Swedish Association of Anthropology and Geography, of the Royal Society of Natural Sciences of Gothenburg, and having received the *David Livingstone Centenary Medal* in 1930 from the American Geographic Society for his relevant contributions to geographical exploration (Marenssi 2007). This distinction established in 1913 by the Hispanic Society of America is awarded each year by the American Geographic Society for *scientific achievements in the field of geography of the Southern Hemisphere.*

In an article published in the newspaper Crítica, Sobral stated in the 1940s that *the South Pole must be an Argentine possession*, later adding that *in the Antarctic no right is more authentic than the Argentine right* (Aceñolaza 2020, p. 31). Despite his Antarctic experience and the international recognition obtained, his offer was not taken into account and he was not able to participate in these expeditions. Apparently, the Navy did not forgive him for resigning as an ensign and choosing to study geology at the University of Uppsala in Sweden.

Pascual Sgrosso (1900–19??), a pioneering geologist in Antarctic research (Fig. 23.3), received a doctorate in Natural Sciences from the University of Buenos Aires in 1938, with a thesis on *Contribution to the knowledge of mineral deposits and the application rocks of the province of Jujuy*. Although he appreciates the support received from Juan Keidel and Pablo Groeber in his doctorate, there is no record of who was his director. He had previously graduated as an industrial engineer from that university. Sgrosso worked at YPF from 1923 to 1930, where he ended up as head of the Campo Vespucio deposits in Salta. He joined the National Directorate of Mines and Geology in 1931, which allowed him to finish his geological studies. Later he was a professor at the then Museum Institute of the National University of La Plata, today the Faculty of Natural Sciences and Museum, where he taught mineralogy classes and became Head of the Mineralogy and Petrology Division between the years 1942 and 1956 in which he retired. In that period, he supervised 43 doctoral theses (see Riccardi 2018). He was also a professor at the National University of Cuyo and the University of Buenos Aires.

On his first trip to Antarctica, which lasted 43 days, Sgrosso toured the South Shetland Islands, especially Deception Island, as well as the Antarctic Peninsula, mainly the Melchior Archipelago, Palmer Island, Margaret Bay, Neko Harbor, and Paradise Bay reaching the island of Alexander I. Sgrosso and his collaborators carried

Fig. 23.3 Pascual Sgrosso, the pioneer geologist in Antarctic expeditions, made the first Argentine geological observations on Deception Island. *Source* Museo de La Plata archives and courtesy of Dora E. Gerez, Buenos Aires

out geological and geophysical investigations, in addition to studying the fish fauna of the region and observing the meteorological conditions of the regions visited, having as a collateral task, replenishing the lighthouse on Primero de Mayo Island, built by the Argentine Navy during the 1942 expedition. On this trip he was able to observe the large geological units of the Antarctic Peninsula, characterizing the plutonic rocks, the metamorphic rocks, the Cretaceous and Tertiary sequences of the Marambio (Seymour) and Cerro Nevado (Snow Hill) islands, and especially the neovolcanic sequences of Deception Island. His observations allow him to correlate the different environments with those of Patagonia, comparing the Patagonian Mountain Range with the Peninsula and the easternmost sector with extra-Andean Patagonia (Sgrosso 1948). He concludes that whatever the interpretation of the geological evolution, he proposes that this Antarctic sector is the southern continuity of the Andes and confirms the correctness of the name Antartandes proposed by Arctowski (1903).

In his second and last visit between December 1948 and January 1949, he expanded the geological observations of the previous expedition (Sgrosso 1958) and presented a complete petrographic description of a great variety of rocks that are treasured in the Antarctic collection of the Museum of La Silver (Carrasquero 2020).

In those years there was a growing awareness of Argentine rights over Antarctica (Sgrosso 1947) as demonstrated by the conference organized by the Secretariat of Culture of the Ministry of Education given on that occasion by Agustín Riggi, director of the National Research Institute of the Natural Sciences. That conference given in 1949 gave rise to a complete publication on *Argentine Antarctica, its geography and its history* where it updates this knowledge, disseminating its geographical features and describing in detail the geology that was known at the time. In the conclusions, it is established that *the geological and physiographic continuity between our Patagonia and the Andean Cordillera, with Graham Land and adjacent islands, is clear and evident. The Antarctic sector between 25° and 74°W, south of the 60° parallel, is unquestionably Argentine.* (Riggi 1950).

23.2 The Argentine Antarctic Institute

This institute, a world pioneer of its kind in scientific research on the white continent, was created in 1951 (Fig. 23.4) under the direction of Colonel Hernán Pujato (1904–2003) to promote and coordinate scientific activities, with the purpose of *continuing to ensure in an inalienable manner for the Argentine Nation the historical, geographical and territorial rights that assist it over the sector of the Antarctic zone that belongs to it.*

The following year, an expedition was organized with twenty-six scientists, including pioneering geologists such as Isaías R. Cordini from the Direction of Mines and Geology, Juan Olsacher from the National University of Córdoba, Horacio H. Camacho from the University of Buenos Aires, Arturo Corte from the National University of Cuyo, Horacio Díaz from YPF, Horacio González Laguinge and

Fig. 23.4 Historic headquarters of the Argentine Antarctic Institute located in front of 9 de Julio Avenue in Buenos Aires and the last photograph before its expropriation. *Source* Dirección Nacional del Antártico, DNA-IA archives

Eduardo González Stegman from Fabricaciones Militares and José Pascual from Banco del Crédito Industrial Argentino.

Furthermore, in those first years, several geologists joined the Institute who, along with others, participated in the Antarctic campaigns (Fig. 23.5). Dr. Otto Schneider (1912–1999), a prestigious German geophysicist who was already in our country, was appointed Head of the Scientific Department of the Institute between 1956 and 1969. During his work, he gave it a marked international imprint, especially through the actions undertaken in 1957 *International Geophysical Year*. His career in research and teaching gave him great prestige and recognition (van Zele 1999). He has been a professor in Buenos Aires and La Plata and a senior researcher at Conicet.

Fig. 23.5 Otto Schneider, Néstor H. Fourcade, and Osvaldo C. Schauer, one of the first Earth Sciences specialists incorporated into the Argentine Antarctic Institute. *Source* Dirección Nacional del Antártico, DNA-IA archives

The stable part of the Antarctic Institute was joined by Néstor H. Fourcade, a graduate of the National University of Córdoba, who was characterized by his petrographic studies of the volcanic sequences of the South Shetland Islands (Fourcade 1960, 1968; Fourcade and Viramonte 1972) and by dozens of publications with numerous colleagues where he contributed both his petrographic knowledge and that acquired in his Antarctic expeditions. Another of the Institute's first geologists was Osvaldo C. Schauer (1925–2010), a graduate of the National University of La Plata who worked until 1961 (see Cingolani 2010) and participated in various Antarctic campaigns (Schauer et al. 1960; Schauer and Fourcade 1964). In those years he was sent to England where he spent a stay with R.J. Adie, who was preparing a text on Antarctic geology.

One of the outstanding personalities who contributed to the Institute's initial Antarctic campaigns from 1950–1951 was Isaías Rafael Cordini (1902–1966) (Fig. 23.6), a doctor in Natural Sciences who produced in 1942 one of the first limnological studies in the country with his thesis on *Laguna La Brava, Province of Buenos Aires: contribution to its limnological knowledge* directed by zoologist José Carbonell (Cordini 1942). As soon as he graduated, he joined the National Direction of Mines and Geology, dedicating himself to geological studies until the rest of his days.

Before finishing his doctorate, the National Cultural Commission awarded him the regional Prize in 1939 for his limnological studies on Lake Nahuel Huapi; The Paraguayan government named him *Commander of the Order of Merit* for his work on *The Pilcomayo Rivers in the Patiño region*; The National Academy of Exact, Physical and Natural Sciences distinguished him with the Eduardo Holmberg Prize for his career, at the same time that he received the *National Science Prize* in 1947. With all these scrolls he participated in five Antarctic campaigns, producing three important geological contributions.

Fig. 23.6 Isaías R. Cordini in his Antarctic campaign, and in his last years when he worked in the National Direction of Geology and Mining. *Source* Dirección Nacional del Antártico, DNA-IA archives

In his extensive review of the Argentine Antarctic sector, Cordini proposed a Precambrian age for the basal complex (Cordini 1959) exposed in the Bertrab nunatak, a southeastern sector of the Weddell Sea region that, years later, was dated at about 1,000 Ma. Described the intense deformation of the basement of Laurie Island and following the hypotheses of the time, attributed it to the Lower Paleozoic (Cordini 1955). In these two works, he makes a complete description of what was known at the time together with the data obtained in his Antarctic campaigns. After his polar studies, he carried out new limnological research and fundamental contributions such as *The Saline Reserves of Argentina* and *The Diatomiferous Deposits of Argentina*, two works carried out in the sixties that are still valid.

Among the Antarctic pioneers is Prof. Juan Olsacher (1903–1964) who participated in the 1952–1953 campaign (Fig. 23.7), one of the first graduates of the National University of Córdoba, a disciple of Roberto Beder, who received his doctorate in 1930 with a strong mineralogical vocation (Leanza 1964). After his Antarctic campaigns of 1952–1953, Olsacher published numerous contributions to his geology in the 1950s, highlighting his petrographic studies of *Deception Island and other sectors of the South Shetlands* (Olsacher 1955, 1956a, b, c, 1957; Olsacher et al. 1957). With the collaboration of Hebe Dina Gay, he designed and made the topographical and geological models of Deception Island at a scale of 1: 25,000 that for many years adorned the entrance hall of the former headquarters of the Institute.

Horacio H. Camacho (1922–2015) (Fig. 23.8) was one of the first to participate alongside Olsacher and Corte in the Antarctic campaigns. He graduated from the

Fig. 23.7 Juan Olsacher as a young professor of Mineralogy at the National University of Córdoba and the topographical and geological models of Decepción Island that he prepared with the collaboration of Dr. Hebe Dina Gay after his 1952–1953 campaign that was exhibited for decades in the main hall of the former headquarters of the Argentine Antarctic Institute. *Source* courtesy of Ricardo Sureda and Dirección Nacional del Antártico, DNA-IA archives

Fig. 23.8 Horacio Camacho in his 1952 Antarctic campaign with doctors Enrique Balech and Antonio Cornejo, and later teaching as a professor. *Source* del Río (2016) and Camacho (1979)

University of Buenos Aires as a doctor in Natural Sciences in 1948 (Aguirre Urreta 2016) where he was a partner with Jorge Villar Fabre (1921–1968). Arturo Corte (1919–2001) carried out geological, petrographic observations and cryopedological studies in Bahía Esperanza, Puerto Paraíso and the South Shetland Islands y Melchior (Camacho and Villar Fabre 1957; Corte and Somoza 1957).

Camacho also made paleontological contributions in Antarctica, the result of his campaigns complemented by the study and complete review of existing international works (Camacho 1959, 1979). In the 1950s and 1960s, the Antarctic Institute Publications series, which includes 27 scientific works, was a vehicle to publicize scientific research in the Argentine Antarctic sector, which lasted until 1994.

It is interesting to note that Eduardo González Stegman of Fabricaciones Militares in his 1953–1954 campaign obtained a collection of samples from Nelson Island in South Shetland that was later studied at the National Direction of Geology and Mining by Quartino (1959).

23.3 National Antarctic Direction

At the end of 1969, the National Antarctic Direction was created under the direction of General Jorge Edgardo Leal (1921–2017), which gave a sustained boost to Antarctic research (Fig. 23.9).

Among his objectives, Leal always stressed that the mere presence on the white continent was not enough if it was not associated with scientific and technical knowledge to assess the possibilities of every order that it contains. General Jorge Edgardo Leal was the one who led the first Argentine expedition to the South Pole in 1965, the only one to arrive from the Weddell Sea.

Starting in 1972, shortly after the inauguration of the Marambio Base, the Geoantar Program began, which involved strong joint participation between the Argentine

Fig. 23.9 Jorge E. Leal led the first Argentine expedition to the South Pole and his arrival at the Pole in December 1965. *Source* Dirección Nacional del Antártico, DNA-IA archives

Antarctic Institute and the Faculty of Exact and Natural Sciences of the University of Buenos Aires. From its beginnings, the program was led by Carlos A. Rinaldi (1930–2017), and starting in 1972, it marked a strong participation of geological studies in scientific activities in the Argentine Antarctic sector. Rinaldi obtained his doctorate in Natural Sciences from the University of Buenos Aires in 1964 and was a professor of Introduction to Geology for several decades at that institution until his retirement (Olivero 2018).

Rinaldi can be considered a true Antarctic geologist, first for his numerous campaigns and works in the Argentine Antarctic sector, and second for his efforts that favored the participation of geologists from different universities and institutions in the campaigns. His actions were reciprocated since in 1985 he was appointed Director of the Argentine Antarctic Institute, a position he held until his retirement. He was the first civilian to be in charge of that responsibility, and later temporarily the National Director of the Antarctic (Rinaldi 2016).

As a geologist, he led a team that, through the 1972/1975 campaigns of the Geoantar program, created the first complete geological map and a pioneering description of the geology of Seymour or Marambio Island in 1978 (Rinaldi et al. 1978). In addition, he facilitated access to Antarctic research for numerous colleagues (Fig. 23.10). Among them, Roberto Caminos (1931–1997) (Caminos 1972; Caminos et al. 1973; Llambías 1997), Eduardo Olivero, Armando Massabie, Juan Spikerman, Rodolfo del Valle, Francisco Medina, among many others, stood out in those early years from the University of Buenos Aires. Numerous contributions were produced in those years, both as undergraduate and doctoral theses. Among the first, of Olivero and Medina stand out, who did their final degree work in 1975 and continued working in the Antarctic sector.

The University of Salta was represented in these campaigns by professors and students, among whom José Viramonte and Antonio Igarzabal stood out for their contributions (Fig. 23.11). Viramonte carried out numerous campaigns with the Argentine Antarctic Institute and with his Spanish colleagues, producing original

Fig. 23.10 Roberto Caminos (first from right) and Carlos Rinaldi (third) in a 1972 Antarctic campaign on the Hurd Peninsula on Livingston Island. *Source* Dirección Nacional del Antártico, DNA-IA archives

contributions to the Quaternary and active volcanism of Deception Island (see Viramonte 2017; Viramonte et al. 1973, 1974). Igárzabal made a couple of contributions to the geomorphological characterization of Decepción Island in 1973 and 1974.

The researchers at the La Plata Museum had an important participation in the study of fossil vertebrates, first classifying the specimens collected by the geologists of the Antarctic Institute and later being part of the Antarctic campaigns. In 1977, an expedition led by Eduardo Tonni and Alberto Cione collected numerous pieces in the Tertiary of Marambio Island (Tonni and Cione 1978).

Starting in 1977 there was almost continuous collaboration of researchers from the La Plata Museum in field work. The first Antarctic Mesozoic reptiles were found in campaigns carried out by the institute and studied by Argentines (del Valle et al. 1977; Gasparini et al. 1984). This is how the first mosasaurs and plesiosaurs were described in the seventies. From those years to the present, numerous specimens have

Fig. 23.11 The researchers José Viramonte at the entrance to Decepción Island and Ricardo Alonso, both from the National University of Salta during their Antarctic campaigns. *Source* courtesy of those photographed

been collected and are stored in the La Plata Museum, which has one of the most complete collections of these Antarctic reptiles in the world in taxonomic variety and quantity (see details in Reguero 2016). Its researchers have made fundamental contributions to terrestrial reptiles, marine and terrestrial mammals, and birds, which together have been an important contribution to biogeography, climate, paleoecology and taphonomy of these forms in the Argentine Antarctic sector with strong paleobiological implications. An historical synthesis of Antarctic paleoherpetology can be consulted by O'Gorman et al. (2022).

The participation of researchers from the University of Buenos Aires together with personnel from the Argentine Antarctic Institute has produced numerous works on the Marambio, Cerro Nevado, Cokburn and James Ross islands, and on the Antarctic Peninsula, both geological exploration and specific studies on different aspects of stratigraphy, sedimentology, geochronology, micropaleontology, paleomagnetism, and volcanology of the different units. In addition, with other researchers they carried out studies and collected samples on the extent of the mineralization observed in the Antarctic Peninsula (del Valle et al. 1974). The reports of these works, as well as the results of the explorations, are treasured in the series Contributions of the Argentine Antarctic Institute along with studies from other disciplines. In the first years, more complete reports of a scientific nature appeared in the Publications series, such as number one of the series with a review of more than 250 pages (Cordini 1955). The series extended from 1955 to 1994 over 27 issues where they were published, not only geological works. This series with 550 contributions from 1956 to the present contains numerous geological works.

Among the first researchers of the program stand out the explorations and geological studies carried out for more than four decades by Rodolfo del Valle (Fig. 23.12), a geologist who graduated in 1972 and completed his doctorate in 1991 at the University of Buenos Aires. His thesis work on *Geological evolution of the Mesozoic post-arc basin of the extreme Northeast of the Antarctic Peninsula* directed by Carlos Rinaldi is a complete update of the retroarc region of the Antarctic Peninsula (del Valle 1991).

Fig. 23.12 Rodolfo del Valle, a reference for Antarctic geologists with several decades of study on the white continent in one of his campaigns on the Antarctic Peninsula. *Source* Dirección Nacional del Antártico, DNA-IA archives

During those years del Valle has planned, coordinated and executed scientific projects in Antarctica on different aspects of geology, geodesy, geophysics, paleontology, and glaciology as part of the Argentine Antarctic Program. He carried out scientific campaigns between 1973 and 2010, most of them in mobile camps.

Another prominent Antarctic geologist is Eduardo Olivero (Fig. 23.13), who has carried out more than twenty campaigns since 1974 when he started as a student. He completed his final bachelor's thesis the following year on the island of Cerro Nevado under the direction of Carlos Rinaldi, graduating that same year from the University of Buenos Aires. In 1976, a visit to James Ross Island sparked an interest that lasted decades in fully understanding the stratigraphy of the marine Cretaceous of Antarctica. His studies were fundamental to understanding the biostratigraphy, sedimentology, and sedimentation environments that allowed him to reconstruct the Mesozoic evolution of the region, which were published in various international journals (Olivero 2020). In his 1986 research on James Ross Island, he produced a discovery of great global impact, the discovery of an armored ornithischian dinosaur, from the ankylosaur group. Until then, Antarctica was the only continent where dinosaurs were not known. Given its current isolation, separated by oceanic depths from the rest of the continents, the presence of large terrestrial vertebrates indicated, beyond any doubt, that Antarctica had been physically connected to other continental masses (Olivero et al. 1991). His Antarctic studies have received numerous recognitions, including being a member of the Academy of Sciences of Buenos Aires and receiving in 2022 the consecration awards in Earth Sciences from the National Academy of Exact, Physical and Natural Sciences and the Pellegrino Strobel Award from the University of Buenos Aires.

It can be stated that the work of Rodolfo del Valle, José Viramonte, and Eduardo Olivero in the seventies, together with numerous geologists, paleontologists, and geophysicists from different institutions, have been fundamental in laying the foundations of geological knowledge of our Argentine Antarctic sector. These works paved the way for an explosion of contributions in recent decades that were made known to the international community through publications, symposia, and Antarctic congresses.

Fig. 23.13 Eduardo Olivero in 1975 during his undergraduate work, and in 1986 recording the discovery of the first dinosaur in Antarctica. *Source* courtesy Eduardo Olivero

In recent decades, studies in different geological disciplines have led to the consolidation of the work of Argentine geologists (Fig. 23.14) such as Rodolfo del Valle, Roberto Scasso, José Viramonte, Sergio Marenssi, Sergio Santillana, Claudio Parica, among several others, in the Antarctic sector. This brief review has undoubtedly left out numerous research carried out in this region in the field of Earth Sciences, but the selection of a few has had the objective of showing the important activity carried out by the Argentine Antarctic Institute, the first world institution dedicated exclusively to Antarctic scientific research in these distant lands of our country.

In recent years, intense and dedicated research into the Argentine Antarctic sector has continued, carried out by new generations who have extended the example of those pioneers. Among them stands out Sergio Marenssi, professor of the Department of Geological Sciences of the University of Buenos Aires, who was Director of the Argentine Antarctic Institute. He has made important contributions to the Tertiary stratigraphy of Marambio Island with the collaboration of Sergio N. Santillana, with whom he has more than 30 years of Antarctic work. Marenssi has made numerous contributions to the climatic deterioration of high latitudes since the Eocene in the

Fig. 23.14 Rodolfo del Valle, José Viramonte, and Sergio Marenssi, three prominent geologists who have made important and pioneering Antarctic contributions. *Source* courtesy of those photographed

Fig. 23.15 Marcelo Reguero from the National University of La Plata, Roberto Scasso, and Andrea Concheyro from the University of Buenos Aires, three active researchers who participated with the Argentine Antarctic Institute in important contributions in the region. *Source* courtesy of those photographed

Antarctic Peninsula, while he has participated in numerous discoveries and studies of Cenozoic fossils made known with Marcelo Reguero.

Marcelo Reguero is a vertebrate paleontologist (Fig. 23.15), with a PhD from the National University of La Plata, who has been working and studying Mesocenezoic Antarctic faunas for more than forty years. His studies have received numerous national and international recognitions.

Andrea Concheyro (Fig. 23.16), a doctor in geology from the University of Buenos Aires, graduated under the direction of Horacio Camacho in her bachelor's degree and Norberto Malumián in her doctorate, has made important contributions to micropaleontology and the study of nannofossils in Antarctica.

From the beginning, she dedicated herself to micropaleontology, especially the study of nannofossils, which has made her an international reference in this group. In addition to her teaching in Buenos Aires and other universities in Argentina, she has given postgraduate courses in Brazil, Ecuador, the United States, Mexico, and Italy. The experience obtained from her is required by both companies and academia and especially for Antarctic studies, where she has been a researcher at the Argentine Antarctic Institute for more than twenty-five years. She has participated in countless Antarctic campaigns and the results of his research have been made known in specialized symposiums, local and international magazines, and through numerous conferences, among which the one given at the Sweden Royal Academy of Sciences on the occasion of the centenary of the Swedish expedition to Antarctica stands out. She is a full member of the Antarctic Academy where upon her induction she spoke on *Geology and Micropaleontology of the James Ross Basin: the science that emerged from the cold.*

Fig. 23.16 Andrea Concheyro using the pickaxe, taking notes, and with her team of collaborators surveying profiles in search of microfossils on Marambio Island in the 2009 campaign. *Source* courtesy Andrea Concheyro

23.4 Final Considerations

Our geological activity in Antarctica began in the twentieth century with a memorable epic involving the first Argentine geologist José María Sobral, followed by an incredible and glorious rescue carried out by the corvette Uruguay under the command of Julián Irizar. Years of sacrificial work followed, more silent but continuous since 1904 on Laurie Island, the oldest permanent base in the entire Antarctic territory, where the glaciological research of Augusto Tapia in 1920, another pioneering geologist in these regions, stands out.

The creation of the National Antarctic Commission in 1940 began Argentine scientific expeditions on the continent, where the geologist Pascual Sgrosso and the efforts of Agustín Riggi stand out to publicize the geographical and geological importance of our Antarctic sector.

We must wait for the establishment of the Argentine Antarctic Institute in 1951 to have a staff of Antarctic geologists who, with the collaboration of institutions such as the Direction of Mines and Geology and with the scientific leadership of Otto Schneider, began to carry out systematic observations and studies. In these investigations, relevant figures stood out, such as geologists Isaías Cordini, Juan

Olsacher, and Horacio Camacho, along with other scientists who worked mainly in the Antarctic Peninsula and adjacent islands.

Almost 20 years later, the founding of the National Antarctic Direction in 1969 made it possible to coordinate the efforts of the different military institutions operating on the white continent. This institution prioritized scientific research and began to fight for the sovereignty of knowledge. When the contributions of the Argentine Antarctic Institute, the scientific arm of the National Direction, are observed, it can be seen how since 1970 the research carried out has more than doubled, which shows a sustained effort over time. In this action, the Geoantar program under the direction of Carlos Rinaldi has been key to geological investigations. Rinaldi's assumption as the first civilian director of the institute starting in 1985, is reflected in the quality and quantity of geological work carried out. At that time numerous geologists intervened, not only making their studies known through the Contributions and Publications of the Institute but also a marked international externalization of the research of our Antarctic sector began with publications in symposia and specific books and journals of high impact from outside.

References

Aceñolaza FG (2020) Plataforma continental argentina, Islas Australes y Antártida Argentina, la geología como construcción de soberanía, fundamentos geológicos y razones histórico-jurídicos sobre la última frontera de Argentina. Instituto Superior de Correlación Geológica, 193 p, Tucumán

Aguirre-Urreta MB (2016) Horacio Homero Camacho (1922–2015). Revista De La Asociación Geológica Argentina 73(2):294–299

Arctowski H (1903) Die Antarktischen Eisverhaltnisse, Auszug aus meinem Tagebuch der Südpolarreise der „Belgica", 1898–1899. Petermanns Mitteilungen. Erganzungsheft 144, 1099 p, Gotha

Camacho HH (1959) Antártida a través de sus faunas y floras fósiles. Instituto Antártico Argentino, Contribuciones 29:1–18, Buenos Aires

Camacho HH (1979) Paleontología Antártica. Antártida 9:18–23, Buenos Aires

Camacho HH, Villar Fabre J (1957) I. Observaciones geológicas y petrográficas en Bahía Esperanza, Puerto Paraíso e islas Shetland del Sur y Melchior. In Algunas observaciones geológicas y criopedológicas en la Antártida. Instituto Antártico Argentino, Publicación 4:1–128, Buenos Aires

Caminos R (1972) Antártida Argentina. In: Leanza AF (ed) Geología Regional Argentina, Academia Nacional de Ciencias. Córdoba, pp 771–796

Caminos R, Marchese H., Massabie A, Morelli J, Rinaldi C, Spikernann JP (1973) Geología del sector noroccidental de la península Hurd, Isla Livingstone, Shetland del Sur, Antártida Argentina. Instituto Antártico Argentino, Contribución 162

Carrasquero S (2020) Apuntes sobre la geología y sus protagonistas: Pascual Sgrosso. Boletín de Novedades de la Asociación Geológica Argentina, Julio 2020:8–9, Buenos Aires

Cingolani C (2010) Osvaldo Carlos Schauer (1925–2010) Necrológica. Revista de la Asociación Geológica Argentina (2009) 6(1):157–159

Cordini IR (1942) Laguna La Brava, Provincia de Buenos Aires: contribución a su conocimiento limnológico. Universidad de Buenos Aires, Facultad de Ciencias Exactas y Naturales, unpublished PhD thesis, 87 p, Buenos Aires

Cordini IR (1955) Contribución al conocimiento del sector antártico argentino. Instituto Antártico Argentino, Publicación 1:1–273, Buenos Aires

Cordini IR (1959) El conocimiento Geológico de la Antártida. Instituto Antártico Argentino, Publicación 6:1–160, Buenos Aires

Corte AE, Somoza AL (1957) II. Observaciones criopedológicas y glaciológicas en las islas Decepción, Media Luna y Melchior. In Algunas observaciones geológicas y criopedológicas en la Antártida. Instituto Antártico Argentino, Publicación 4:129–131, Buenos Aires

del Río C (2016) Horacio Homero Camacho 1922–2015. Ameghiniana 53(1):74–76

del Valle RA (1991) Evolución geológica de la cuenca mesozoica de pos-arco del extremo Noreste de la Península Antártica. Universidad de Buenos Aires, Facultad de Ciencias Exactas y Naturales, unpublished PhD thesis, 321 p, Buenos Aires

del Valle RA, Morelli J, Rinaldi CA (1974) Manifestación cupro-plumbífera "Don Bernabé" Instituto Antártico Argentino, Contribución 175:1–33, Buenos Aires

del Valle RA, Medina F, Gasparini Z (1977) Nota preliminar sobre el hallazgo de reptiles fósiles marinos del suborden Plesiosauria en las islas James Ross y Vega, Antártida. Instituto Antártico Argentino, Contribuciones 212:1–13, Buenos Aires

Fourcade NH (1960) Estudio Geológico-petrográfico de Caleta Potter, Isla 25 de Mayo, Islas Shetland del Sur. Instituto Antártico Argentino, Publicación 8:1–120, Buenos Aires

Fourcade NH (1968) Geología y petrografía del Monte Irizar y adyacencias, Isla Decepción, Isla Shetland del Sur. 3° Jornadas Geológicas Argentinas, (1966). Actas 3:117–122

Fourcade NH, Viramonte JG (1972) Present Situation of Volcanic Activity in Deception Island, Antártida, Smithsonian Institute, Center for Short Lived Phenomena, 7 p, Washington

Gasparini Z, del Valle R, Goñi R (1984) An Elasmosaurus (Reptilia, Plesiosauria) of the Upper Cretaceous in the Antarctic. Instituto Antártico Argentino, Contribuciones 305:1–24, Buenos Aires

Leanza AF (1964) Prof Dr Juan Augusto Olsacher (1903–1964). Revista de la Asociación Geológica Argentina19(4):195–200

Llambías EJ (1997) Roberto Caminos (1931–1997), Nota Necrológica. Revista De La Asociación Geológica Argentina 52(3):410–414

Marenssi S (2007) Doctor José María Sobral (1880–1961): de los hielos antárticos al olvido argentino. Revista De La Asociación Geológica Argentina 62(4):495–497

Olivero EB (2018) Carlos A Rinaldi (1930–2017), Necrológica. Revista de la Asociación Geológica Argentina 75 (2):303–308

Olivero EB (2020) Rocas, fósiles y sucesos en espacio y tiempo: Armando un rompecabezas antártico-fueguino. Asociación Argentina para el Progreso de las Ciencias, Revista Ciencia e Investigación Reseñas 8(3):74–87, Buenos Aires

Olivero EB, Gasparini Z, Rinaldi CA, Scasso R (1991) First record of dinosaurs in Antarctica (Upper Cretaceous, James Ross Island): paleogeographic implications. In: Thomson MRA, Crame JA, Thomson JW (eds) Geological evolution of Antarctica. Cambridge University Press, pp 617–622

Olsacher JA (1955) Informe geológico preliminar sobre las relaciones entre el Arco de las Antillas del Sur y los océanos Atlántico y Pacífico. Revista de la Facultad de Ciencias Exactas y Naturales (Córdoba) 17(3–4):2–14, Córdoba

Olsacher JA (1956a) Estudio Geológico y Petrográfico de la Isla Decepción (Antártida Argentina). Instituto Antártico Argentino, Publicación 2:1–76, Buenos Aires

Olsacher JA (1956b) Observaciones geológicas en Almirante Brown (Antártida Argentina). Revista de la Facultad de Ciencias Exactas y Naturales (Córdoba) 18(1–2):165–170, Córdoba

Olsacher JA (1956c) Breves observaciones geológicas en Puerto Paraíso, Península Antártica. Revista de la Facultad de Ciencias Exactas y Naturales (Córdoba) 18(1–2):165–170, Córdoba (Instituto Antártico Argentino, Contribución 9)

Olsacher JA (1957) Observaciones geológicas en Caleta Armonía, Isla Nelson, Shetlands del Sur, Antártida Argentina. Revista de la Facultad de Ciencias Exactas y Naturales (Córdoba) 19(1–2):199–221, Córdoba (Instituto Antártico Argentino, Contribución 20, 1958)

Olsacher JA, Díaz H, Teruggi M (1957) Contribución a la Geología de la Antártida Occidental. I.- Contribución al conocimiento geológico de la Isla Decepción. Instituto Antártico Argentino, Publicación 2:1–152, Buenos Aires

O'Gorman JP, Acosta Hospitaleche C, Reguero MA, Gasparini Z (2022) Antecedentes históricos de la paleoherpetología argentina en Antártida. Publicación Electrónica De La Asociación Paleontológica Argentina 22(1):399–410

Quartino BJ (1959) Rocas eruptivas de Caleta Armonía, Isla Nelson, Shetland del Sur. Instituto Antártico Argentino, Contribución 25:1–26, Buenos Aires

Reguero M (2016) Investigaciones de vertebrados en la cuenca de James Ross. In: Rinaldi C Mitos y verdades en el desierto Blanco, Ediciones Aql, 258–281, Villa Martelli

Riccardi AC (2018) Origen y desarrollo de la enseñanza de la Geología en la Universidad Nacional de La Plata. Revista Facultad de Ciencias Exactas, Físicas y Naturales 5(Suplemento 1):33–47, Córdoba

Riggi A (1950) La Antártida Argentina, su geografía y su geología. Publicación del Instituto Cultural y Didáctica, Museo Argentino de Ciencias Naturales Bernardino Rivadavia 4:32 p, Buenos Aires

Rinaldi C (2016) Mitos y verdades en el desierto Blanco. Ediciones Aql, 2da Edición corregida y ampliada, 464 p, Villa Martelli

Rinaldi CA, Massabie A, Morelli J, Rosenman LH, del Valle RA (1978) Geología de la isla Vicecomodoro Marambio, Antártida. Instituto Antártico Argentino, Contribuciones 217:1–37, Buenos Aires

Schauer O, Fourcade N (1964) Estudio geológico - petrográfico del extremo occidental de la isla 25 de Mayo, islas Shetland del Sur. Instituto Antártico Argentino, Publicación 23:1–11, Buenos Aires

Schauer O, Fourcade NH, Dalinger RE (1960) Contribución al conocimiento geológico del extremos occidental de la Isla 25 de Mayo y Península Ardley, Islas Shetland del Sur. Instituto Antártico Argentino, Contribución 22:1–16, Buenos Aires

Sgrosso P (1947) La Península Antártica o Tierra de Graham y sus Archipiélagos adyacentes. Revista Industria Minera 66, Buenos Aires

Sgrosso P (1948) La Antártida Argentina. Revista del Museo de La Plata (Nueva Serie). Geología 4(23):178–197

Sgrosso, P. 1958. Contribución al conocimiento geológico y petrográfico de la Antártida Argentina. Revista del Museo de La Plata (Nueva Serie). Geología 4(29):3711–384

Tonni EP, Cione A (1978) Una nueva colección de vertebrados del Terciario inferior de la isla Vicecomodoro Marambio (Seymour Island), Antártida. Obra Del Centenario Del Museo De La Plata 5:73–79

van Zele MA (1999) Otto Schneider (1912–1999). Revista De La Asociación Geológica Argentina 54(1):93–94

Viramonte J (2017) La ardiente pasión por los volcanes. Asociación Argentina para el Progreso de las Ciencias, Revista Ciencia e Investigación Reseñas 5(2):55–75, Buenos Aires

Viramonte JG, Bossi GE, Fourcade NR (1973) Estudio Preliminar de los Sublimados de la Isla Decepción, Antártida Argentina. 5° Congreso Geológico Argentino (Córdoba). Actas 1:271–282

Viramonte JG, Sureda RJ, Fourcade NR (1974) Estado Volcánico de la Isla Decepción, Islas Shetland del Sur, Antártida Argentina. Contribución Del Instituto Antártico Argentino 174:1–15

Chapter 24
The Founding of New Universities

Abstract For many years, the universities of Buenos Aires, Córdoba, and La Plata were the only ones teaching geology until the 1950s, when the universities of Tucumán, Salta, and the University of the South in Bahía Blanca were founded. At the end of the 1960s and beginning of the 1970s, the universities of San Juan, San Luis, Patagonia, Jujuy, La Pampa, Catamarca, and Río Cuarto were created. Soon, geology courses began to be offered in all of them. Finally, geology courses were created in the universities of Río Negro, Comahue, La Rioja, Tierra del Fuego and Cuyo. All of these universities where geology was taught were public institutions, except the University of San Juan Bosco, which was private and run by the Salesians. However, a few years later it merged with the National University of Patagonia. All of these new universities began to graduate geologists and with the support of the National State they consolidated to train new professionals with different orientations following regional needs. The study of geology grew steadily, incorporating local students, who perhaps would not have had another opportunity to dedicate themselves to the study of geosciences.

Keywords Taquini plan · Jujuy · Cuyo region · Bahía Blanca · Patagonia · Tierra del Fuego

For many years the universities of Buenos Aires, La Plata and Córdoba were the only ones where geology was taught and the geologists that the country needed were trained. In the 1950s, the national universities of Tucumán, Salta, and the South in Bahía Blanca began the geology degree program almost at the same time.

In isolation, a degree program was created at the University of San Juan Bosco in Comodoro Rivadavia in 1963, the only private university in the country founded by the Salesians that a few years later merged into a single university with the National University of Patagonia.

A second batch of universities as the San Juan, San Luis, Patagonia, Jujuy, La Pampa, Catamarca, and Río Cuarto began training geologists between 1968 and 1975. Part of these universities were founded in the 70s by the Taquini Plan, by

V. A. Ramos, *History of the Geological Sciences in Argentina*, The Latin American Studies Book Series, https://doi.org/10.1007/978-3-031-83139-3_24

which 14 national institutions were created between 1971 and 1973, which led to the doubling of university students in Argentina in less than 4 years (see Taquini 2022).

Finally, geology courses were created at the national universities of Río Negro, Comahue, La Rioja, Tierra del Fuego and Cuyo from 2010 to 2018.

Currently, there are eighteen universities where geologists are trained. The three oldest have been described in Chap. 5 and those of Tucumán and Salta in Chap. 17. Below will be a synthesis of those that have begun during the last century and have made important contributions to our geology.

24.1 National University of the South

The geology degree began in 1956 with a body of local professors (Fig. 24.1) such as Augusto Calmels, Oscar Bernasconi, César Prozzi, Oscar Edgardo Gelós, Héctor Orlando and Alfredo Losada who taught a large part of the subjects. The teaching was reinforced by traveling professors from the National University of La Plata such as doctors Ángel Borrello, Alfredo Cuerda, Guillermo Furque and Osvaldo Schauer. The geology department was created in 1961, and its first director was Augusto Pablo Calmels (Mariño and Melchor 2017). The first graduates Norma Pezzutti, Haroldo Laya and Melba Guerstein were received in mid-1963, Shortly after, professors Arturo Corte (Trombotto and Brown 2001), Kitaro Hayase, Simón Burmistrov, and Raúl Dessanti joined, the latter as a traveler professor.

Since its founding, the Department of Geology has been characterized by its contributions in the field of mineralogy and reservoir geology that were consolidated with the arrival of Kitaro Hayase from Waseda University, Tokyo. The first doctors Jorge Dristas, Pedro Maiza and Teresa Manera, trained by Hayase (Mas 2007), received their degrees in 1972. At the end of the sixties, Eduardo Llambías

Fig. 24.1 Augusto Calmels (1924–2017) and Arturo E. Corte (1919–2001) the first two directors of the Department and Kitaro Hayase (1911–1977) a master of mineralogy, director of the first doctoral theses. *Source* Dimieri and Frisicale (2018)

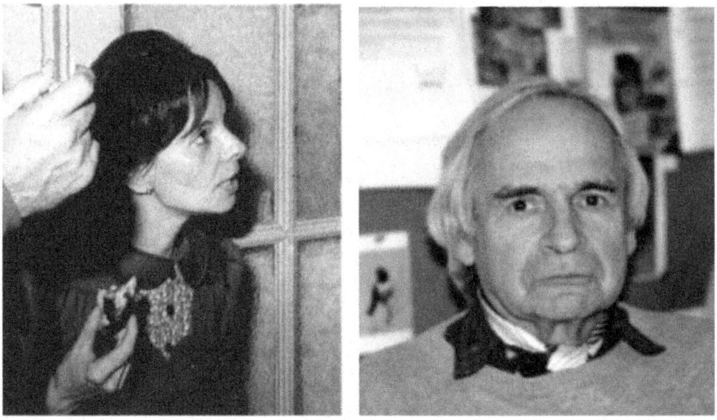

Fig. 24.2 Lidia Malvicini in mineral deposits and Eduardo Llambías in petrology, two excellent professors who had to leave teaching in Bahía Blanca in 1975. *Source* Dimieri and Frisicale (2018)

and Lidia Malvicini joined as professors, reinforcing the areas of petrology and deposits.

The new professors, together with the purchase of equipment for the laboratories and the research carried out, consolidated the department. All this activity was unfortunately cut short in 1975 during the intervention of the National University of the South, which unified the departments of Geology and Biology. The following year, under the military dictatorship, several professors (Fig. 24.2) such as Calmels, Losada, Prozzi, Llambías and Malvicini, among others, resigned along with their supervisors. Three students were kidnapped and found murdered in December 1975 and several more who left the city marked those years of dictatorship (see more details in Dimieri and Frisicale 2018).

With the return to democracy, the department was consolidated, becoming a recognized research center in petrology, mineralogy, and structural geology (Fig. 24.3). Noted for their contributions were Mirta Quattrocchio in palynology, Eduardo Domínguez in mineral deposits, José Kostadinoff in geophysics, Ernesto Bjerg in the study of ultramafic rocks, among many others who currently carry out teaching and research. A separate paragraph deserves the studies carried out in Ventania by Werner von Gosen, professor at the University of Erlangen, who together with local researchers contributed to the understanding of the complex thrust structure of these mountain ranges.

Since the organization of the 6th. Argentine Geological Congress in 1975, the department sponsored numerous conferences and symposia, both local and international, that allowed it to show the progress made by its researchers on different topics.

Fig. 24.3 Emblematic building of the National University of the South where the Department of Geology operated for some years and the new facilities from 1985. *Source* Dimieri and Frisicale (2018)

24.2 National University of Patagonia San Juan Bosco

The teaching of geology in Patagonia began in 1963 with the University of Patagonia San Juan Bosco. This private institution created by the Salesians, organized the first geology course with teachers mostly contributed by YPF, and its first graduates graduated in 1970. Various conflicts in the university environment demanding better training and a higher academic level led at the end of 1973, coinciding with a national policy that promoted the opening of new universities, to the creation of a national university.

At the beginning of 1973, the National University of Patagonia was founded, which began teaching the following year with a large part of the teaching staff made up of YPF professionals and geologists from the Patagonia Plan of the Argentine Mining Geological Survey. Its first graduates were received in 1977. In those years it did not have full-time professors, so research was carried out in the institutions to which the professors belonged.

A few years later, the National University of Patagonia San Juan Bosco was created in 1980 through the unification of the two previous universities, initiating a process of improvement in teaching and research development. Headquarters were opened in the cities of Esquel, Trelew, Puerto Madryn and the former headquarters of Ushuaia (see more details in Paredes et al. 2018).

In the first years, Roberto Ferello, Félix R. Ugarte and Ernesto Lombard, professionals from YPF, were professors there, along with others. Among them, Félix R. Ugarte (1927–1982) stood out, during his stay in the San Jorge basin, taught and published on the coal-Jurassic basin of Patagonia (Ugarte 1966a, b). These contributions were in the first issue of the Geological Sciences Series of the Magazine of the University of Patagonia San Juan Bosco, a short-lived journal.

At the beginning of the 90 s, full-time professors such as Eduardo Musacchio, Juan C. Sciutto and Adriana M. Nillni joined the University and were able to start geological research. The doctoral courses began in 1989 with Raúl E. Giacosa being the first doctor. Professors at this stage were Hugo Arbe, Arturo Canero, Adolfo Genini, Marcelo Márquez, Hugo D. Pezzuchi, Juan C. Sciutto and Raúl Giacosa, among others (Fig. 24.4).

Arturo E. Canero (1941–2017) stands out, as a professor of Geochemistry, elected dean of the Faculty of Natural Sciences, vice-rector and rector between 1982 and 1995. Similarly, Adolfo Genini, professor of mineralogy, fulfilled the roles of dean and rector of the Faculty of Natural Sciences of the university until 2013.

Among the research topics developed are studies on the active Hudson volcano, the crystalline basement of Patagonia, the hydrothermal Au–Ag deposits of the Deseado massif, and the tectonostratigraphic, paleontological and paleoenvironmental aspects of the Golfo San Jorge and Cañadón Asfalto basins, together with the study of its volcanic basement (Fig. 24.5).

The most notable researchers in these years have been Eduardo A. Musacchio (1940–2011), professor of Paleontology and Philosophy of Sciences and international specialist in micropaleontology (Vallati 2012) and Miguel J. Haller (Fig. 24.6), first professor emeritus of the university (Fig. 24.7) with an important career at the Puerto Madryn campus as a full professor and principal researcher at Conicet, president of the Argentine Geological Association and president of the 15th. Argentine Geological Congress of 2002 (Haller 2020).

Fig. 24.4 Professors Arturo Canero in geochemistry and Adolfo Genini in mineralogy have had outstanding University Management responsibilities for consolidating the studies. *Source* Paredes et al. (2018)

Fig. 24.5 Researchers Eduardo A. Musacchio and Miguel J. Haller made important contributions to their disciplines. *Source* Paredes et al. (2018)

Fig. 24.6 Current headquarters of the Department of Geology of the National University of Patagonia San Juan Bosco in Comodoro Rivadavia. *Source* Paredes et al. (2018)

In recent years, coinciding with a generational change, a quality teaching group has been consolidated with research dedicated to the main problems of the region, where the studies carried out in the *Golfo de San Jorge* Basin stand out, as well as the volcanic and geological risk in general. This group organized the 21st. Argentine Geological Congress in Puerto Madryn in 2022.

24.3 National University of San Juan

The geology major was created in the city of San Juan at the initiative of Emiliano P. Aparicio (1921–1988) in 1968 as a Department of Natural Sciences of the Faculty of Engineering and Exact, Physical and Natural Sciences of the National University of Cuyo, whose main headquarters were in Mendoza. A group of teachers from that faculty, doctors Florian Wetten, Jorge C. Oliveri, and Julio A. Llano, headed by Aparicio, proposed the course plan considering the proximity of the Andes and the excellent geology and structure of the adjacent regions. In its beginnings, it contributed to the development of the Cordillerano Plan in the Cuyo region, which needed geologists to implement the exploration programs of the Fabricaciones Militares (see Chap. 15).

Aparicio had participated as president of the Organizing Commission in the First Argentine Geological Conference in the city of San Juan in 1960, with which the geological congresses began.

With the creation of the National University of San Juan in 1973, as part of the Taquini plan, the consolidation of the degree was further promoted, now as part of the Faculty of Exact and Natural Sciences, independently of Engineering. In its initial teaching staff (Fig. 24.7), Emiliano Aparicio (Zambrano 1989), Jorge C. Oliveri (Castro de Machuca 2009), Apolo Ortiz (Salfity 2016), Julio Llano and Florián Wetten stood out, as well as engineers Juan C. Perucca and Fernando Volponi (see more details in Castro de Machuca 2018).

New professors contributed to the consolidation of teaching such as Alberto Regairaz, Felisa Bercowski, Juvenal Zambrano, Otto Wilko Simon, and Alfredo Monetta, the latter in charge of the Museum of Natural Sciences that Aparicio had created in 1964. The first graduates were received in 1975. and the first doctor was Osvaldo Bordonaro in 1983, a renowned biostratigrapher of the early Paleozoic.

A considerable part of San Juan graduates work in mining exploration and exploitation and have made important contributions to the knowledge of new mineral occurrences in the province. Among these, the discoveries of world-class gold deposits in the Veladero mine and porphyry copper in Los Azules, both in the Principal Cordillera of San Juan, and the silver-gold deposits of the Casposo mine in

Fig. 24.7 Four professors marked the development of Geology in San Juan: Emiliano Aparicio, Jorge Oliveri, Juvenal Zambrano, and Bruno Baldis in the 70s. *Source* Castro de Machuca (2018)

Fig. 24.8 Headquarters of the Department of Geology at the Malvinas University Campus in Rivadavia. *Source* Castro de Machuca (2018)

the Frontal Cordillera, which showed the success of creating a school of geology at the foot of the Andes. In the initial discussions about the approval of the degree at the new National University of San Juan, Arturo J. Amos, at the time dean of the Faculty of Natural Sciences of the University of La Plata, had stated that *San Juan is unquestionably the place most suitable in our country for the location of a Geology school* (Castro de Machuca 2018).

Its current building was inaugurated in 1989 when the faculty and the degree program moved to the Islas Malvinas University Campus (Fig. 24.8).

The arrival of Bruno Baldis (1937–1997) in 1976 to San Juan gave his teachings and studies on the early Paleozoic of the Precordillera a new rhythm and a more international perspective with his research (Aceñolaza 1998). Baldis graduated from the University of Buenos Aires in 1964 and dedicated more than thirty years to the study of the San Juan Precordillera, which made it known globally for his research. His work with his disciples on the Precordillera published in Episodes had a high international impact due to the importance of its geological characteristics (Baldis et al. 1984). He directed several international geological correlation projects sponsored by UNESCO and organized the 11th Argentine Geological Congress in the city of San Juan in 1990. On that occasion, with the support of the local municipality, they named the main square of Zonda after Ricardo F. Stappenbeck, in homage to that young German geologist and pioneer who studied the Precordillera at the beginning of the twentieth century (Fig. 24.9).

Another professor who introduced solid training in sedimentology and historical and regional geology was Juvenal Zambrano (1927–2010). Graduated from the National University of Córdoba where he received his doctorate, he arrived in 1974 with extensive oil experience. He made important contributions to the knowledge of the aquifers of the Ullum Valley and other sectors of the province, as well as in other regions of the country through his time at the Regional Groundwater Center

Fig. 24.9 Bruno Baldis, president of the congress, with Bernabé Quartino, unveiling in 1990 the plaque in homage to Ricardo F. Stappenbeck in the Zonda main square that bears his name. *Source* Ramos (2016)

based in San Juan. He was a pioneer along with Carlos María Urien in the knowledge of the offshore sedimentary basins of our continental shelf, a topic where he made important scientific contributions. As a teacher, he is remembered for his low profile, sober, prudent in his actions and his generosity in the delivery of his knowledge that he did not stop perfecting until the last moment (Suvires 2010).

24.4 The National University of San Luis

It is another of the universities created by the Taquini Plan at the beginning of 1973. The bachelor's degree in geology, with a specialty in mineral deposits, was created that same year with courses beginning the following year. One of the promoters was Horacio Perino who had graduated from the National University of Córdoba as a geologist, and later as a doctor in Natural Sciences in 1950. He was until his retirement a professor of general geology.

The initial teachers were traveling professors from La Plata or Buenos Aires and professionals who worked in the institutions of the province. Among the first (Fig. 24.10), Pedro Criado Roque (1921–1988) stood out, who joined the university in 1977. He graduated from the National University of La Plata where he received his doctorate in 1944 and worked at YPF for many years. He toured much of the country with his surveys until 1960 when he retired from YPF and went to work in Cuba for two years (Rolleri 1989). The Cuban revolution had triumphed in 1959, and he was part of a group of idealistic professionals who went to offer their services to the revolution. He returned to the country in 1962, and shortly after rejoined YPF where he reached the position of Director of Exploration. He made fundamental contributions to Argentine regional geology that have been recorded in the first symposia organized by the National Academy of Sciences. As a professor, he taught structural, historical, and Argentine geology, promoting regional field trips with his students. He was president of the Organizing Commission of the 8th Argentine

Fig. 24.10 Horacio Merino, Pedro Criado Roque, and Carlos Costa, three prominent professors at the National University of San Luis. *Source* Ortiz Suárez and Morla (2018)

Geological Congress in 1981 held in the city of San Luis, together with Horacio Perino. On that occasion, they gave the name Dr. Franco Pastore to a street in the city, for being the first Argentine geologist, a native of San Luis and with important geological contributions made in the province.

The arrival in San Luis of Conicet researchers, such as Eduardo Llambías, Lidia Malvicini, César Prozzi and Elda Di Paola, marked the development of new doctorates and scholarships, with a renewed push for research (Ortiz Suárez and Morla 2018). These new teachers allowed the first doctoral thesis to be completed in 1990 and many students to begin their doctorate.

In those years Carlos Costa, a graduate of San Luis with postgraduate training in photogeology in Colombia, undertook the study of the neotectonics of San Luis, until then unknown. Costa began as a professor of structural geology in 1987 and completed his thesis almost self-taught in 1992 under my direction. In a few years, he became a referent in terms of neotectonics in our country, studying the paleoseismology of different *fossil earthquakes*. His studies transcended the local ambit, being an obligatory reference in these studies in South America and beyond our borders (see Costa 2020). Costa provided his knowledge of structural geology as a professor until his retirement in San Luis and is widely known for his neotectonics courses throughout the country. He is one of the most internationally prominent teachers of that house of studies.

The level reached by the Department of Geology is currently in the hands of its graduates, almost all of them with postgraduate degrees, who have led through teaching and research to far surpass the teaching received from their first professors.

24.5 National University of Río Cuarto

This university is another of those created by the Taquini Plan in 1971, but the teaching of geology began in 1975 at the request of the geologist Hugo Tomassino (Fig. 24.11), a graduate of the National University of Córdoba. In their first years, they had a series of vicissitudes that led to almost closing their geology degree (Pinotti 2018). Its consolidation began with the arrival of Mario P. Cantú in 1979, who was the first full-time professor. Cantú, a graduate of Buenos Aires, had developed his professional activity at INTA, which allowed the teaching of Quaternary geology to be consolidated in Río Cuarto, especially the genesis of soils, geomorphology, and environmental geology, turning this university into a national reference in these areas disciplines. A few years later, the first graduate was received in 1982. Among the teachers, Ana María Esparza in charge of the petrology area stands out. She was a graduate of the National University of the South, who had started, together with Julio Llano in 1972, the teaching of petrography at the University of San Juan.

Years later, with the incorporation of Wolfgang Volkheimer, the first Conicet scholarships were obtained in 1983, which allowed the development of the postgraduate course, receiving the first doctors, Mercedes Prámparo and Ana María Zavattieri (Fig. 24.12), with outstanding performance in the field of palynology. New graduates such as Juan Otamendi in 1995, Ana María Combina in 1996, and Lucio Pinotti in 1998, were the professors of the department to obtain their doctorate in Río Cuarto.

With the valuable contribution in the supervision of postgraduate research by Francisco Nullo, Teresa Sánchez, Eduardo Llambías, Eugenio Aragón, Martín Iriondo, Miguel Haller, Daniel Poiré, Luis del Río, Mario Hernández, Luis Benedetto, Victor A. Ramos and Carlos Costa's graduates have had sustained training in the human resources that currently make up their teaching staff.

The petroleum geology group led by María Lidia Sánchez had the support of the industry through courses and internships. Among them, the support provided by Gualter Chebli, Daniel Kokogian, Gustavo Vergani and Facundo Fuentes stands out.

Fig. 24.11 Hugo Tomassino, creator of the Geology degree, Ana M. Esparza, professor of petrography, and Juan Otamendi, professor of geochemistry. *Source* Pinotti (2018)

Fig. 24.12 Mario Cantú and the first female doctors of Río Cuarto, Mercedes Prámparo and Ana María Zavattieri. *Source* Pinotti (2018) and courtesy Mercedes Prámparo and Ana M. Zavattieri

Among his professors, Juan Otamendi and his disciples stood out for their regional work on the petrology of the Sierras Pampeanas with contributions to the best international journals in their specialty. Professor Milka Kronegold de Brodtkorb, named honorary doctor of the University of Río Cuarto, dedicated her last years as a professor to teaching opaque mineral and metallogenesis courses at this institution (Pinotti 2018).

Currently, with a solid teaching staff, the department was consolidated (Fig. 24.13) with its research in Quaternary geology and in its petrological and geochemical studies of the Sierras Pampeanas.

Fig. 24.13 New building of the Department of Geology inaugurated in 2018. *Source* Pinotti (2018)

24.6 National University of La Pampa

The geology degree began with the creation of the National University of La Pampa in 1975, based in the city of Santa Rosa. The organization of the career was in charge of Augusto Pablo Calmels (1924–2017) (Fig. 24.14), who had been laid off from the National University of the South that year (Mariño and Melchor 2017). Initially teaching several subjects, and then concentrating on the first general geology courses, Calmels had a long career in the new Department of Geology.

He had management tasks as department director and dean of the faculty. He contributed to the organization of the Professional Council of Natural Sciences of the province of La Pampa and promoted the current Pampas Natural Sciences Conference since 1980. The first years were difficult but starting in 1990 the first graduates began to participate in teaching while completing their doctorate. Among them, Silvio Casadío (Fig. 24.15) stands out for his initial studies in La Pampa in the Cretaceous-Paleogene interval, Ricardo Melchor in the Paleozoic sedimentary rocks of the province, and Hugo Tickyj for the structure and petrology of the Paleozoic crystalline basement.

Years later, as a result of the University Quality Improvement program, the establishment of human resources with recognized careers was achieved, such as Edsel Brussa (Benedetto, 2008) and Marcelo Zárate (Fig. 24.16), who began to teach paleontology and Argentine geology classes, respectively. The first of them participated very actively in various academic aspects until he died in 2008.

Of utmost importance for the development of geological studies was the management of Marcelo Zárate, which culminated in the creation of the Institute of Earth

Fig. 24.14 Augusto Pablo Calmels at the ceremony of his appointment as Honorary Professor of the University in 2007. *Source* Mariño and Melchor (2017)

Fig. 24.15 Silvio Casadío, Ricardo Melchor and Hugo Tickyj during fieldwork. *Source* Courtesy Silvio Casadío

Fig. 24.16 Two professors, Edsel Brussa and Marcelo Zárate promoted teaching and research starting in the 1990s. *Source* Benedetto (2008) and courtesy Marcelo Zárate

and Environmental Sciences of La Pampa (Incitap), dependent on Conicet and the university (Zárate 2023). This institute has two major central themes, geology and soil science. In the first, aspects linked to sedimentology, petrology, tectonics, paleontology, stratigraphy and geochronology are studied in the geographical area of La Pampa and adjacent sectors (Bertotto and Umazano 2018).

24.7 National University of Comahue

The Provincial University of Neuquén was nationalized and the National University of Comahue was created in 1971, starting courses the following year. The first geology courses were taught in 1966 as support for engineers in the Faculty of Engineering of the provincial university at its Challacó headquarters. Renowned industry geologists such as Oscar Galante, Fernando Ramón, Luis Cazau, Jorge Cangini, Roberto Gazzani, Carlos Devizia, Fortunato Valenti, Omar González, and Antonino Salleo participated as teachers at this headquarters. These courses were moved to Neuquén starting in 1972, but the geology degree did not yet exist.

After a dark period of interventions during the entire military regime, with the return of democracy came the normalization of the university between 1984 and 1986. This period was led by the geologist Arístides Romero (1926–1995) as a normalizing rector. This prestigious geologist graduated from the University of Buenos Aires in 1955 and was also *Docteur de l'Université Paris-Sorbonne* with a thesis on clay mineralogy (Ramos 2021).

In those years Romero promoted research and teaching contexts and supported the creation of the Clay Mineral Research Center (CIMAR) in 1986, one of the first research centers of the National University of Comahue under Jorge Vallés (Fig. 24.17). In the 90s, the Patagonian Center for Metallogenetic Studies was created, directed by Marta B. Franchini, as well as the Dino Project Geo-Paleontological Natural Park under the direction of Jorge Calvo (Fig. 24.18), which is located in Lake Barreales north of the city of Neuquén. Despite all this activity and the geology courses that were given to the Faculty of Engineering, the degree was still not created (Vallés 2018).

Fig. 24.17 Professors Jorge Vallés and Marta Franchini, directors of the Clay Mineral Research Center and the Patagonian Center for Metallogenetic Studies. *Source* Vallés (2018)

Fig. 24.18 The Geo-Paleontological Natural Park in the vicinity of Lake Barreales and the paleontologist director Jorge Calvo (1961–2022). *Source* Centro Paleontológico Lago Barreales archives

The first initiatives began in 2005, but the geology degree program was finally created in 2010. This degree program was widely accepted in the region, with hundreds of students interested in pursuing a degree in geology. The first geologists graduated between 2016 and 2017.

Jorge Vallés organized the 18th. Argentine Geological Congress in the city of Neuquén in 2011, with the support of the Department of Geology of the National University of Comahue.

24.8 The Geology Career in the New Universities

In recent years, geology courses were established in the new universities created in this century. Among them, the National University of Río Negro stands out, created in 2007 and whose geology degree began in 2010 at the General Roca headquarters.

The degree program had the initial collaboration of doctors Raúl Giacosa and Eugenio Aragón (Fig. 24.19) and the course was led by Silvio Casadío.

At the same time, the degree in Paleontology was created, which was led by Rodolfo Coria and the collaboration of Leonardo Salgado. It currently has a modern building for teaching classrooms and extensive research facilities at the Paleobiology and Geology Research Institute dependent on the National University of Río Negro and Conicet (Fig. 24.20).

This career stands out for a modern pedagogical program based on learning about phenomena complemented by research at all levels (Diez and Casadío 2018).

Another new university is the National University of Tierra del Fuego, Antarctica, and the South Atlantic Islands, which was created in 2009, based on the old Ushuaia headquarters of the National University of Patagonia San Juan Bosco. The geology degree began in 2012 at the Institute of Polar Sciences, Environment and Natural Resources (Fig. 24.21). It began with geologists residing in Ushuaia, and researchers from the Austral Center for Scientific Research (CADIC). Later, professionals from

Fig. 24.19 Raúl Giacosa and Eugenio Aragón with Pablo González, current professor of the degree in Geology. *Source* Diez and Casadío (2018)

Fig. 24.20 Classroom building for Teaching and Research Institute in Paleobiology and Geology of the general Roca headquarters of the National University of Río Negro. *Source* Diez and Casadío (2018)

other parts of the country joined, as well as from Ushuaia, both from the Centro Austral and other provincial entities. The formation of the teaching staff is still in training (González Guillot 2018).

Among his first professors, Jorge Rabassa (Fig. 24.22) stands out, with extensive experience in Quaternary and glacial geology (Rabassa 2023), and Eduardo Olivero, stratigrapher, paleontologist, and great expert on the Cretaceous sequences of Tierra del Fuego and the Antarctic Peninsula (Olivero, 2020), both of them coming from the Austral Center of Research and residing in Ushuaia. The new professor Mónica Escayola is responsible for the petrological studies.

The teaching staff is in full development and is growing in experience with its own graduates. A notable characteristic of this university is that it is located in the port (Fig. 24.23), which is the closest route to the Antarctic continent.

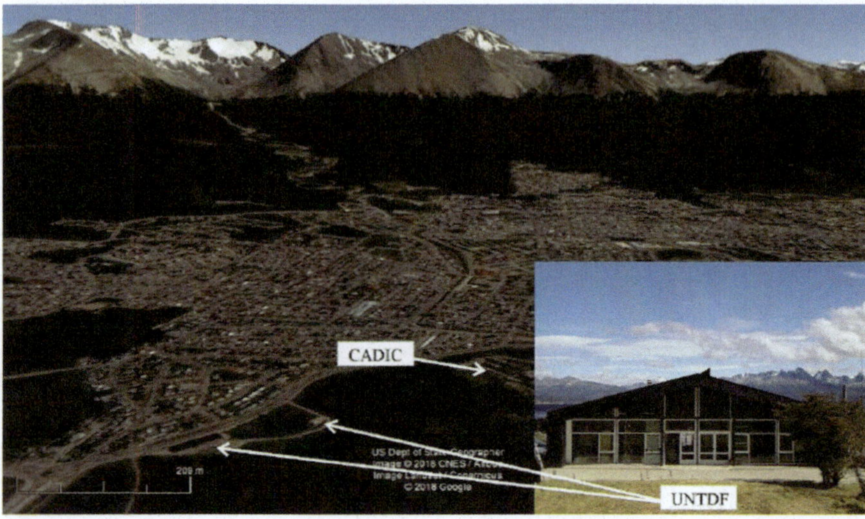

Fig. 24.21 Location of the National University of Tierra del Fuego and the Austral Center for Scientific Research (CADIC). *Source* González Guillot (2018)

Fig. 24.22 Professors Jorge Rabassa, Eduardo Olivero, and Mónica Escayola. *Source* Courtesy of those photographed

The latest geology degree program created is that of the National University of Cuyo in Mendoza. Although this university was founded in 1939, the Faculty of Exact and Natural Sciences was established as such in 2014. In that year, a higher cycle of geology was incorporated into the degree in basic sciences. The first director of the course is Bernardo González Riga, Conicet researcher and first professor of geology. In 2018, the 3rd year began at the Malargüe headquarters under the coordination of Adriana Blesa (González Riga et al. 2018).

Fig. 24.23 Classroom extension of the National University of Cuyo at the Malargüe headquarters where the degree in Geology is taught. *Source* Courtesy Universidad Nacional de Cuyo

24.9 Final Considerations

The brief synthesis carried out shows that the first three universities where the teaching of geology began, Buenos Aires, Córdoba, and La Plata, were followed after almost half a century during the end of the Peronist government by three others, Salta, Tucumán, and del Sur in Bahía Blanca, created in the 50 s. We must wait for the Taquini Plan in the 1970s for the teaching of geology courses to become generalized in numerous universities such as the National University of Patagonia, San Juan, San Luis, Río Cuarto, Jujuy, La Pampa, and Comahue. However, the latter was created in 1971, and the teaching of the degree in geology began much later.

It is interesting to note that of all the universities that teach geology, the only one that was private in nature was that of San Juan Bosco in Comodoro Rivadavia from 1963 to 1980. The necessary investment to set up laboratories and have a good level of research and quality of teaching led to merging with the national university to form the National University of Patagonia San Juan Bosco in 1980.

The last wave of creation of geology courses at the beginning of this century was concentrated in the national universities of Río Negro, Tierra del Fuego, La Rioja, and Cuyo. The last geology degree was formed in 2016 at the National University of Cuyo, an institution with a long history founded in 1939.

A common characteristic that all these initiatives had is that after a few initial years of indecision, the vocation to study geology grew steadily, incorporating local students, who perhaps would not have had another opportunity to dedicate themselves to the study of geosciences.

References

Aceñolaza F (1998) Bruno Alberto Juan Baldis (1937–1997). Revista De La Asociación Geológica Argentina 53(3):141–142

Baldis B, Beresi M, Bordonaro O, Vaca A (1984) The Argentine Precordillera as a key to Andean Structure. Episodes 7(3):14–19

Benedetto JL (2008) Dr Edsel Daniel Brussa (1961–2008). Revista de la Asociación Geológica Argentina 63(4):776–778

Bertotto GW, Umazano AM (2018) La enseñanza de la Geología en la Universidad Nacional de la Pampa. In: Martino R, Ramos VA (eds) El nacimiento de la geología en la enseñanza universitaria, Revista Facultad de Ciencias Exactas, Físicas y Naturales 5(Suplemento 1):123–129, Córdoba

Castro de Machuca B (2009) Dr Jorge Carlos Oliveri: 1921–2008. Revista de la Asociación Geológica Argentina 64(2):357–359

Castro de Machuca B (2018) Orígenes de la carrera de Geología en la Universidad Nacional de San Juan. In: Martino R, Ramos VA (eds) El nacimiento de la geología en la enseñanza universitaria, Revista Facultad de Ciencias Exactas, Físicas y Naturales 5(Suplemento 1):98–105, Córdoba

Costa C (2020) Buscando terremotos fósiles. Asociación Argentina para el Progreso de las Ciencias, Revista Ciencia e Investigación Reseñas 8(4):42–55, Buenos Aires

Diez MA, Casadío S (2018) Historia en tiempo presente de la enseñanza e investigación en Ciencias de la Tierra en la Universidad Nacional de Río Negro. In: Martino R, Ramos VA (eds) El nacimiento de la geología en la enseñanza universitaria, Revista Facultad de Ciencias Exactas, Físicas y Naturales 5(Suplemento 1):141–151, Córdoba

Dimieri LV, Frisicale MC (2018) Síntesis histórica de la enseñanza de la Geología en la Universidad Nacional del Sur. In: Martino R, Ramos VA (eds) El nacimiento de la geología en la enseñanza universitaria, Revista Facultad de Ciencias Exactas, Físicas y Naturales 5(Suplemento 1):80–88, Córdoba

González Guillot M (2018) La Carrera de Licenciatura en Geología en la Universidad Nacional de Tierra del Fuego, Antártida e Islas del Atlántico Sur. In: Martino R, Ramos VA (eds) El nacimiento de la geología en la enseñanza universitaria, Revista Facultad de Ciencias Exactas, Físicas y Naturales 5(Suplemento 1):152–157, Córdoba

González Riga BJ, Tovar M, Tarabelli MF, Blesa AP, Mescua JF (2018) La carrera de geología más joven del país: Licenciatura en Geología de la Universidad Nacional de Cuyo. In: Martino R, Ramos VA (eds) El nacimiento de la geología en la enseñanza universitaria, Revista Facultad de Ciencias Exactas, Físicas y Naturales 5(Suplemento 1):158–1637, Córdoba

Haller MJ (2020) De Suiza a la Patagonia, 46 años de practicar geología por todo el país. Asociación Argentina para el Progreso de las Ciencias, Revista Ciencia e Investigación Reseñas 8(2):62–76, Buenos Aires

Mariño E, Melchor R (2017) Augusto Pablo Calmels: 1924–2017. Revista De La Asociación Geológica Argentina 74(4):583–586

Mas GR (2007) Dr Kitaro Hayase: Un Maestro. En Historia de la Geología Argentina, INSUGEO, Miscelánea 2016:73–74, Tucumán

Olivero EB (2020) Rocas, fósiles y sucesos en espacio y tiempo: armando un rompecabezas antártico-fueguino. Asociación Argentina para el Progreso de las Ciencias, Revista Ciencia e Investigación Reseñas 8(3):74–87, Buenos Aires

Ortiz Suárez AE, Morla PN (2018) Nacimiento y evolución de la Licenciatura en Ciencias Geológicas, Universidad Nacional de San Luis. In: Martino R, Ramos VA (eds) El nacimiento de la geología en la enseñanza universitaria, Revista Facultad de Ciencias Exactas, Físicas y Naturales 5(Suplemento 1):106–112, Córdoba

Paredes JM, Giacosa RE, Foix N, Allard JO (2018) La enseñanza de la Geología en Comodoro Rivadavia. In: Martino R, Ramos VA (eds) El nacimiento de la geología en la enseñanza universitaria, Revista Facultad de Ciencias Exactas, Físicas y Naturales 5(Suplemento 1):89–97, Córdoba

Pinotti LP (2018) Historia y evolución del Departamento de Geología de la Universidad Nacional de Río Cuarto. In: Martino R, Ramos VA (eds) El nacimiento de la geología en la enseñanza universitaria, Revista Facultad de Ciencias Exactas, Físicas y Naturales 5(Suplemento 1):113–122, Córdoba

Rabassa J (2023) Cincuenta años con geoformas y paisajes, glaciares y suelos congelados. Asociación Argentina para el Progreso de las Ciencias, Revista Ciencia e Investigación Reseñas 11(2):67–94, Buenos Aires

Ramos VA (2016) Historia del conocimiento geológico de la Precordillera. 3° Jornadas de Geología de Precordillera, Acta Geológica Lilloana 28 (Suplemento):154–162, Tucumán

Ramos VA (2021) El geólogo Arístides Romero: memorias de un luchador. Fundación Azara, Universidad Maimónides, Revista De Historia Natural 11(1):5–35

Rolleri EO (1989) Pedro Criado Roque 1921–1988. Revista De La Asociación Geológica Argentina 43(3):428–431

Salfity J (2016) Apolo Ortiz (1936–2015), Necrológica. Revista De La Asociación Geológica Argentina 73(1):156–159

Suvires GM (2010) Juvenal Jorge Zambrano (1927–2010), Necrológica. Revista De La Asociación Geológica Argentina 66(1):296–299

Taquini A (compiler) (2022) Universidad y cambio social: Plan Taquini: pasado, presente y futuro. EDUNTREF, 336 p, Buenos Aires

Trombotto D, Brown J (2001) Arturo Eduardo Corte 1919–2001. Arct Antarct Alp Res 33(3):378–379

Vallati P (2012) Dr Eduardo Aldo Musacchio (1940–2011) Nota necrológica. Revista de la Asociación Geológica Argentina 68(2):306–309

Vallés JM (2018) El nacimiento de la enseñanza de la Geología en la Universidad Nacional del Comahue. In: Martino R, Ramos VA (eds) El nacimiento de la geología en la enseñanza universitaria, Revista Facultad de Ciencias Exactas, Físicas y Naturales 5(Suplemento 1):130–140, Córdoba

Zambrano JJ (1989) Emiliano Pedro Aparicio 1921–1988. Revista De La Asociación Geológica Argentina 43(3):425–427

Zárate MA (2023) La lupa geológica en la llanura y los suelos. Asociación Argentina para el Progreso de las Ciencias, Revista Ciencia e Investigación Reseñas 11(1):91–104, Buenos Aires

Chapter 25
A Geological Revolution

Abstract This chapter summarizes the facts related to the important changes that occurred in the geological sciences due to the modifications in the basic paradigms throughout the nineteenth century. These changes affected different institutions, although some responded quicker than others to the new situations. The *Continental Drift* found geologists in Argentina in an advantageous position that allowed them to participate substantially in the development of the new hypotheses. An example of this is the influence that Keidel had on Wegener's ideas regarding the relations between South Africa and South America. Our geologists quickly accepted the displacement of the continents in contrast to those of the northern hemisphere. However, the revolution in *Plate Tectonics* was not quickly accepted. In this revolution, the paleomagnetic studies of Eng. Valencio were at the frontier of knowledge. Geologists were slower to accept the new hypotheses, but finally recognized them, and played an important role in understanding the mechanisms that led to the uplift of the Andes and in the analysis of processes such as flat-slab subduction, accretion of allochthonous terranes, and crustal delamination and mantle removal.

Keywords Continental drift · Paleomagnetism · Plate tectonics · Flat-slab subduction · Terranes

In the first chapters, the development of Argentine geology is analyzed through historical records. As institutions were created, emphasis was placed on the evolution of knowledge and the contributions made in each of them, highlighting their most active periods. However, events and advances in that knowledge were transversal to all of them, even though in some they had more impact than in others.

The first of these has been the theory of *Continental Drift*, which had a strong consensus in our country in the first decades of the twentieth century, almost from the beginning, unlike in the northern hemisphere.

The second of these events took place at the end of the 60s with the advance of knowledge of the dynamics of the ocean floor, which produced great changes in the understanding of geological processes. These changes were interpreted as a scientific revolution since important premises had to be left aside and the data reinterpreted

V. A. Ramos, *History of the Geological Sciences in Argentina*, The Latin American Studies Book Series, https://doi.org/10.1007/978-3-031-83139-3_25

according to the new evidence. The most important change in these premises was that the continents were moving, and this was very far from the concept that had been accepted since the dawn of geology, in which their position was fixed over time since continental drift had not been universally accepted. This revolution was called *Plate Tectonics*, which caused dizzying changes in many concepts and strong controversies until it became established as a widely accepted theory. In this chapter, a universal development of the theory will not be made, since our objective will be to describe the response that the Argentine geological community gave to these new ideas. The background and ideas that prevailed in previous periods in our country will be analyzed, as well as the strong resistance that there was in different areas to their acceptance, which in some cases significantly delayed the understanding of the new paradigms.

25.1 Continental Drift

Since the dawn of the theory of *Continental Drift*, which some consider the first scientific revolution in geology, until the beginnings of *Plate Tectonics*, numerous works have studied the history of these two great revolutions in Earth sciences. In the history of continental drift, there is a consensus in accepting that Alfred Wegener (1880–1930) in his various works from 1912 (Wegener 1912, 1915, 1920, 1922) until 1922, crystallized in his last work of 1929 (Wegener 1929), was the father of the theory that proposed the mobility of the continents. This recognition is based on the fact that he was the first to comprehensively and multidisciplinary analyze the correlation between the different continents and their displacement, a work to which he dedicated a large part of his life, dying in 1930 in Greenland while trying to measure their displacement (Ramos 2013).

Among these analyses of his history, the famous 1928 symposium on *Continental Drift* organized by the American Association of Petroleum Geologists (van der Gracht 1928) stands out. It is evident from reading the conclusions of the symposium that there was consensus among geologists in the northern hemisphere and especially in the Anglo-Saxon community, with honorable exceptions, that the continents could not move.

However, especially those who worked in the Sierras de la Ventana in the province of Buenos Aires did not offer resistance to these theories. In those years, Eduard Suess (1831–1914) had asked Gustav Steinmann (1856–1929) whether the Sierra de la Ventana was a branch of the Andes, a question that Steinmann referred to Juan (Hans) Keidel, a young disciple who had just joined the Geology and Mines Division of Argentina. Keidel replied in a long letter in 1906 saying *It is a Paleozoic Mountain range developed on the margin of Gondwana and has no relations with the Andes* (see details in Chap. 9). The answer impressed Suess so much that he had it published in the proceedings of the Imperial Academy of Sciences, Mathematics and Natural Sciences of Austria under the title *On the construction of the Argentine Andes* (Keidel 1907).

Keidel had arrived in our country that year and did not yet know these mountains but based on the explanations given to him by Rodolfo Hauthal, who had already worked in the Sierras de la Ventana, he responded to Steinmann. From there, his interest in studying the region was awakened and he decided to visit the Sierra de Pillahuincó for the first time in 1912. During this visit, he found strong evidence of its glacial origin, such as clasts with pentagonal shapes and strong striations. He carried out several works on this subject, the first of which he presented in 1913 at the 12° International Geological Congress in Toronto (Keidel 1913). In this Congress, he not only made known the geology of the mountains but also correlated their glacial deposits with the Lafonia Tillite in Malvinas and those of Dwyka in South Africa, justifying their similarities with the Cape System, a proposal that aroused much interest (Fig. 25.1). His teacher Steinmann, in his report published in Germany the following year on the international congress, made complimentary comments on Keidel's correlation, welcoming a new *continental bridge* between Africa and South America (Steinmann 1914). This interpretation of the *continental bridges* was used by the *fixists* who did not accept the movement of the continents.

However, when Keidel published his second and most complete work in 1916 on *The Geology of the Sierras of the Province of Buenos Aires and their Relations with the Mountains of South Africa* he did not postulate the presence of a continental bridge. In this work, he concluded that *The Sierras of Buenos Aires are interposed between the Paleozoic structures of the Andean region and the mountains of the Cape, linking,* ***in truth in an unknown way****, the movements of the Upper Paleozoic of South America and South Africa* (Keidel 1916, p. 63).

When Wegener, who had not attended the international congress in Toronto, learned of Keidel's descriptions and correlations through Steinmann's summary of the supposed continental bridges, he contacted Keidel. Wegener had been a strong critic of the existence of bridges due to the impossibility of sinking lighter continental crust into the denser oceanic crust (Wegener 1912), so he quickly rejected the proposal for a new bridge. Furthermore, upon learning of Keidel's second work, he decides to modify his first edition of 1915 of *The Formation of Continents and Oceans* and publishes a new version (*Die Entstehung der Kontinente und Ozeane*, Wegener

Fig. 25.1 Striated clasts: **a** and **b** from the Keidel Sauce Grande Tillite and **c** from the Dwyka Tillite (Keidel 1913; du Toit 1926)

Fig. 25.2 Second and third editions (1920 and 1922) of Alfred Wegener's classic work *The Formation of Continents and Oceans*, whose first geological evidence was based on Keidel's correlation work

1920). In these and successive editions he places the observations obtained by *Professor Keidel* as the first and most compelling geological evidence to demonstrate continental drift (Fig. 25.2).

Keidel continued to work on these Upper Paleozoic movements, now in northern Argentina, calling the resulting structures the Gondwanides (Keidel 1922). Keidel's work caught the attention of Alexander du Toit (1878–1948), a South African geologist and enthusiastic defender of the ideas of *Continental Drift*, one of the few who had defended Wegener's ideas at the AAPG symposium in 1928. Du Toit decided to visit Brazil, Uruguay, and Argentina (Fig. 25.3) to verify the nature of the supposed late Paleozoic glacial deposits and to be able to discuss his ideas with Keidel (see details in Ramos 2013, 2023). It is important to highlight that based on du Toit's publications, the Anglo-Saxon world has considered him as the precursor of the correlation between the Ventania and Cape orogenic systems and the development of the Gondwanides. Although in his works of 1927 and 1937, especially in the latter work *Our wandering Continents*, he confirms and gives credit to Keidel's observations and uses his name of the Gondwanides, in English literature Keidel has disappeared both as the author of the first continental correlations and as the father of the definition of the Gondwanides (du toit 1927, 1937). It was Keidel who objectively described the correlations, and it was left to du Toit to follow in his footsteps and learn from his contributions to reconstruct the Gondwanides.

Juan Keidel taught continental drift to several generations of future geologists, first as head of geology at the Mining Department and later as a professor in both Buenos Aires and La Plata. The famous AAPG symposium of 1928 did not affect the evidence available in the southern hemisphere. In du Toit's publications, one

Fig. 25.3 Alex du Toit and the geologist Juan José Nágera (1887–1966) from the Mining Directorate who accompanied him in the small tills of the Precordillera of San Juan. *Source* Nágera (1938)

can see the continuous exchange with Argentine geologists who provided him with numerous observations in favor of the correlation and continental drift.

Among those who contributed to the dissemination and acceptance of continental drift in our country, Anselmo Windhausen stands out. In his masterpiece on Argentine Geology, he describes with a detailed analysis the separation of the different continental blocks, the paleontological and tectonic evidence, and how the Gondwanides allow the reconstruction of the supercontinent. However, he rightly points out that although the Patagonian-Antarctic massif can be recognized among other visible blocks, today there is no idea of how *they were grouped together in those diastrophic events at the end of the Paleozoic Era* (Windhausen 1931, p. 532). In those years, collision processes were unknown, although Windhausen had recognized that Patagonia, for example, had been an independent block (see Ramos et al. 2020).

A good example to illustrate the degree of acceptance of the results of this symposium is given by Víctor Benavides Cáceres, an eminent Peruvian geologist, who in early 1950 was doing his doctorate under the direction of Professor Marshall Kay at Columbia University in New York. Professor Kay informed his students that the following Friday afternoon they were going to have a seminar to be given by a young Argentine geologist, Dr. Horacio J. Harrington. This young man, who had completed his doctorate at the University of Buenos Aires with Keidel, lectured on the geological and paleontological evidence that allowed him to affirm that the Ventania system in Argentina and the Cape system in South Africa were adjacent during the Upper Paleozoic. Based on the presence of the *Glossopteris* flora, as he had discovered in his doctoral thesis in Buenos Aires, he postulated that the two continents were joined together as part of Gondwana. Furthermore, his later research at Oxford on the *Eurydesma* fauna, allowed him to reconstruct a shallow sea during the Early Permian that had developed between adjacent South America and Africa. At the end of the seminar, everyone applauded enthusiastically and went to celebrate. The following Monday, Professor Kay told his students: *Please forget what this young geologist said; out of politeness and respect for our guest, I did not want to contradict him; you should know that continents cannot move and that continental drift is absurd* (Ramos 2020a, b).

25.2 Geosyncline Theory

In the 1960s, tectonics was dominated by the geosyncline theory in Argentina. Its great promoter was the professor and researcher Ángel V. Borrello (1918–1971) of the *Universidad Nacional de La Plata* (see biographical details in Chap. 19), who after an outstanding work in *Yacimientos Carboníferos Fiscales*, dedicated himself from 1961 to research and teaching as head of the Geology Division of the La Plata Museum. In those ten years, he created a school of geotectonics with numerous disciples, through more than forty study trips throughout the Argentine territory, which allowed him to analyze the different geological environments.

His first steps in geosyncline theory were taken under the influence of his thesis supervisor Juan Keidel, who instilled in him his admiration for Hans Stille (1876–1966), whom Borrello considered the father of *a new Geology, in full development by the great Germanic master.* He was impressed by Stille's magnum opus of (Stille1940) on *the attractive subject of the introduction to the tectonic structure of the Americas, whose validity in many passages … is still clearly preserved. I confess that I found in H. Stille's Einführung in den Bau Amerikas (Introduction to the construction of America) an understanding of various geotectonic and geomagmatic characteristics of the Andes* (Borrello 1969a). Despite his enthusiasm for the German school, he couldn´t obtain a scholarship to study in Germany due to the development of the Second World War. His admiration for Stille continued until his later years, as highlighted by his dedication to the Geosynclines of Argentina's book (Borrello 1969a, b) where he expressly wrote: *To the memory of Prof. Dr. Hans Stille, ecumenical teacher of contemporary Geotectonics* (Fig. 25.4).

In addition to Stille, Jean Aubouin, professor emeritus of the University of Paris, and Marshall Kay, professor at Columbia University in New York, were particularly influential in his training as a tectonician. An invitation from Aubouin from the *Laboratoire de Géologie of the Faculté des Sciences* of the Sorbonne Paris allowed him to visit the Alps, the Pyrenees, and other tectonic units of Europe in 1965. He thus established a strong relationship with Auboin, who that year had published *Geosynclines* (Aubouin 1965), which was one of the first modern books in English on geosynclines that had a great impact in the United States. Together they made a

Fig. 25.4 Hans Stille (1876–1966), Marshall Kay (1904–1975), and Jean Aubouin (1928–2020), three great tectonicians of the first half of the twentieth century who marked the Borrello geosyncline formation. *Source* Lotze (1941), Dott (1976), and Gaudant (2021)

field trip to the Andes whose results were published the following year in a classic work on the Andean geosyncline (Aubouin and Borrello 1966).

On his return from the 22° International Geological Congress in New Delhi, India, he went to Columbia University to exchange ideas with Marshall Kay, who had published two classic works years earlier, *Geosyncline Nomenclature and the Craton* in 1947, and *North American geosynclines* in 1951, works that influenced Borrello's later studies.

In New Delhi, he had the opportunity to exchange ideas with Vladimir Belousov, professor at Lomonosov University and Director of the Lithosphere Institute, influential and powerful structural geologist, father of an entire school of vertical tectonics and tectonic inversion that dominated Soviet geotectonics for several decades. The result of Beloúsov's incursions in the Andes in 1960 can be read in his *Outline of the Tectonics of the Andes* of 1963, which presents a little-known structural section between the city of Mendoza and Aconcagua (Fig. 25.5) (Beloúsov 1963).

Two of Beloúsov's textbooks, *Structural Geology* and *Basic Problems of Geotectonics* (Beloúsov 1971, 1975), translated into Spanish, were popular in the 1970s. He was president of the IUGS Upper Mantle Program, which had its final symposium in Buenos Aires in 1970. It was organized by the American Geophysical Union, coordinated by Dr. Vladimir Belousov, one of the greatest Soviet tectonician from the Moscow Lithosphere Institute (Fig. 25.6), and by engineer Daniel Valencio, a brilliant Argentine geophysicist, professor at the University of Buenos Aires and specialist in paleomagnetism. The meeting was full of lively discussions where Frederick Vine and Drum Matthews spoke about magnetic belts and their symmetry on both sides of the oceanic ridges, and John Sclater explained the subsidence curve due to the cooling of the ocean floor. However, Beloúsov had the last word, claiming that the studies carried out in the oceans did not take into account what was known about continental orogens, and he destroyed the hypotheses presented by these young people. I was writing the last chapter of my thesis, and his arguments were so convincing

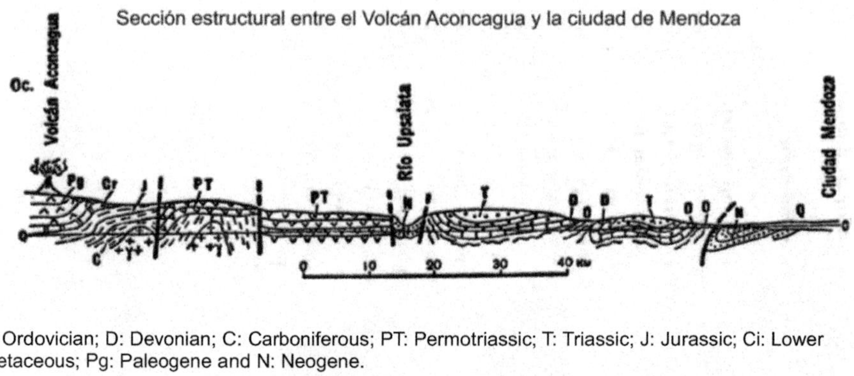

Fig. 25.5 Structural section surveyed by Belousov in 1960 between the city of Mendoza and the Aconcagua Volcano (sic) where the proposed classic structural style stands out, dominated by high-angle faults that become vertical at depth. Note that the thrust belt east of Aconcagua is resolved with a sub-vertical fault. *Source* Beloúsov (1963)

Fig. 25.6 Vladímir Vladímirovich Beloúsov (1907–1990) and Viktor Khain (1914–2009) dos grandes tectonistas rusos que tuvieron relación con Borrello. *Source* Shnyukov and Yanko-Hombach (2020) and Drachev et al. (2010)

that I modified the structural sections in depth, which went from a Mingramm-style tangential shortening (see Ramos, 1996 for Mingramm's contributions) to subvertical structures of the Beloúsov-type (see Ramos 1970).

After the Buenos Aires symposium, I met Beloúsov again at the International Geological Congress in Paris in 1980, where he lectured on the oceanification of the continents, which gave the apparent impression of the movement of continental margins that were fixed. At the next congress in Moscow in 1984, he continued to deny plate tectonics and continental displacement, but his popularity had fallen substantially, and he was no longer the arrogant professor of 1970. He had already published his *Why do I not accept plate tectonics?* (Belousov 1979).

In Moscow, I met Viktor Khain (1914–2009), one of the greatest Russian tectonicians who pushed plate tectonics against all odds in the USSR at the time when Beloúsov was president of the Lithosphere Institute. Due to disagreements with the authorities, he was sent into exile in Siberia. In 1984 I was back in Moscow, and we had just met. When he found out I was Argentine, he first asked me: *What do we know about Professor Borrello's Paleozoic ophiolites?* When I explained to him that they limited the accretion of allochthonous terrain, he excitedly asked me to send him a bibliography so he could learn more about the results. Years later, with more than a thousand published works, he received the highest distinctions in the Western world and was named a member of the main European academies of science, receiving the Gustav Steinmann Medal of the German Geological Society. The last time we were able to talk to him was in 1991 in Campinas, São Paulo. With his eternal enthusiasm, he was impressed by the progress of Plate Tectonics in the Andes. Meanwhile, Borrello, throughout the sixties, in addition to his work on the biostratigraphy of the Cambrian of the Precordillera, published numerous contributions to geosyncline theory from 1963 onwards, with special emphasis on the Precordillera system (Borello 1967, 1968a, b, c, 1969a, b: Borrello and Cuerda 1967). In 1969 he published his masterwork, *Los geosinclinales de la Argentina*, which became a very popular textbook in the geological community in those years. On October 28, 1971, he gave a closing lecture at the First National Symposium on Economic Geology, held in the city of San Juan (Cuerda 1972), on the tectonic evolution of the Precordillera. It was the first time I heard him give a lecture, and I was impressed with the fervor and passion he put into his presentation, and even more so when he was applauded for several minutes by an enthusiastic audience. Members of the Geological Association of Mendoza brought him to this city to repeat the lecture at night, but his heart could not resist and he died in a hotel in Mendoza shortly before his new lecture. After his death, works by him continued to appear, such as that of Aubouin et al. (1973) and subsequent ones.

The impact and general acceptance of the geosyncline theory in Argentina are undoubtedly due to the teachings of Borrello, strongly followed by his disciples, which had its epicenter at the National University of La Plata, but followers in almost all institutions.

25.3 Megashears and Basement Tectonics

With the widespread access to satellite images from Landsat programs in the early 1970s, a series of linear features became visible on a regional scale, which led to a revival of the old concepts of basement tectonics (Hobbs 1912). This tectonics, which some called new basement tectonics, had focused on topics such as fault reactivation, the influence of basement faults on metallogeny, and hydrocarbon accumulation. They used geophysics and remote sensing techniques to interpret subsurface and surface geology. They were based on work such as that of Sales (1968), which introduced concepts of crustal mechanics to analyze deformation in the foreland (Sales 1968). These concepts were added to the evaluation of the ERTS-1 images that enhanced these guidelines for mining and oil exploration (Saunders et al. 1973), using the classic concepts of John D. Moody (Fig. 25.7) on wrench-fault tectonics (Moody and Hill 1956; Moody 1973).

These concepts were popularized by the works on basic wrench tectonics for oil exploration (Wilcox et al. 1973). In our country, they were introduced by Tod P. Harding in 1973, in his analysis of the Huincul Ridge (Harding 1973, 1974) and generalized together with Jim D. Lowell in 1979 in his classic work on *Structural styles, their plate-tectonic habitats, and hydrocarbon traps in petroleum provinces* (Harding and Lowell 1979). In this work, they describe different structural styles involving the basement with wrench faults in compressional and extensional blocks and reverse and extensional basement faults in a wide variety of geological environments. Both Harding and Lowell were in Buenos Aires giving courses for YPF and other oil institutions. The results can be seen in the paradigm shift in the structure of the Huincul Ridge produced by the work of Ismael Orchuela and Victor Ploskiewicz shortly after (Orchuela et al. 1981; Ploskiewicz et al. 1984 and later works).

Among the first works in its academic application, the one by Beatriz Coira and collaborators in 1975 stands out. Using these techniques, they defined the importance of large lineaments such as the Gastre fault in northern Patagonia (Coira et al. 1975). In the central region of the Sierras Pampeanas, several works highlight the presence of large lineaments in the control of mineralization associated with porphyry copper in districts such as Bajo de la Alumbrera in Catamarca (Mon 1976; Ramos 1977).

Fig. 25.7 John D. Moody (1918–1883) the father of wrench tectonics. *Source* Halbouty (1985)

A geological correlation project that began in 1983, the IGCP 202 *Megafaults of South America* led by Francisco Hervé, strengthened the identification and importance of these faults in the tectonic evolution of South America. Its final international meeting was held in San Juan led by Bruno A. Baldis and its results were published the following year in a special issue of the *Revista Geológica de Chile*, number 31 of 1987. In those years the geological and tectonic evolution of a region was interpreted as controlled by these megashears which by transpression or transtension produced uplifts and basins.

The greatest exponent of this school was Bruno A. Baldis (Figs. 25.8 and 25.9), in those years a professor at the National University of San Juan. Numerous works produced by Baldis with his disciples found an explanation in almost any geological environment in these regional megashears that moved land for hundreds of kilometers, controlled sedimentation over time, and were responsible for the final structure of a region.

The excessive importance given to megashears during the 1970s and early 1980s was diminished with the application of new concepts with a strong weight in geophysical, geochemical, and paleomagnetic data. However, there are still isolated cases where attempts are made to use megashears in complex areas to elucidate the tectonic evolution of a region (see Japas et al. 2016; González et al. 2021, among others).

Fig. 25.8 Bruno A. Baldis (1937–1997) and Luis Dalla Salda (1942–2015), two drivers of megashear tectonics. *Source* Castro de Machuca (2018) and Cingolani (2015)

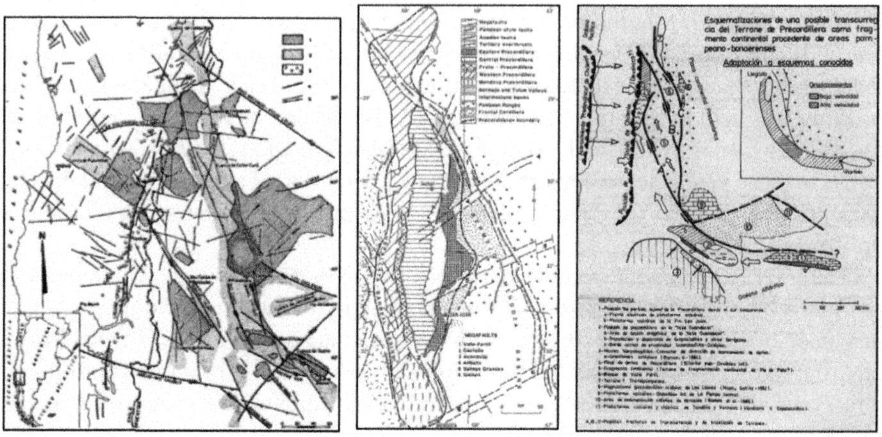

Fig. 25.9 Examples of how large megashears were interpreted as controlling the development of pull-apart basins, volcanic centers, and tectonic evolution of the region in northern Patagonia, between the Precordillera and Sierras Pampeanas, and in the translation of the Precordillera (after Dalla Salda and Francese 1987 and Baldis et al. 1984, 1989)

25.4 The Beginnings of *Plate Tectonics*

The emergence of new geophysical evidence on the movement of continents and oceans produced a total effervescence among geologists and geophysicists in the late sixties and marked the beginning of the scientific revolution that has become *Plate Tectonics*. An important milestone in this revolution was the 1967 American Geophysical Union symposium on seafloor spreading held in Washington, which brought together Dan McKenzie, Jason Morgan, Xavier Le Pichon and Jack Oliver, among many others, to discuss John Tuzo Wilson's proposal (1965). This symposium was followed by many works that are now considered classics on the subject. This new knowledge and ideas explained the processes that led to continental drift and the interaction between rigid plates (see Ramos 2020a, b).

Regarding the Andes, there were initially two competing models. One proposed by John Dewey explained magmatism, orogenesis, and the formation of sedimentary basins through a simple subduction of oceanic crust (Dewey 1969; Dewey and Bird 1970). In contrast to this model, Robert Dietz suggested that the main orogeny of the Upper Cretaceous along the Andes was formed by the collision of South America against an island arc and the closure of an ocean to give rise to the current subduction with a continental volcanic arc (Dietz and Holden 1970). With time and new geological and geophysical data, Dewey's proposal prevailed, which is known as the Andean tectonic model.

While this was happening in the northern hemisphere, in our country we were in the midst of the boom of the geosyncline theory, applauding the emotional conference of 1971 by Dr. Borrello. A small group participated in the final symposium of the Upper Mantle Project in Buenos Aires, listening to the closing conference of Dr. Belousov at the end of 1970, which destroyed the evidence presented by young geophysicists and geologists obtained during the duration of the project. The only voice that taught *Plate Tectonics* was that of Pedro Criado Roque (Fig. 25.10), who, invited by several universities, gave interesting conferences on the subject on a traveling basis.

Unlike the theory of *Continental Drift*, where our geologists were pioneers, *Plate Tectonics* took a while to be accepted in Argentina. In those years it was spoken of as a fascinating story, but it had no relation to geological observations, petrographic studies, or even structural studies. When I returned to the country in mid-1974 and working at the National Geological Survey, there was little talk or analysis of *Plate Tectonics*. At the University of Buenos Aires, there was an exception, generated by engineer Daniel Valencio (1928–1987) and his working group (Fig. 25.11). At the request of Félix González Bonorino, the first Paleomagnetic Laboratory in South America was founded in 1964. The hiring of Eng. Valencio had a prominent role in the studies of continental drift between Africa and South America. These studies determined for the first time, the separation path between the two continents, which made Eng. Valencio an international referent in Paleomagnetism (Borches 2008; Sinito and Vilas 1987). Eng. Valencio was a Professor of Geophysics and senior researcher at Conicet until his early death at the age of 59.

The analysis of what was published in 1976, as illustrated by the proceedings of the Argentine Geological Congress in Bahía Blanca, shows that the discussions ranged from basement tectonics with megashears to works on geosynclinal evolution. Geotectonics was reduced to analyzing the large megashears and their control on sedimentation and structure as shown by Baldis and collaborators in northwestern Argentina. The excellent balanced structural sections of Alberto Mingramm were intervened to show the supposed effects of the megashears (Baldis et al. 1976). Other works tried to explain how the geosyncline cycle guided volcanism in the Chilean

Fig. 25.10 Pedro Criado Roque (1921–1988), professor at the National University of San Luis with decades of experience at YPF. *Source* Rolleri (1989)

Fig. 25.11 Ken Creer, making magnetic measurements, and Daniel Valencio, taking notes in 1970, on a field trip in the Karoo in South Africa, and teaching geophysics at the Manzana de Las Luces. *Source* Biblioteca Digital F.C.E.N

sector. The references to *Plate Tectonics* were to show that it did not work, as Hubert Miller (1936–2020) argued when describing the variations in metamorphism in the Pacific margin, which had nothing to do with the paired belts of Miyashiro, and even less with the *simplistic speculations* of Ian Dalziel and collaborators published in Nature in 1974. In this classic work, Dalziel was the first to demonstrate how the closure of a previously unknown fossil marginal basin occurred, which Miller did not accept (Miller 1976).

An important change occurred in 1976, when in Denver, Colorado, I took a course on Plate Tectonics with Dr. William Dickinson, then a professor at Stanford University, an undisputed reference in the geological processes associated with the new global tectonics. On that occasion, Dr. Dietrich Roederer, a professor at Pennsylvania State University, taught us how fold and thrust belts work and the fundamentals of structural section balancing. Upon my return from the course, I applied these teachings to the structure of the Neuquén basin (Fig. 25.12), presenting the first structural sketches within the framework of *Plate Tectonics* of the Neuquén Andes with new approaches that began with the report on the province of Neuquén (Ramos 1978). One of the first works published upon my return was that of the fold and thrust belt of Lake Belgrano, where I analyzed the epidermal style of deformation, a novel for that sector of the Patagonian Cordillera, which contrasted with that of the Lake Fontana segment (Ramos 1976, 1979).

At the Argentine Geological Congress in Neuquén held in 1978, except for some incipient interpretations of Lower Cretaceous volcanism in the northern Patagonian Cordillera and some new data on Paleozoic paleomagnetism, there are almost no works related to *Plate Tectonics*.

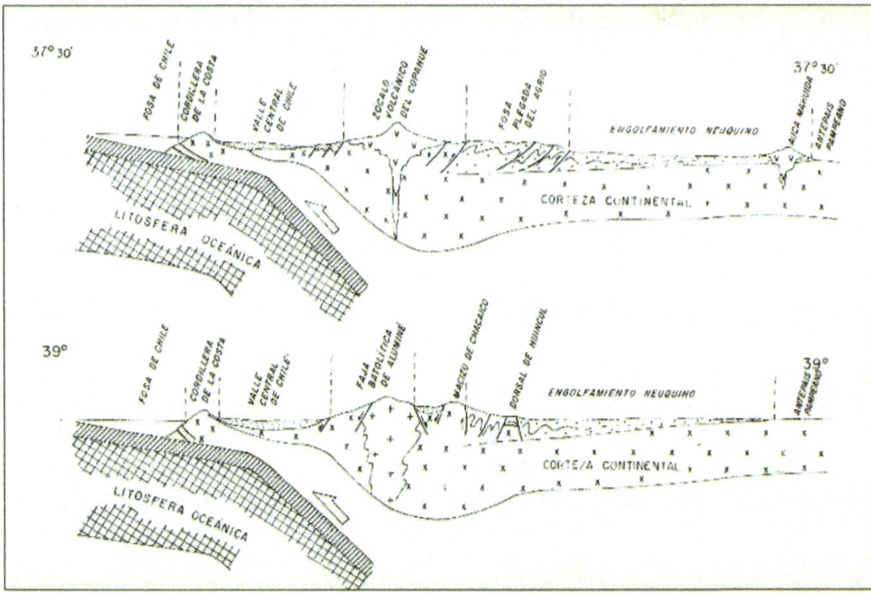

Fig. 25.12 One of the first attempts in the framework of *Plate Tectonics* is published in the Report of the province of Neuquén where an attempt is made to explain the different response to the subduction of the central sector of the Neuquén basin with its volcanic arc regarding the southern sector where the magmatic arc exposes granites (Ramos 1978)

In 1980, I was asked by the National University of Patagonia San Juan Bosco to give a course in Comodoro Rivadavia on Geotectonics to introduce the new concepts of global tectonics. This was followed by numerous courses at the universities of Córdoba and Tucumán in 1981 and Comahue in 1982, which continued with some courses organized by the Argentine Geological Association in Buenos Aires and Mendoza. From these initial courses, dozens of courses followed in almost all the universities of the country and in neighboring countries such as Chile, Uruguay, Paraguay, etc. These new ideas combined with the experiences of months of fieldwork in various sectors of the Andes led me to reinterpret the Andean evolution of different sectors.

An important role in the advancement of knowledge in those years has been played by the IGCP 120 Project of the UNESCO Geological Correlation Program on Magmatic Evolution of the Andes under the direction of Dr. Umberto Cordani of the University of São Paulo. The project began in 1976 and continued until 1985 (Ramos 2020a, b and works cited here). This project allowed Argentine researchers to present the geological evolution of the Andes in numerous international meetings, sparking discussions and exchanges with different specialists that enriched Argentine Geology with their publications (Linares et al. 1982). The results of this project began to be seen at the Argentine Geological Congress in San Luis in 1981. The contributions of Dalla Salda and collaborators characterized the Tertiary volcanic arc of the Patagonian Cordillera; the eastern extension of the granitic plutons of Monte San Lorenzo; those of Caminos and others on the tectonic evolution of the Fuegian Cordillera and the ophiolites of its marginal basin (Dalla Salda et al. 1981; Ramos and Palma 1981, Caminos et al. 1981).

In 1981 the Symposium on *Accretionary Terranes and Cordilleran Tectonics* in Ithaca, United States, within the deep seismic program COCORP of Cornell University produced a paradigm shift in the basic concepts of how continents were formed. The opportunity to discuss with Peter Coney, Jim Monger, David L. Jones, Alan Cox, Pete Palmer, and Jack Oliver, among many others, allowed us to understand the modern concept of terrane and its diagnosis based on the first work published in Nature at the end of 1980 on the subject (Coney et al. 1980). At this symposium, the hypothesis on the allochthony of the Precordillera was presented together with Terry Jordan and Rick Allmendinger.

The following year, in 1982, the first course on Plate Tectonics was given at the Argentine Geological Association, which was jointly taught with engineer Daniel Valencio, Jorge Vásquez, and José María Urien and had a well-attended audience.

The success obtained and the need to offer it to a greater number of interested people led to a new course being held that same year in the large hall of the Undersecretariat of Mining. This course had a crowded audience, and Juan V. Ploszkiewicz was also added as a guest professor (Fig. 25.13).

25.5 Horizontal Subduction in the Andes

In those years, through the courses and dissemination carried out, plus the internal tectonic schools given at YPF by top-notch professors who came from abroad, the geological community began to be introduced to new tectonic ideas. Controversies arose at Argentine geological congresses, where there were frequent disputes, especially in the analysis of folded belts, between *verticalist* geologists and the so-called *skateboard* geologists who were in favor of tangential tectonics. The first discussions took place at the Argentine Geological Congress in 1984.

Another important milestone in knowledge was the Latin American Geology Congress held in Buenos Aires in 1982. On that occasion, we presented the first work on horizontal subduction in the Andes with special emphasis on the Pampas sector (Isacks et al. 1982). The discussions and controversies that arose and the geological consequences caused important debates. Due to the recent Malvinas War were no participants from the United States or the United Kingdom. This work was followed by others (Fig. 25.14) where an active horizontal subduction zone in the Pampas was compared with the Laramide flat slab, which had a great international impact (Jordan et al. 1983a, b).

These works began to demonstrate how *Plate Tectonics* could explain the distribution, structure, and other characteristics of the main Argentine geological provinces, recognized since the early works of Groeber in 1938. These facts helped to accept Plate *Tectonics* as an important theory to understand our geology. At the beginning

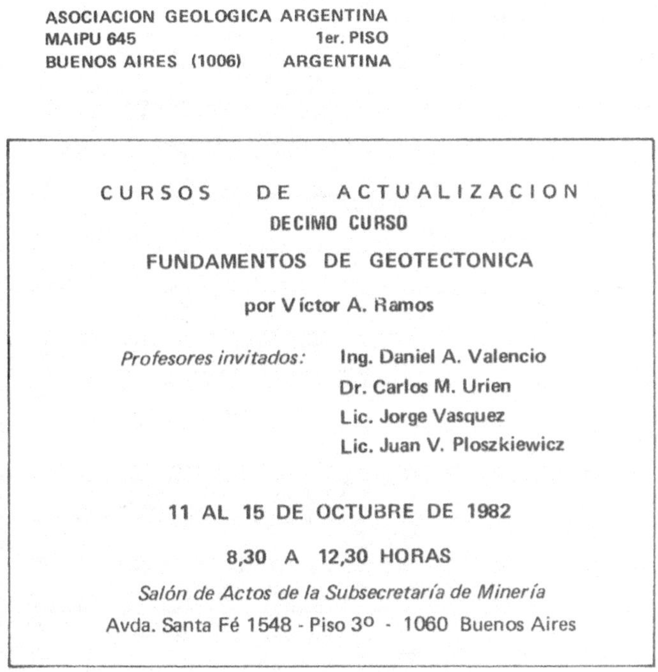

ASOCIACION GEOLOGICA ARGENTINA
MAIPU 645 1er. PISO
BUENOS AIRES (1006) ARGENTINA

CURSOS DE ACTUALIZACION
DECIMO CURSO
FUNDAMENTOS DE GEOTECTONICA
por Víctor A. Ramos

Profesores invitados: Ing. Daniel A. Valencio
Dr. Carlos M. Urien
Lic. Jorge Vasquez
Lic. Juan V. Ploszkiewicz

11 AL 15 DE OCTUBRE DE 1982
8,30 A 12,30 HORAS
Salón de Actos de la Subsecretaría de Minería
Avda. Santa Fé 1548 - Piso 3º - 1060 Buenos Aires

Fig. 25.13 Brochure of the tenth course of the Argentine Geological Association given by Victor A. Ramos and with the participation of guest professors in 1982

of 1984, the first regular and intensive courses on *Plate Tectonics* were started at the University of Buenos Aires, taught jointly by Eng. Daniel Valencio and Juan F. Vilas, with a large attendance of students from the last years and graduates of the UBA and other universities. Another course that had a strong impact and was well attended was the one by Suzanne M. Kay in 1985, who spoke to us about Petrology and Plate Tectonics in the Argentine Geological Association. These courses and the fieldwork associated with students from different universities, in addition to those from Buenos Aires, contributed to its dissemination among several generations.

25.6 The Terranes Accreted to Gondwana

In 1984 at the Argentine Geological Congress in Bariloche, before a large international audience, not only of geologists from Chile and Brazil but also from Europe and the United States, the first public discussions on the processes of *Plate Tectonics* were held before an attendance of hundreds of colleagues and students. Concepts such

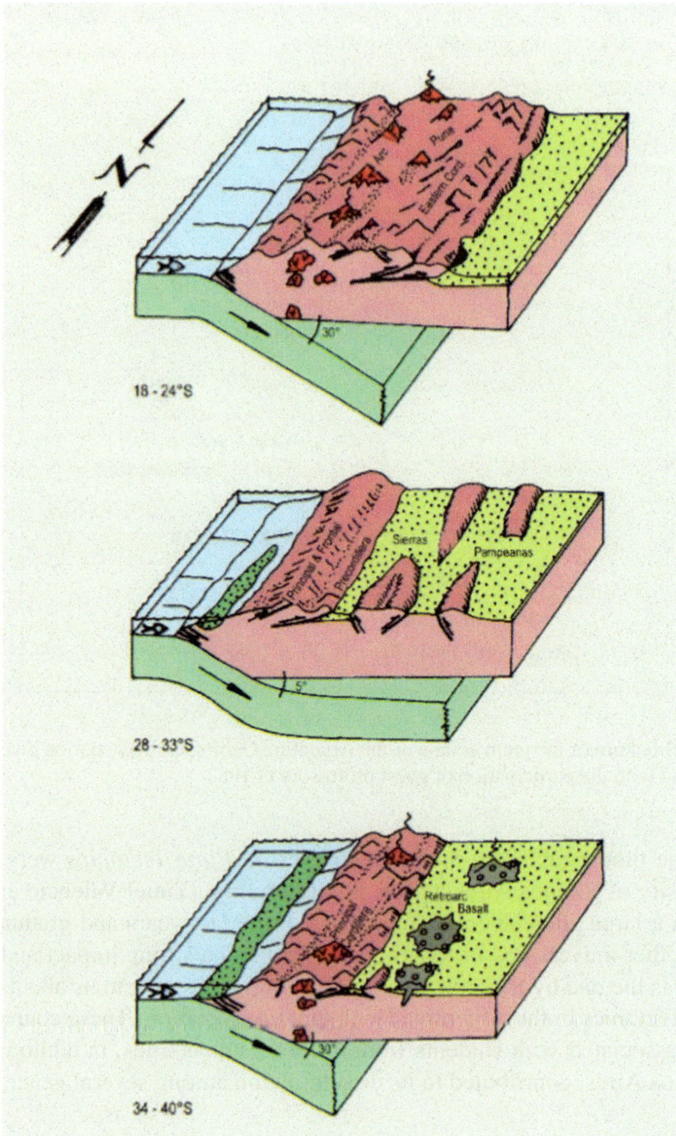

Fig. 25.14 Classic structural diagram by Terry Jordan and collaborators showing the different geological provinces controlled by the dip of the Benioff zone and its geological characteristics (Jordan et al. 1983a, b)

as what an ophiolite is, collision of terranes, and how they are identified, occupied a large part of the heated debates in the sessions and subsequent meetings.

The greatest expectations at this 1984 Bariloche congress were in the Symposium on *Tectonic Evolution of the Andes* that we organized through the Commission on Tectonics (IUGS) and the Argentine Committee of the Lithosphere Program. Some of its 12 works introduced novel concepts that ranged from the characterization of ophiolites, allochthonous terranes in the Andes such as Chilenia, and various structural styles of fold and thrust belts (Fig. 25.15). It can be said that this symposium, together with the sessions on Structural Geology and Geotectonics, marked the full entry of *Plate Tectonics* into the Argentine geological community.

However, there was no consensus on the hypotheses presented on the occurrence of allochthonous terranes such as Chilenia or Patagonia; rather, there was a strong rejection of these proposals that quickly generated two antagonistic groups that for more than a decade confronted each other in various meetings. The group headed by Dr. Bruno Baldis, a National University of San Juan professor, led the discussions, which reached their climax at the Horco Molle meeting in 1986 (Fig. 25.16).

The meeting was organized by Professor Gilberto Aceñolaza at the Horco Molle experimental reserve, recently created at the National University of Tucumán. During the several days of meetings, the different specialists did not reach any agreement; the students and graduates from Buenos Aires were confronted with the strong rejection of the allochthonist ideas expressed by a Tucumán-San Juan front.

Little by little, through publications in high-impact specialized journals such as *Tectonics*, it began to be accepted that the Precordillera was part of Cuyania, an allochthonous terrane originating from Laurentia (Ramos et al. 1986 and subsequent ones). However, at the Symposium on Geology of the Andes and its relation to Hydrocarbon and Mineral Resources held in Santiago de Chile in 1985, where Contantino Mpodozis and I proposed again that Chilenia was not part of Gondwana,

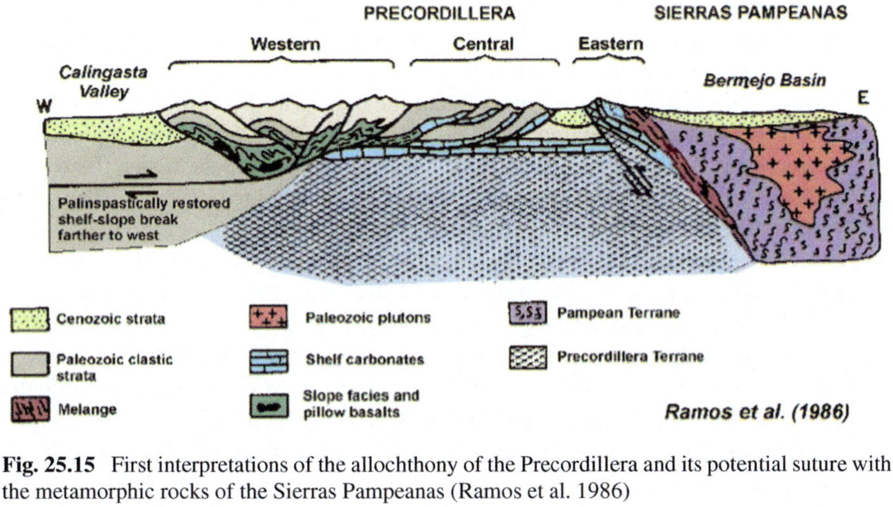

Fig. 25.15 First interpretations of the allochthony of the Precordillera and its potential suture with the metamorphic rocks of the Sierras Pampeanas (Ramos et al. 1986)

Lower line: Miguel A. Galiski, Ricardo Mon, Luisa Villar, Magda Koukharsky, Francisco Nullo, Victor Ploszkiewicz, Beatriz Coira, José Viramonte, Carlos Rapela, Carlos Cingolani, Luis Dalla Salda, N.N., López Ferretti, Victor Ramos, Hugo Fournier and his partner, Julio C. Ávila, Hugo Posse, N.N., and N.N.; Upper line: José M. Cortés, N.N., Carlos Beros, Augusto Rapalini, José Sellés Martínez, Enrique Triep, Antonio Introcaso, Luis Spalletti, José M. Cabanillas, Ricardo Omarini, Juana Toseli, Bruno Baldis, and Carlos Lisi.

Fig. 25.16 Tectonic meeting in Horco Molle organized in Tucumán by Gilberto Aceñolaza in November 1986

but an allochthonous terrane, the reaction against it was even greater. Meanwhile, other alternatives were proposed for this allochthon, associated with a great continent–continent collision between Laurentia and Gondwana. They proposed a microcontinent Occidentalia that extended from Peru to southern Chile which left behind Laurentia when separated from Gondwana in the Middle Ordovician. This hypothesis was led by Prof. Luis Dalla Salda (1942–2015) of the National University of La Plata and published in the journal *Geology*, which received strong attention in the United States (Dalla Salda et al. 1992). This new hypothesis led to the organization of a Penrose Conference in San Juan in 1995, the first one held in South America by the Geological Society of America, coordinated by Ian Dalziel, Pete Palmer, Luis Dalla Salda and Carlos Cingolani under the name *The Argentine Precordillera: A Laurentian Terrane?* (Cingolani 2015). At this meeting, which was attended by numerous North American colleagues specializing in the Appalachians, there were lively discussions both in the presentations and on the field trips (Fig. 25.17).

That same year, an improved alternative to the Precordillera allochthony proposed by Astini and collaborators was presented, which instead of coming from the northern Appalachians came from the Ouachitas Bay (Astini et al. 1995). This new alternative presented a complete sedimentological description and new paleontological analyses of the different units that were in contrast to the Occidentalia proposal.

The discussion continued through various meetings and publications until finally the proposal of Ricardo Astini and collaborators reached consensus, where the Precordillera as part of the Cuyania microcontinent with more than 1000 km of length collided with the Gondwanan margin. The paleomagnetic evidence obtained

Fig. 25.17 Ian Dalziel, Eldridge Moore, and Luis Dalla Salda at the 1995 Penrose Conference looking at the angular unconformity between the Ordovician limestones of the San Juan Formation, and the Silurian shales of the Cuesta del Tambolar, San Juan River valley

by Augusto Rapalini and Ricardo Astini in 1998 (Fig. 25.18 and 25.19) apparently put an end to the controversy (see review in Ramos 2004).

Fig. 25.18 Augusto Rapalini from the Daniel Valencio Laboratory at the University of Buenos Aires who calculated the first paleopoles that allowed the establishment of the allochthony of Cuyania

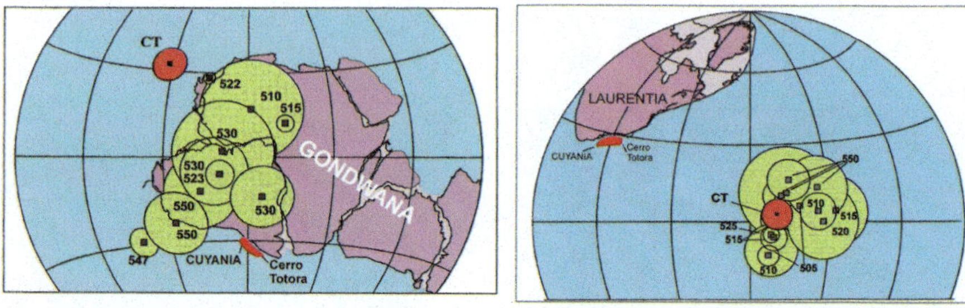

Fig. 25.19 Paleomagnetic evidence showing at 530 Ma where the Cuyania terrane came from. Its polar wandering curve did not agree with that of Gondwana but did coincide with that of Laurentia (modified from Rapalini and Astini 1998)

25.7 The Precordillera More Autochthonous Than the "Flor de Ceibo"

The geological community had left behind the proposals of Horco Molle with a Gondwanan Precordillera arrived through a continental-scale megashear from another sector of Gondwana (Baldis et al. 1989) and there was consensus that Cuyania had been an allochthonous block derived from Laurentia. A novel work by Stan Finney in 2003 (Fig. 25.20), again proposed its autochthonous origin based on new geochronological data of zircons from the Precordillera of northern San Juan. This work coincided with the *IX International Symposium of the Ordovician System* being held in San Juan, which had Stan Finney as a keynote speaker at the opening conference, who again presented his novel data. In those days, a local newspaper that interviewed him coined the title *The Precordillera more autochthonous than the Flor de ceibo.* A later symposium held that same year in Chile, again had him as the main speaker, putting him at the height of his popularity.

Part of the geological community disagreed with that interpretation and promoted a special issue of the journal *Gondwana Research* edited by Graciela Vujovich and collaborators on *Cuyania: an exotic block of Gondwana* that was dedicated to analyzing its Laurentian origin. The issue had 14 papers, with an initial one by Stan Finney presenting Cuyania as an undisputed part of Gondwana with the new data, and all the remaining ones confirming its Laurentian origin. When the issue was going to press, Finney's request was received to withdraw it from the volume to be printed. The Laser Chron Center laboratory at the University of Arizona under George Gehrels had misplaced the labels of one of the samples. Finney's 2003 paper had samples from the basement of Cuyania and others from the central sector of the Sierras Pampeanas.

Ricardo Astini, a sedimentologist from the National University of Córdoba, could not believe the new results and went with the same *baquiano* who had accompanied Finney to sample the dated unit in the same location. He sent the samples to the Laser Chron Center in Arizona and when Gehrels saw the new results, very different from the previous ones, he reviewed the chain of analysis of the previous samples

Fig. 25.20 Stan Finney, George Gehrels, and Ricardo Astini, protagonists of the different vicissitudes of the recognition of the allochthony of the Precordillera as part of Cuyania. *Source* courtesy I.C.S.-I.U.G.S. and the University of Arizona

and discovered an error in the labeling. Gehrels himself, in a heartfelt message that he sent us, assumes responsibility for the error, given that the supposed Gondwanan sample from the Precordillera was actually from the Sierras Pampeanas. This episode definitively closed the discussion on the allochthony of Cuyania.

25.8 The Allochthony of Patagonia

The original proposal made in 1984 on the allochthonous origin of Patagonia generated heated debates for decades (Ramos 1984). However, Windhausen had already outlined in 1924 that Patagonia was an independent block of South America in pre-Cretaceous times, associating it with the development of the Gondwanides (Fig. 25.21). This proposal was unknown for many years since it was published in the weekly supplement of a little-known newspaper (Storni 1946). This proposal is reinforced by the statements of Keidel who maintained a year before that Patagonia was a region that remained independent from the rest of South America, as a remnant of a larger continent that included Antarctica (Keidel 1925, pp. 299–300).

The initial resistance to the allochthonous origin of Patagonia gave rise to new alternatives, among which stand out those of Pankhurst et al. (2006), Chernicoff et al. (2013), and Ramos and Naipauer (2014). A turning point in this controversy was the discovery of archeocyatids of Antarctic origin in northern Patagonia by González et al. (2011 and 2013) and the studies by Pángaro et al. (2016), who identified an extensive peripheral foreland basin, the Hespérides Basin, associated with the collision of Patagonia. This late Palaeozoic basin covered a large part of the province of Buenos Aires, including the Tandilia system, and was connected with the South African basins. This allowed for the first time to have a structural, magmatic, and sedimentary framework related to the first-order collision between Patagonia

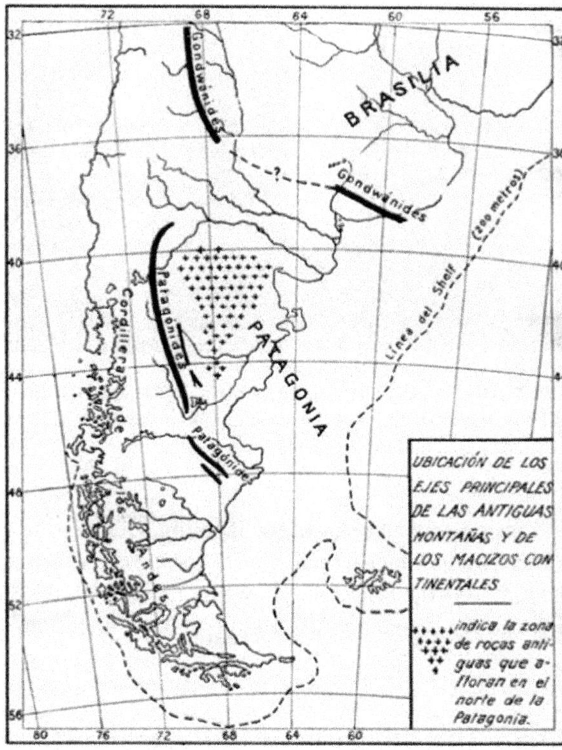

Fig. 25.21 Diagram of the Brasilia massifs as part of Gondwana and Patagonia, separated by the Gondwanides, an orogen formed in the Permian that continued into the Cape System in southern Africa (Windhausen 1924). Note his proposal of the *Patagónides* as a Mesozoic chain

associated with the Antarctic continents and the continent of Gondwana (see Ramos et al. 2020, and citations therein). Despite some new proposals of local origin and very restrictive approaches, there is finally consensus on the proposed collision of Patagonia (Fig. 25.22).

25.9 Final Considerations

Despite a late acceptance of *Plate Tectonics* in our country, from the 1980s onwards a firm consensus began to emerge regarding its importance for understanding, particularly the geological evolution of the Andes. The development of its magmatism, its sub-Andean basins, and the resulting structures were successfully explained by applying the fundamental principles of global tectonics. In particular, the morphostructural characteristics of the different geological provinces, recognized since the beginning of our geology, have for the first time an adequate geological explanation. These processes in continuous evolution have allowed, through numerous specific studies, the current knowledge of our geology.

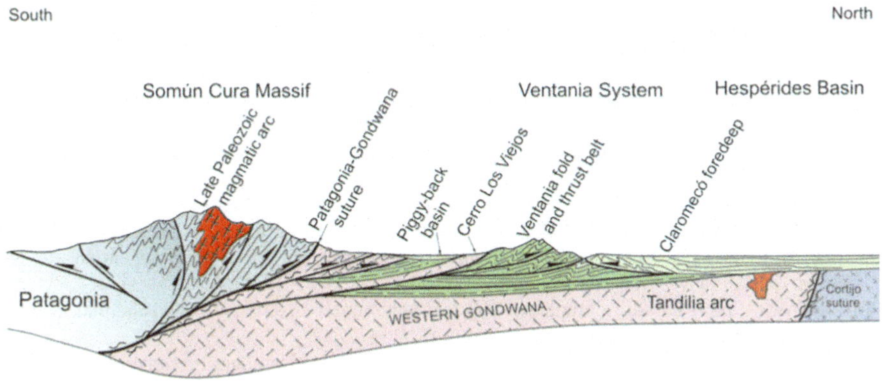

Fig. 25.22 Structural section illustrating the tectonic relationships between Patagonia and Gondwana with the development of the peripheral foreland basin of the Hesperides overlying the Tandilia system (Ramos et al. 2020)

Despite having left behind the initial discussions on the variations in the geometry of the subduction zones analyzed in space and time, new questions arise that make it increasingly difficult to obtain a complete understanding of the processes. Today the allochthonous nature of different crustal massifs is widely accepted, but there are still residual discussions on their limits and the precise times of accretion. There is no doubt that the discussions, some of which were heated and enthusiastic, have allowed for a rapid advance in knowledge. However, the understanding of new subcrustal processes located in the mantle promises new times of discussion and analysis for new, more holistic understandings of our geology.

References

Astini RA, Benedetto JL, Vaccari NE (1995) The early Paleozoic evolution of the Argentina Precordillera as a Laurentian rifted, drifted, and collided terrane: a geodynamic model. Bull Geolog Soc Am 107:253–273

Aubouin J (1965) Geosynclines. Developments in Geotectonics 1. Elsevier, Amsterdam, 335 p

Aubouin J, Borrello AV (1966) Châines alpines et chaines Andines: regard sur la géologie de la cordillere des Andes au parallele de l'Argentine moyenne. Bulletin de la Societé Geologique de France, 7° Serie, Paris, 8:1050–1070

Aubouin JA, Borrello AV, Cecione G, Charrier R, Chotin P, Frutos J, Thiele R, Vicente JC (1973) Esquisse paleogeographique et structurale des Andes Meridionales. Revue de Géographie Physique et de Geologie Dynamique, Paris 15(1–2):11–71

Baldis BAJ, Gorroño R, Ploskiewicz V, Sarudiannsky R (1976) Geotectónica de la Cordillera Oriental, Sierras Subandinas y comarcas adyacentes. 6° Congreso Geológico Argentino, Actas, Buenos Aires 1:3–22

Baldis B, Beresi M, Bordonaro O, Vaca A (1984) The Argentine Precordillera as a key to Andean Structure. Episodes 7(3):14–19

Baldis B, Peralta S, Villegas R (1989) Esquematizaciones de una posible transcurrencia del terreno de Precordillera como fragmento continental procedente de áreas pampeano-bonaerenses. InsuGeo, Serie Correlación Geológica, Tucumán 5:81–100
Beloúsov VV (1963) Esbozo de la tectónica de los Andes (translated by Schumacher DW). Boletín de la Sociedad Moscovita de Ciencias Naturales, Sección Geología, Moscú 2:6–24
Beloúsov VV (1971) Problemas Básicos de Geotectónica. Ediciones Omega, Barcelona, 854 p
Beloúsov VV (1975) Geología Estructural. Editorial Mir, Moscú, 304 p
Beloúsov VV (1979) Why I do not accept plate tectonics? Eos 60(7):207–211
Borches C (2008) Daniel Valencio y el Paleomagnetismo. La Ménsula, Año, Buenos Aires 2(4):1–4
Borrello AV (1967) Estado actual del conocimiento geológico del flysch en la Argentina. Revista Museo de La Plata (NS) Sección Geología 6(44):125–153
Borrello AV (1968a) Estudio comparado del tectonotema en el desarrollo geosinclinal. Revista Museo de La Plata (NS) Sección Geología 6(47):209–220
Borrello AV (1968b) Los pisos estructurales de la Precordillera-Cordillera Frontal en su evolución orogénica. 3° Jornadas Geológicas Argentinas (Comodoro Rivadavia, 1966), Actas, Buenos Aires 1:227–232
Borrello AV (1968c) Sobre el vulcanismo embrionario neoídico de los Andes Patagónicos Fueguinos. Boletín de la Academia Nacional de Ciencias, Córdoba 47(I):21–26
Borrello AV (1969a) Los geosinclinales de la Argentina. Dirección Nacional de Geología y Minería, Anales, Buenos Aires 14:1–188
Borrello AV (1969b) El flysch paleoídico de Mendoza. 4° Jornadas Geológicas Argentinas (Mendoza), Actas, Buenos Aires 1:75 89
Borrello AV, Cuerda AJ (1967) Sobre la evolución estructural del geosinclinal Paleozoico de la Precordillera. (Tramo montañoso Jejenes-Jáchal, provincia de San Juan). Revista Museo de La Plata (NS) Sección Geología 6(43):111–124
Caminos R, Haller M, Lapido O, Lizuaín A, Page R, Ramos VA (1981) Reconocimiento geológico de los Andes Fueguinos, Territorio Nacional de Tierra del Fuego. 8° Congreso Geológico Argentino (San Luis). Actas 3:759–786
Castro de Machuca B (2018) Orígenes de la carrera de Geología en la Universidad Nacional de San Juan. In: Martino R, Ramos VA (eds) El nacimiento de la geología en la enseñanza universitaria, Revista Facultad de Ciencias Exactas, Físicas y Naturales, Córdoba 5(Suplemento 1):98–105
Chernicoff CJ, Zappettini EO, Santos JOS, McNaughton NJ, Belousova E (2013) Combined U-Pb SHRIMP and Hf isotope study of the Late Paleozoic Yaminué Complex, Rio Negro Province, Argentina: Implications for the origin and evolution of the Patagonia composite terrane. Geosci Front 4:37–56
Cingolani C (2015) Prof Dr Luis Hugo Dalla Salda (1942–2015). Revista de la Asociación Geológica Argentina 72(4):578–586
Coira BL, Nullo FE, Proserpio C, Ramos VA (1975) Tectónica de basamento de la región occidental del Macizo Nordpatagónico, provincias de Río Negro y Chubut. Revista De La Asociación Geológica Argentina 30(4):361–383
Coney PJ, Jones DL, Monger JWH (1980) Cordilleran suspect terranes. Nature 288:329–333
Cuerda A (1972) Profesor Dr Angel V Borrello. Revista de la Asociación Geológica Argentina 26(4):529–530
Dalla Salda L, Francese J (1987) Las megaestructuras del Macizo y Cordillera Norpatagónica argentina y la génesis de las cuencas volcano-sedimentarias terciarias. Revista Geológica De Chile 31:3–13
Dalla Salda L, Lequizamón M, Mazzoni M, Merodio J, Rapela C, Spalletti L (1981) Características del vulcanismo paleógeno en la Cordillera Patagónica entre las latitudes 39°30' y 41°20' S. 8° Congreso Geológico Argentino (San Luis), Actas, Buenos Aires, 1:629–657
Dalla Salda LH, Cingolani CA, Varela R (1992) Early Paleozoic orogenic belt of the Andes in southwestern South America: result of Laurentia-Gondwana collision? Geology 20(7):617–620
Dewey J (1969) Continental margins: a model for conversion of Atlantic type to Andean Type. Earth Planet Sci Lett 6(3):189–197

Dewey J, Bird JM (1970) Mountain Belts and the New Global Tectonics. J Geophys Res 75:2625–2647
Dietz RS, Holden JC (1970) Reconstruction of Pangaea: breakup and dispersion of Continents, Permian to Present. J Geophys Res 75:4939–4955
Dott RH Jr (1976) Memorial to Marshall Kay 1904–1975. Memor Geol Soc Am 07:1–9
Drachev SS, Nikishin AM, Pushcharovsky DY, Ryabukhin AG (2010) Memorial: Victor Efimovich Khain (1914–2009). Bull Am Assoc Pet Geol 94(10):1639–1640
du Toit AL (1926) The geology of South Africa. Oliver and Boyd, Edinburgh, 463 p
du Toit AL (1927) A geological comparison of South America with South Africa, vol 381. Publications Carnegie Institute, 157 p
du Toit AL (1937) Our wandering continents. Oliver and Boyd, London, p 366
González PD, Tortello F, Damborenea S (2011) Early Cambrian archaeocyathan limestone blocks in low-grade metaconglomerate from El Jagüelito Formation (Sierra Grande, Río Negro, Argentina). Geologica Acta 9:159–173
González PD, Tortello MF, Damborenea SE, Naipauer M, Sato AM, Varela R (2013) The Archaeocyaths from South America: review and a new record. Geological Journal 48(2–3):114–125
Gaudant J (2021) Jean-Armand Aubouin. President 1976. Société Géologique de France. https://www.geosoc.fr/historique/depuis-2012-la-nouvelle-sgf/64-propos/tous-les-presidents/presidents-sgf/99-jean-armand-aubouin.html
González PD, Naipauer M, Sato AM, Varela R, Basei M, Cábana MC, Vlach S, Arce M, Parada M (2021) Early Paleozoic structural and metamorphic evolution of the Transpatagonian Orogen related to Gondwana assembly. Int J Earth Sci 110:81–111
Halbouty MT (1985) John D Moody (1918–1983) Bull Am Assoc Petroleum Geol 69(1):135–136
Harding TP (1973) Áreas potencialmente explotables para hidrocarburos resultantes de deformaciones producidas por fallas de desplazamiento lateral. Instituto Argentino del Petróleo, Resumen, Buenos Aires, 1 p
Harding TP (1974) Acumulaciones importantes de hidrocarburos originadas por deformaciones causadas por fallas laterales. Petrotecnia 2(17):12–19
Harding TP, Lowel JD (1979) Structural styles, their plate-tectonic habitats, and hydrocarbon traps in petroleum provinces. Bull Am Assoc Pet Geol 63(7):1016–1058
Hobbs WH (1912) Repeating patterns in the relief and in the structure of the land. Bull Geolog Soc Am 22:123–176
Isacks B, Jordan T, Allmendinger R, Ramos VA (1982) La segmentación tectónica de los Andes Centrales y su relación con la geometría de la placa de Nazca subductada. 5° Congreso Latinoamericano Geología (Buenos Aires). Actas 3:587–606
Japas S, Ré GH, Oriolo S, Vilas JF (2016) Basement-involved deformation overprinting thin-skinned deformation in the Pampean flat-slab segment of the southern Central Andes, Argentina. Geolog Mag 153(5/6):1042–1065
Jordan T, Isacks B, Ramos VA, Allmendinger RW (1983a) Mountain building model: the Central Andes. Episodes 1983(3):20–26
Jordan T, Isacks B, Allmendinger R, Brewer J, Ando C, Ramos VA (1983b) Andean tectonics related to geometry of subducted plates. Bull Geolog Soc Am 94(3):341–361
Keidel H (1907) Ueber den Bau der argentinischen Anden. Sitzungsberichte der kaiserlichen Akademie der Wissenschaft, Mathematisch-Naturwissenschaftliche Classe, Wien 114(1):649–674
Keidel J (1913) Über das Alter, die Verbreitung und die gegenseitigen Beziehungen der verschiedenen tektonischen strukturen in den argentinischen Gebirgen. 12° Session du Congrés Géologique International, Compte Rendu, Toronto 1:671–687
Keidel J (1916) La geología de las Sierras de la Provincia de Buenos Aires y sus relaciones con las montañas de Sudáfrica y Los Andes. Ministerio de Agricultura de La Nación, Sección Geología, Mineralogía y Minería, Anales, Buenos Aires 11(3):1–78

Keidel J (1922) Sobre la distribución de los depósitos glaciares del Pérmico conocidos en la Argentina y su significación para la estratigrafía de la serie de Gondwana y la paleogeografía del hemisferio austral. Boletín de la Academia Nacional de Ciencias, Córdoba, vol 25, pp 239–368

Keidel J (1925) Sobre el desarrollo paleogeográfico de las grandes unidades geológicas de la Argentina. Sociedad Argentina de Estudios Geográficos, GAEA, Anales, Buenos Aires 4:251–365

Linares E, Cordani UG, Munizaga F (eds) (1982) Magmatic evolution of the Andes. Earth-Sci Rev 18:199-450

Lotze F (1941) Hans Stille wissenschaftliches Werk. Naturwissenschaften 40:593–595

Miller H (1976) El basamento de la Provincia de Aysén (Chile) y sus correlaciones con las rocas premesozoicas de la Patagonia argentina. 6° Congreso Geológico Argentino. Actas 1:5–142

Mon R (1976) La tectónica del borde oriental de Los Andes en las Provincias de Salta, Tucumán y Catamarca, Argentina. Revista De La Asociación Geológica Argentina 31(1):65–72

Moody JD (1973) Petroleum explorations aspects wrench-fault tectonics. Bull Am Assoc Pet Geol 57:449–476

Moody JD, Hill MJ (1956) Wrench fault tectonics. Bull Geol Soc Am 67:1207–1246

Nágera JJ (1938) Geografía física de la República Argentina, Editorial A, Kapelusz, Buenos Aires, 232 p

Orchuela IA, Ploszkiewicz JV, Viñes RF (1981) Reinterpretación estructural de la denominada dorsal Neuquina. 8° Congreso Geológico Argentino (San Luis). Actas 3:281–293

Pankhurst RJ, Rapela CW, Fanning CM, Márquez M (2006) Gondwanide continental collision and the origin of Patagonia. Earth-Sci Rev 76:235–257

Pángaro F, Ramos VA, Pazos PJ (2016) The Hesperides basin: a continental-scale upper Palaeozoic to Triassic basin in southern Gondwana. Basin Research 28(5):685–711

Ploszkiewicz JV, Orchuela IA, Vaillard JC, Viñes RF (1984) Compresión y desplazamiento lateral en la zona de Falla Huincul: Estructuras asociadas, provincia del Neuquén. 9° Congreso Geológico Argentino (Bariloche), Actas, Buenos Aires 2:163–169

Ramos VA (1970) Geología de los primeros contrafuertes de la Puna saltojujeña entre San Antonio de los Cobres y el Moreno. Universidad de Buenos Aires. Facultad de Ciencias Exactas y Naturales, unpublished PhD thesis, Buenos Aires, 94 p

Ramos VA (1976) Estratigrafía de los lagos Fontana y La Plata, provincia del Chubut, República Argentina. 1° Congreso Geológico Chileno (Santiago de Chile), Actas 1(A):43 64

Ramos VA (1977) Basement tectonics from Landsat imagery in mining exploration. Geologie en Mijnbow, Amsterdam 56(3):243–256

Ramos VA (1978) Estructura. In: Rolleri EO (ed) Geología y recursos naturales de la Provincia del Neuquén, 7° Congreso Geológico Argentino (Neuquén), Relatorio. Buenos Aires, pp 99–118

Ramos VA (1979) Tectónica de la región del Río y Lago Belgrano, Cordillera Patagónica Argentina. 2° Congreso Geológico Chileno (Arica), Actas I(B), vol 1, p 32

Ramos VA (1984) Patagonia: ¿Un continente paleozoico a la deriva? 9° Congreso Geológico Argentino (SC Bariloche). Actas 2:311–325

Ramos VA (1996) Las primeras secciones estructurales balanceadas de Argentina: obra de un pionero. Revista De La Asociación Geológica Argentina 51(1):92–95

Ramos VA (2004) Cuyania, an exotic block to Gondwana: Review of a historical success and the present problems. Gondwana Research 7(4):1009–1026

Ramos VA (2013) Keidel (1877–1954) y du Toit (1878–1948), dos pioneros de la deriva continental. 3º Congreso Argentino de la Historia, Actas, Salta, pp 159–168

Ramos VA (2020a) Cincuenta años de Tectónica de Placas en los Andes. In: Hamity L (ed) Conmemoración del sesquicentenario de la Academia Nacional de Ciencias, 9 al 13 de septiembre de 2019, Academia Nacional de Ciencias. Córdoba, pp 13–40

Ramos VA (2020) Magmatic evolution of the Andes: the contributions of the IGCP Project 120. In: Bartorelli A, Teixeira W, Brito-Neves BB (eds) Geocronologia e tectônica do continente sul americano: a contribuição de Umberto Giuseppe Cordani. Sociedade Brasileira de Geologia, São Paulo, 4 th, Cap 7, 579–592

RamosVA (2023) Hans Keidel and Alexander du Toit's relationship and its impact on Wegener's continental drift hypothesis. In: Hynes AJ, Murphy B (eds) The consummate geoscientist: a celebration of the career of Maarten de Wit, vol 531. Geological Society, Special Publications, London. https://doi.org/10.1144/SP531-2022-1

Ramos VA, Naipauer M (2014) Patagonia: where does it come from? J Iber Geol 40(2):367–379

Ramos VA, Palma MA (1981) El batolito granítico del Monte San Lorenzo, Cordillera Patagónica, provincia de Santa Cruz. 8° Congreso Geológico Argentino (San Luis), Actas 3:257 280

Ramos VA, Jordan TE, Allmendinger RW, Mpodozis C, Kay SM, Cortés JM, Palma MA (1986) Paleozoic terranes of the central argentine chilean andes. Tectonics 5(6):855–880

Ramos VA, Lovecchio P, Naipauer M, Pángaro F (2020) The collision of Patagonia: Geological facts and speculative interpretations. Ameghiniana 57(5):464–479

Rapalini AE, Astini RA (1998) Paleomagnetic confirmation of the Laurentian origin of the Argentine Precordillera. Earth Planet Sci Lett 155:1–14

Rolleri EO (1989) Pedro Criado Roque 1921–1988. Revista De La Asociación Geológica Argentina 43(3):428–431

Sales J (1968) Crustal mechanics of Cordilleran foreland deformation: a regional and a scale model approach. Bull Am Assoc Pet Geol 52:2016–2044

Saunders DF, Thomas GE, Kindsman FE, Beatty DF (1973) Evaluation of the commercial utility of Erts-1 imagery in structural reconnaissance for minerals and petroleum. NASA, Report VI-702700F-152

Shnyukov E, Yanko-Hombach V (2020) History of the geological study of mud volcanoes in the black sea region. In: Mud volcanoes of the black sea region and their environmental significance. Springer, 494 p

Sinito AM, Vilas JF (1987) Daniel Alberto Valencio (1928–1987), Nota Necrológica. Revista De La Asociación Geológica Argentina 42(1–2):218–223

Steinmann G (1914) Vom internationalen Geologenkongress in Toronto. XIII Geologischer Unterricht. Geologische Rundschau 5:683–684

Stille H (1940) Einführung in den Bau Amerikas. Gebrüder Borntraeger, Berlin, 717 p

Storni CD (1946) Doctor Anselmo Windhausen en el 15° aniversario de su muerte. Revista De La Sociedad Geológica Argentina 2(2):89–112

Tuzo Wilson J (1965) A new class of faults and their bearing upon continental drift. Nature 207:343–347

van der Gracht WAJM (ed) (1928) Theory on continental drift. In: A symposium on the origin and movement of land masses both intercontinental and intracontinental, as proposed by Alfred Wegener (New York, 1926). American Association of Petroleum Geologists, 230 p

Wegener A (1912) Die Entstehung der Kontinente. Petermanns Geograph Mitteilungen 58(I):185–195, 253–256, 305–309. Transl from German by Jacoby WR (2001) J Geodyn 32:29–63

Wegener A (1915) Die Entstehung der Kontinente und Ozeane, 1st ed. Friedrich Vieweg and Sohn, Braunschweig. 2nd ed. 1920, 3rd ed. 1922; 4th ed. 1924; 4th revised ed. 1929

Wegener A (1920) Die Entstehung der Kontinente und Ozeane, 2nd ed. Friedrich Vieweg & Sohn, Braunschweig

Wegener A (1922) Die Entstehung der Kontinente und Ozeane, 3rd ed. Friedrich Vieweg & Sohn, Braunschweig

Wegener A (1929) Die Entstehung der Kontinente und Ozeane, revised ed. Friedrich Vieweg & Sohn, Braunschweig

Windhausen (1924) Die Entstehung der Kontinente und Ozeane, 4th ed. Friedrich Vieweg & Sohn, Braunschweig

Wilcox RE, Harding TP, Seely DR (1973) Basic wrench tectonics. Bull Am Assoc Pet Geol 57:74–96

Windhausen A (1931) Geología Argentina, segunda parte: Geología Histórica y Regional del Territorio Argentino. Ed Jacobo Peuser, 646 p, Buenos Aires

Chapter 26
Almost Three Hundred Years of Geology

Abstract In these almost three centuries of advances in knowledge, there have been notable events. From the obscure interpretations of the "tombs of giants" and the fragmentary geological knowledge of the Traveling Naturalists, we move on to the general descriptions of our mineral resources made after 1852 during the Argentine Confederation based in Paraná. Finally, the teaching of Geology began in 1865 at the University of Buenos Aires and geological research at the National Academy of Sciences in Córdoba, which was complemented by the explorations of the La Plata Museum in Patagonia. These initial surveys were the basis for the organization of the first National Geological Survey in 1904 with a legion of young and well-prepared German geologists who surveyed the Central and Northern regions of the country. This allowed the first Argentine geologists to begin their contributions and studies that in a few years allowed the consolidation of the regular geological survey of the country. At the same time, and with the need to explore the potential of oil and the creation of *Yacimientos Petrolíferos Fiscales*, it was necessary to train a large number of geologists who turned to the industry. Those years were followed by the creation of institutions such as *Fabricaciones Militares*, the National Atomic Energy Commission, and *Yacimientos Carboníferos Fiscales*, where geologists mobilized metalliferous and radioactive resources and the exploitation of coal. At the same time, new universities were opened that permitted an expansion of the educational offer in Geological Sciences that was distributed throughout the country. The creation of Conicet reinforced research both in universities and in different institutes where our geologists work. A comprehensive observation of the stages that have led to the present situation of geology shows us that with some ups and downs, economic and political crises, and strong vicissitudes, we have reached the twenty-first century with a well-consolidated geological community and with a promising future for the Geological Sciences activity.

Keywords Main accomplishments · Early history · Mapping · Teaching · Present situation

V. A. Ramos, *History of the Geological Sciences in Argentina*, The Latin American Studies Book Series, https://doi.org/10.1007/978-3-031-83139-3_26

Almost three hundred years have passed since the first observations made in these lands by the Jesuit fathers who attributed to *tombs of giants* the remains of large fossil mammals buried in the Santa Fe plain. In the missions, other Jesuits made contributions to the mineral kingdom and its fossils, which can be considered incipient contributions to the knowledge of our geology. Brief descriptions by enlightened travelers followed these observations on their way from Buenos Aires to Potosí, which were complemented by the studies of naturalists such as Tadeo Haenke who settled in the viceroyalty and made contributions that had remained partially unpublished in their time. It can be said that in the eighteenth century, the contributions to our geology were mostly basic and incomplete.

26.1 United Provinces of the Río de la Plata

It was not until the nineteenth century that naturalists from the Río de la Plata, such as Father Dámaso Larrañaga, made the first geological descriptions and interpreted the geology of the cliffs on both banks of the Río de la Plata. It can be said that this work, carried out in 1819, was the first geological study of the United Provinces of the Río de la Plata.

These studies were followed by the observations made by the traveling naturalist Alcide d'Orbigny between 1826 and 1833, as part of a project of the *Muséum d'Histoire Naturelle* in Paris, whose scientific journey was recorded in nine volumes with important contributions to the Natural History of South America. Partially superimposed on the previous one, there is the voyage of Charles Darwin who between 1832 and 1835, made fundamental geological observations on the Atlantic coast between Montevideo and Tierra del Fuego and in the Andes Mountain range that allowed him years later to support his theory on the Evolution of Species.

The history of Geology in the first half of the nineteenth century is marked by these illustrious naturalist travelers whose contributions were made known years later in Paris and London, respectively. With them this period of geological observations closes, and we must wait a few decades for the resurgence of new studies that are executed during the national organization after the battle of Caseros in 1852.

26.2 The National Organization

The profound change in geological studies was the product of the Argentine Confederation in 1852 by the hand of Justo J. de Urquiza. Naturalists and geologists such as Alfred Marbais Du Graty, Auguste Bravard, Victor Martín de Moussy and Germán Burmeister made important contributions, preparing samples for international exhibitions and working in the first Natural History museums, both in Paraná and later in Buenos Aires. Their contributions allowed the country's geological and mining resources to be shown to the world through their publications.

Thanks to the efforts of Juan María Gutiérrez, the new rector of the University of Buenos Aires, the teaching of Geology began in 1865 by hiring Italian professors, where Pellegrino Strobel stood out for his contributions not only to Earth Sciences but also for his promotion of the Argentine Scientific Society and other academic activities. Not only did he give the first Geology class in the country, but he is also considered the father of Argentine Archaeology for his first publications in this discipline.

A few years later, on the initiative of Domingo F. Sarmiento and Germán Burmeister, was founded the National Academy of Sciences of Córdoba in 1869. In 1873, the first German scientists arrived, among whom Alfred Stelzner stood out. It can be said that with the actions of these geologists, both in Córdoba and in Buenos Aires, not only did the teaching of Mineralogy and Geology finally crystallize, but regional surveys and the first geological maps of the country began to be carried out. In those years, the beginning of the teaching of Eduardo Aguirre in 1878, the first Argentine professor of Geology at the university, was an important milestone.

The foundation of the Museum of Natural Sciences of La Plata in 1884 by Francisco P. Moreno, together with the need to demarcate the boundaries between Argentina and Chile along the Andes Mountain range, allowed explorations to be carried out mainly in the Patagonian mountain range and its eastern foothills. Once again, with the participation of Swiss and German geologists, a broad push was given to geological research in Patagonia.

With the close of the nineteenth century, a pioneering era of the beginnings of Geology in our country came to an end. The foundations were laid for a sustained growth of geological activity in the coming years.

26.3 The Consolidation of Geology

The economic boom in Argentina at the beginning of the twentieth century, together with the restructuring of the Ministry of Agriculture, laid the foundations for the creation of the Geology and Mining Department. In 1904, a series of previously separate departments were brought together to create what can be considered the first National Geological Survey. The decision to put it in the hands of Enrique Hermitte was an extraordinary one to guarantee the success obtained in the following decades.

A host of young German geologists, as well as surveyors and chemists, systematically surveyed large regions of the country for 20 years. These geological studies, accompanied by petrological and mineralogical analyses carried out by more than fifteen geologists directed by Juan Keidel, produced an important leap in the quality and extent of the surveys carried out. Not only were extensive areas incorporated into the knowledge, but the basic principles of stratigraphy and structural geology of the territory were established with important scientific findings. These studies were the basis for finding new economically exploitable oil deposits in Comodoro Rivadavia in 1907 and in Neuquén in 1917, initiating a successful period of hydrocarbon discovery. The Geology and Mining Direction was not only dedicated to exploring mineral deposits, but also, through the use of modern drilling rigs, provided drinking water to numerous towns in the interior. Under the supervision of Ricardo Stappenbeck, hydrogeological studies were initiated, revealing numerous aquifers that were previously unknown in vast semi-arid regions of the interior of the country.

Those years saw the first Argentine geologists in action: José María Sobral, who graduated from the University of Uppsala in 1912, and Franco Pastore and Juan José Nágera, the first graduates of the University of Buenos Aires in 1914 and 1915. All of them started in the Geology and Mining Direction, which in 1922 was conducted by Sobral. Many of the German geologists returned to Germany, or retired, or dispersed among different institutions. A few remained in the Geology and Mining Direction from 1922, a decade marked by budgetary problems that did not allow the previous pace of surveys to continue or new foreign geologists to be hired. The crisis of 1930 was fatal and in 1931 this institution reduced its number of geologists to four, insufficient for the tasks to be carried out.

26.4 The Beginning of New Argentine Institutions

From this group of pioneering German geologists, Juan Keidel and later Pablo Groeber joined the teaching of Geology at the universities of Buenos Aires and La Plata, training a legion of young Argentines. These new geologists were quickly incorporated into oil exploration, which in the 1930s and 1940s had an explosive boom. The efforts of Hipólito Yrigoyen to founding a state company for the prospecting and exploitation of hydrocarbons in 1919 were crystallized in the presidency of Marcelo T. de Alvear, who created *Yacimientos Petrolíferos Fiscales* in 1922 under the direction of colonel and civil engineer Enrique Mosconi, a graduate of the University of Buenos Aires. These were years of rapid growth and the consolidation of a team of oil geologists, initially led by Guido Bonarelli and later led by Egidio Feruglio, two brilliant Italian geologists. These two, together with other Italian geologists who arrived in the country at that time, created an excellent Italian school of geology in the state company that allowed the training of several generations of Argentine geologists in oil exploration and exploitation. Shortly after, the General Direction of *Fabricaciones Militares* was created in 1941, initially under the supervision of engineer Victorio Angelelli, who began an iron ore exploration program that allowed the development of the Argentine steel industry. In the following years, it played a major role in the regional exploration plans that brought to light the first deposits of disseminated copper that marked the beginning of large-scale mining in Argentina in the following years.

Another institution that made important contributions to national development was the National Atomic Energy Commission created in 1952. Its geologists under the supervision of Victorio Angelelli and Pedro N. Stipanicic found numerous deposits of radioactive minerals in different provinces of the country, concentrating in recent years on the extra-Andean Patagonia where reserves were confirmed in important uranium deposits that guarantee resources for several decades of mining exploitation of radioactive minerals.

The National Direction of Mining and Geology, from the fifties onwards, carried out various regional geological survey programs, which in a few years covered a large part of Argentina with its 1:200,000 scale sheets. A large group of geologists such as Félix González Bonorino, Juan C.M. Turner, Guillermo Furque, Jorge Polanski, Emilio González Díaz, among many others, produced a systematic renovation of our geology.

In parallel with the development of these institutions, geology was consolidated first in the universities of Buenos Aires and La Plata and later in Córdoba. New universities incorporated the teaching of geology in Salta, Tucumán, San Juan, Comodoro Rivadavia and Bahía Blanca, followed by others in various parts of the country. The creation of the National Council for Scientific and Technical Research (Conicet) in 1958 by Bernardo Houssay, allowed the development of researchers with full-time dedication in universities and new institutes created by Conicet.

It can be objectively stated that in the second half of the twentieth century, research and development of teaching took root in numerous institutions, marking a period of important achievements in incorporating geological knowledge into the country's progress.

26.5 The *Plate Tectonics* Revolution

We have left certain facts and advances in geological knowledge that were transversal to all institutions for the last part of the book, even though they had more impact and development in some than in others. Unlike the theory of *Continental Drift*, in which our geologists have been pioneers worldwide, *Plate Tectonics* was not so easily accepted by the local geological community. It took almost two decades for its acceptance, even though the advanced paleomagnetic studies carried out by Daniel Valencio and his collaborators were at the forefront of knowledge in those years. The analysis of the different lines of research and the various tectonic theories in force in those years shows us the difficulties of insertion that the new ideas on global tectonics had in the geological community. A not entirely objective review of what happened at that time tries to show the conflicts that were experienced in those years, trying to tell a story in the present time to leave a testimony of how this knowledge has entered our country. Despite putting the utmost honesty in the story of the development of the activities, it is a story lived in a personal way and may be colored by the experiences of the moment, but I consider that despite these shortcomings it is important to leave this record for future generations. Time will allow later to make a more objective and starker version of the facts.

While I was writing these pages throughout the different chapters, not only the personal experiences received during more than sixty years of geological activity began to emerge from my memory, but also the vivid stories and testimonies of my teachers. They belonged to a transitional generation formed by the German professors who taught them and the foreign bosses they had in the different institutions, whom they came to know directly. Personalities such as Juan Keidel, Pablo Groeber, and Egidio Feruglio, among others, were illuminated by the stories received from these first geologists. The long conversations held with I. Osvaldo Bracaccini, Marcelo Yrigoyen, Edgardo Rolleri, Pedro Criado Roque and Juan C.M. Turner, among many others, have allowed me to learn firsthand many facts that I have tried to include in this history.

In these pages, I have tried to rescue from oblivion many geologists who have made transcendent contributions to our geology. I have also treated with indulgence others whose errors have led their disciples away from the path of truth. Although I have tried to balance the contributions in the different disciplines, I notice when reading this history that there is a bias marked by tectonics and I ask my readers to understand and to forgive me.

There are many things left to relate and I hope that in an eventual second edition, I can cover these omissions. Furthermore, I have had to finish before the end of the twentieth century due to the explosion of knowledge and contributions that have been made in recent years. I can only thank those who have encouraged me to write this story since without their support I would never have finished it.

Buenos Aires, October 15, 2024.

The manufacturer's authorised representative in the EU is Springer Nature Customer Service Centre GmbH, Europaplatz 3, 69115 Heidelberg, Germany. If you have any concerns regarding our products, please contact ProductSafety@springernature.com

Printed and bound by CPI Group (UK) Ltd, Croydon, CR0 4YY
30/09/2026
02225593-0012